MICROMAGNETISM AND THE MICROSTRUCTURE OF FERROMAGNETIC SOLIDS

The main topic of this book is micromagnetism and microstructure as well as the analysis of the relations between characteristic properties of the hysteresis loop and microstructure. Also presented is an analysis of the role of microstructure in the fundamental magnetic properties (for example magnetostriction or critical behaviour) of crystalline and amorphous alloys. The authors apply the theory of micromagnetism to all aspects of advanced magnetic materials including domain patterns and magnetization processes under the influence of defect structures. Coverage includes modern developments in computational micromagnetism and its application to spin structures of small particles and platelets.

Based on the continuum theory of micromagnetism, the physical principles of modern permanent and soft magnetic materials are covered comprehensively. Magnetization processes in small particles are outlined on the basis of the Landau–Lifshitz–Gilbert equation including the effects of thermal fluctuations. Magnetic aspects of intermetallic compounds, nanocrystalline and amorphous alloys are considered in detail within the framework of nucleation and pinning phenomena. Measurements of high-field susceptibility in the approach to ferromagnetic saturation are shown to be an appropriate method for the analysis of magnetically active microstructures in ferromagnets. To demonstrate the power of the theory of micromagnetism, the authors present many examples showing that theoretical predictions are supported by experimental results.

This book will be of interest to researchers and graduate students in condensed matter physics, electrical engineering and materials science, and to industrial researchers working in the electrotechnical and recording industries.

HELMUT KRONMÜLLER received the Dr. rer. nat. degree in 1958 from the Faculty of Physics at the Technische Hochschule, Stuttgart, Germany. In 1968 he became a Lecturer and in 1974 a Professor in Physics at the University of Stuttgart. In 1970 he became a Scientific Member of the Max Planck Society and a Member of the Board at the Max-Planck-Institut für Metallforschung, Stuttgart. Since 1987 he has been Director at the Institute and from 1992 to 1995 he acted in the temporary position of Managing Director. From 1993 to 2000 he was Chairman of the AG Magnetism of the German Physical Society.

The international reputation of the author is revealed by many nominations to organizing committees of conferences, memberships of scientific organizations and co-editorships of various scientific journals. His scientific work is contained in over 1000 publications including books and numerous review articles.

MANFRED FÄHNLE was born in 1951. He was awarded his PhD with distinction from the University of Stuttgart in 1977. Since 1977 he has been a member of the scientific staff at the Max-Planck-Institut für Metallforschung, Stuttgart. In 1990 he became Associate Professor for Theoretical Physics at the Institute for Pure and Applied Physics at the University of Stuttgart. He has been a member of several national and international scientific committees.

MICROMAGNETISM AND THE MICROSTRUCTURE OF FERROMAGNETIC SOLIDS

HELMUT KRONMÜLLER
MANFRED FÄHNLE

Max-Planck-Institut für Metallforschung, Stuttgart, Germany

CAMBRIDGE
UNIVERSITY PRESS

CAMBRIDGE UNIVERSITY PRESS
Cambridge, New York, Melbourne, Madrid, Cape Town, Singapore, São Paulo, Delhi

Cambridge University Press
The Edinburgh Building, Cambridge CB2 8RU, UK

Published in the United States of America by Cambridge University Press, New York

www.cambridge.org
Information on this title: www.cambridge.org/9780521120470

First published 2003
This digitally printed version 2009

A catalogue record for this publication is available from the British Library

Library of Congress Cataloguing in Publication data
Kronmüller, Helmut.
Micromagnetism and the microstructure of ferromagnetic solids / Helmut Kronmüller,
Manfred Fähnle.
p. cm. – (Cambridge studies in magnetism)
Includes bibliographical references and index.
ISBN 0 521 33135 8
1. Magnetic structure. 2. Ferromagnetic materials – Structure. 3. Microstructure.
I. Fähnle, Manfred, 1951- II. Title. III. Series.
QC754.2.M336 K76 2003
538′.44–dc21 2002036832

ISBN 978-0-521-33135-7 hardback
ISBN 978-0-521-12047-0 paperback

Contents

Acknowledgements

This book has been realized on the basis of research activities over several decades in the field of micromagnetic and microstructural problems of modern magnetic materials.

First of all the authors would like to express their heartfelt thanks to the Max Planck Society, which enabled research work over many years in the interdisciplinary fields of micromagnetism and microstructures.

The authors acknowledge gratefully the work of several generations of numerous diploma and Ph.D. students as well as postdocs, engaged on experimental and theoretical topics, which has contributed to the wide spectrum of this book.

The activities in the exciting field of micromagnetism and microstructures at the Max Planck Institute for Metals Research started with the thesis of one of the authors under his advisors, Ulrich Dehlinger[†] and Alfred Seeger. Later on, intense discussions with colleagues active in the field of microstructures showed the importance of combined investigations of magnetic, mechanical and diffusional properties of ferromagnetic materials.

It has been the privilege of the authors to have had numerous discussions with the pioneers of micromagnetism, W.F. Brown, Jr.[†] and W. Döring. The modern concepts of computational micromagnetism and of magnetoelastic interactions were initiated in discussions during a sabbatical year by W.F. Brown, Jr. at the Max Planck Institute for Metals Research in Stuttgart. Well-known results of these discussions are also in the early papers of La Bonte and Brown and of Alex Hubert[†] and Arno Holz[†].

The authors are very grateful to those friends and colleagues with whom they had many stimulating discussions at conferences and during their stays as guest scientists at the Max Planck Institute for Metals Research in Stuttgart. In particular the authors would like to express their gratitude to A. Aharoni[†], A. Arrott, H. Blythe, J.H.V. Brabers, V.A.M. Brabers, K.H.J. Buschow, H.P. Chang, R.W. Chantrell, S. Chikazumi, R. Coehoorn, J.M.D. Coey, H.A. Davies, T. Egami, J. Fidler, J.J.M. Franse, F.J. Friedländer, H. Fujimori, M. Gibbs,

D. Givord, J.M. González, C.D. Graham, H.-Q. Guo, G.C. Hadjipanayis, I.R. Harris, R. Hasegawa, A. Hernando, G. Herzer, H.-R. Hilzinger, S. Hirosawa, M. Homma, A. Hubert[†], F.B. Humphrey, J.P. Jacubovics[†], D. Jiles, S.N. Kaul, N.S. Kazama, H.R. Kirchmayr, E. Kneller[†], U. Krey, H.K. Lachowicz, J.D. Livingston, F.E. Luborsky, A.P. Malozemoff, T. Masumoto, P.G. McCormick, R.D. McMichael, J. Miltat, K.-H. Müller, M.W. Muller, R.C. O'Handley, M. Okuda, D.I. Paul, J. Pastuchenkov, C. Ross, M. Sagawa, R. Schäfer, T. Schrefl, L. Schultz, D.J. Sellmeyr, R. Skomski, K.J. Strnat[†], T. Suzuki, S. Takahashi, T. Takahashi, Y. Umakoshi, M. Vázquez, D. Wagner, F. Walz, Y.J. Wang, E.P. Wohlfarth[†], H. Zijlstra.

The authors gratefully acknowledge stimulating discussions with Dr Dagmar Goll and her commitment to improving and eliminating errors in the manuscript.

Special thanks are expressed to Mrs Monika Kotz and Mrs Inge Schemminger who wrote the manuscript in LaTeX version, accepting with patience many corrections of the manuscript. The tremendous work of Mrs Therese Dragon and Dr Dagmar Goll who implemented the figures and made many proposals to improve their presentation is gratefully acknowledged.

Finally, the authors would like to thank their wives Sonja and Elke for their patience and continuous support during the writing of this book.

The following figures are published by kind permission of Elsevier Ltd. Global Rights Department, Oxford OX 5, 1DX, UK:

Fig. 1.1 reprinted from H. Kronmüller, 'Recent developments in high-tech magnetic materials', *J. Magn. Magn. Mater.* **140/144**, 971 (1995), 25.

Fig. 5.6 reprinted from H. Kronmüller, 'Micromagnetism in hard magnetic materials', *J. Magn. Magn. Mater.* **7** (1978), 341.

Figs. 6.8, 6.18 reprinted from H. Kronmüller, K.D. Durst and G. Martinek, 'Angular dependence of the coercive field in sintered $Nd_{15}Fe_{77}B_8$ magnets', *J. Magn. Magn. Mater.* **69** (1987), 149.

Figs. 6.25, 6.30, 6.31, 6.48 reprinted from H. Kronmüller, K.D. Durst and M. Sagawa, 'Analysis of the magnetic hardening mechanisms in RE-FeB permanent magnets', *J. Magn. Magn. Mater.* **74** (1988), 291.

Figs. 6.32, 6.35, 6.36 reprinted from D. Goll, M. Seeger and H. Kronmüller, 'Magnetic and microstructural properties of nanocrystalline exchange coupled PrFeB permanent magnets', *J. Magn. Magn. Mater.* **185** (1998), 49.

Figs. 7.6, 8.9, 8.22, 8.26, 8.27, 8.28, 8.31 reprinted from H. Kronmüller, *et al.*, 'Magnetic properties of amorphous ferromagnetic alloys', *J. Magn. Magn. Mater.* **13** (1979), 83.

Figs. 7.4, 7.5, 7.7, 7.10, 7.11 reprinted from H. Kronmüller, 'Theory of the coercive field in amorphous ferromagnetic alloys', *J. Magn. Magn. Mater.* **24** (1981), 159.

Figs. 7.8, 7.18 reprinted from H. Kronmüller and T. Reininger, 'Micromagnetic background of magnetization processes in inhomogeneous ferromagnetic alloys', *J. Magn. Magn. Mater.* **112** (1992), 1. Reprinted from W. Fernengel and H. Kronmüller, 'Magnetization processes within the narrow domain structure of amorphous ferromagnetic alloys', *J. Magn. Magn. Mater.* **37** (1983), 167.

Fig. 7.20 reprinted from N. Murillo and J. Gonzalez, 'Effect of the annealing conditions and grain size', *J. Magn. Magn. Mater.* **218** (2000), 53.

Fig. 8.16 reprinted from H. Kronmüller and J. Ulner, 'Micromagnetic theory of amorphous ferromagnets', *J. Magn. Magn. Mater.* **6** (1977), 52.

Fig. 8.30 reprinted from M. Domann, H. Grimm and H. Kronmüller, 'The high-field magnetization curve of amorphous ferromagnetic alloys', *J. Magn. Magn. Mater.* **13** (1979), 81.

Figs. 9.5, 9.8 reprinted from J. Pastushenkov, A. Forkl and H. Kronmüller, 'Temperature dependence of the domain structure in $Fe_{14}Nd_2B$ single crystals during the spin-reorientation transition', *J. Magn. Magn. Mater.* **174** (1997), 278.

Fig. 9.11 reprinted from J. Pastushenkov, A. Forkl and H. Kronmüller, 'Magnetic domain structure of sintered Fe-Nd-B-type permanent magnets and magnetostatic grain interaction', *J. Magn. Magn. Mater.* **101** (1991), 363.

Fig. 9.7 reprinted from K.D. Durst, and H. Kronmüller, 'Determination of intrinsic magnetic material parameters of $Nd_2Fe_{14}B$ from magnetic measurements of sintered $Nd_{15}Fe_{77}B_8$ magnets', *J. Magn. Magn. Mater.* **59** (1986), 86.

Figs. 9.24b, 9.25 reprinted from B. Hofmann and H. Kronmüller, 'Stress-induced magnetic anisotropy in nanocrystalline FeCuNbSiB alloy', *J. Magn. Magn. Mater.* **152** (1995), 91.

Fig. 9.28 reprinted from H. Kronmüller *et al.*, 'Magnetic properties of amorphous ferromagnetic alloys', *J. Magn. Magn. Mater.* **13** (1979), 53.

Figs. 13.1, 13.2, 13.3, 13.4 reprinted from H.F. Schmidts and H. Kronmüller, 'Size dependence of the nucleation field of rectangular parallelepipeds', *J. Magn. Magn. Mater.* **94** (1991), 220.

Figs. 13.10, 14.6 reprinted from H. Kronmüller *et al.*, 'Micromagnetism and the microstructure in nanocrystalline materials', *J. Magn. Magn. Mater.* **175** (1997), 177.

Figs. 13.24, 13.25 reprinted from T. Schrefl, J. Fidler and H. Kronmüller, 'Remanence and coercivity in isotopic nanocrystalline permanent magnets', *J. Magn. Magn. Mater.* **138** (1994), 15.

Figs. 13.12, 13.22 reprinted from R. Hertel and H. Kronmüller, 'Finite element calculations on the single domain limit of a ferromagnetic cube: a solution to μMAG standard problem no. 3', *J. Magn. Magn. Mater.* **238** (2002), 185.

Figs. 13.14, 13.15, 13.16, 13.17, 13.18, 13.19, 13.20 reprinted from H. Kronmüller and R. Hertel, 'Computational micromagnetism of magnetic structures and magnetization processes in small particles', *J. Magn. Magn. Mater.* **215/216** (2000), 11.

Fig. 13.35 reprinted from T. Schrefl *et al.*, 'The role of exchange and dipolar coupling at grain boundaries in hard magnetic materials', *J. Magn. Magn. Mater.* **124** (1993), 251.

Fig. 13.38 reprinted from R. Fischer and H. Kronmüller, 'The role of grain boundaries in nanoscaled high-performance permanent magnets', *J. Magn. Magn. Mater.* **184** (1998), 166.

Figs. 14.5, 14.9, 14.12, 14.13 reprinted from T. Leineweber and H. Kronmüller, 'Dynamic of magnetization states', *J. Magn. Magn. Mater.* **192** (1994), 575.

Figs. 13.5, 13.6, 13.7 reprinted from H. F. Schmidts, G. Martinek and H. Kronmüller, 'Recent progress in the interpretation of nucleation fields of hard magnetic particles', *J. Magn. Magn. Mater.* **104/107** (1992), 1119.

Fig. 6.33 reprinted from H. Kronmüller *et al.*, 'Magnetization processes in small particles and nanocrystalline materials', *J. Magn. Magn. Mater.* **203** (1999), 12.

Fig. 11.2 reprinted from M. Fähnle and J. Furthmüller, 'Various contributions to magnetostriction in amorphous and polycrystalline ferromagnets', *J. Magn. Magn. Mater.* **72** (1988), 6–12.

Fig. 12.1 reprinted from M. Fähnle and G. Herzer, 'On the correlated molecular field theory of amorphous ferromagnets', *J. Magn. Magn. Mater.* **44** (1983), 274.

Fig. 12.2 reprinted from M. Fähnle, 'Monte Carlo study of phase transitions in bond and site-disordered Ising and classical Heisenberg ferromagnets', *J. Magn. Magn. Mater.* **45** (1984), 279.

Figs. 12.3, 12.4 reprinted from M. Fähnle *et al.*, 'The magnetic phase transition in amorphous ferromagnets and in spin glasses', *J. Magn. Magn. Mater.* **38** (1983), 240.

Fig. 12.8 reprinted from M. Fähnle and G. Herzer, 'On the correlated molecular field theory of amorphous ferromagnets', *J. Magn. Magn. Mater.* **44** (1984), 274.

Fig. 5.8 reprinted from H. Kronmüller and D. Goll, 'Micromagnetic theory of the pinning of domain walls at phase boundaries', *Physica B* **319** (2002), 122.

Fig. 14.8 reprinted from J. Fidler *et al.*, 'Micromagnetic simulations of magnetization reversal in rotational magnetic fields', *Physica B* **306** (2001), 112.

Figs. 14.10, 14.11 reprinted from T. Leineweber and H. Kronmüller, 'Influence of the intrinsic dynamics of the magnetization on the hysteresis loop', *Physica B* **275** (2000), 5.

1

Introduction

In Europe it was the Greek philosophers in the first millennium BC who developed the first ideas about the fascinating properties of the so-called loadstone found as the mineral magnetite, Fe_3O_4, at several places. It was Lucretius Carus [1.1] in the first century BC who attributed the origin of the name 'magnet' to the province of Magnesia where the loadstone mineral was found. Another hypothesis was given by William Gilbert in his famous book entitled *De Magnete* [1.2]. He refers to Pliny [1.2] who derived the name magnet from a shepherd named Magnes who was considered one of the discoverers because the nails of his shoes were covered with splinters of magnetite.

In spite of the early discovery of magnetic materials it took nearly three millenniums until a basic understanding of magnetism was possible. The history of magnetism has been treated by Mattis [1.3] up to the quantum theory of the last century. Here a brief review is given of the main steps leading to our present understanding of magnetism and the topics of this book, which considers the relations between magnetism and the microstructure of solids.

It is to the merit of the Greek philosophers that they added to the four ancient elements, earth, fire, water and air, as a fifth element the magnetic materials. From the beginning of its discovery, magnetism has been discussed controversially: the animists such as Thales and Anaxagoras believed in the divine origin of magnets, assuming loadstones to possess a soul. Another school, the mechanists around Diogenes (460 BC), suggested that the humidity of iron feeds the dryness of the loadstone. A third hypothesis, which we consider as the most advanced one, was developed by Empedocles, Epicurus and Democritus by introducing the effluvia as a type of dynamical field that is responsible for the attracting and repelling forces between magnets.

The darkness of two millenniums was lightened by the experimental activities of monks in the thirteenth and sixteenth centuries. In an early experiment, Petrus Peregrinus (1269) showed that on a spherical loadstone lancette-type pieces of iron

1

aligned parallel to the meridians of the sphere that cross at two points denoted as
the poles of the sphere. Later in the sixteenth century, John Baptista Porta [1.4]
disproved the old idea that there exists a humidity exchange between iron and
loadstone. After a lengthy experiment where iron and loadstone were close together
no significant difference in their weight could be found by Porta.

A big step forward in the knowledge of magnetism was the famous book entitled
De Magnete by William Gilbert published in 1600. In this book Gilbert (1544–
1603) [1.5] summarized the total knowledge of his day, employing the experimental techniques available at this time. He also had close contacts with the famous
astronomers of this era, Galileo and Kepler. His criticism of scientists 'who have
both lost their oil and their pains' because they adopted theories from other authors
without testing them by experiments is also valid in our times, where sometimes
careless experiments induce world-wide activities based on nonreproducible results. William Gilbert was the first magnetician – actually he called himself an
electrician – to describe carefully the industrial production of compass needles
by a blacksmith. Here it should be noted that the use of magnets as compasses
was also known in the Chinese empire from at least the second millennium BC.
A woodcut figure published by Gilbert shows a blacksmith hammering a piece of
glowing iron on an anvil. From an empirical point of view, Gilbert correctly described the entire procedure of obtaining a high-quality compass needle. Today
with our knowledge of magnetism we are able to understand the complex magnetic
and microstructural processes occurring during the production of the needle. These
phenomena in a wider sense will be the topic of this book. The micromagnetism–
microstructure problems to be considered during the compass needle's production
are the following:

1. By hammering the piece of iron, pinning centres for domain walls are introduced by the
   formation of dislocations and a fine-grained microstructure.
2. By aligning the piece of iron parallel to a meridian from South to North, the earth's field
   induces a magnetization when cooling below the Curie point.
3. The induced remanence is rather large due to the Hopkinson maximum of the susceptibility just below the Curie temperature.
4. Pinning of domain walls and stabilization of the induced remanence takes place at lower
   temperatures due to the increased magnetoelastic interactions.

It is of interest that Gilbert used a knowledge of astronomy in order to orient
the hammered piece of iron with respect to the Pole Star in the constellation of the
Little Bear, named Septembrio in the woodcut picture.

With Gilbert's ideas the door was thrown wide open for theoreticians such as
Descartes (1596–1660) [1.6] who took over Gilbert's proposal that the terrella
of Peregrinus was a model of the earth's field. The thinking of Descartes about

science may be considered as the transition from metaphysical thinking to that of the modern natural sciences. With the separation of body and soul he propagated the independent study of nature – the age of enlightenment, the renaissance – that had been initiated by scientists such as Gilbert (1544–1603), Galileo (1564–1642), Kepler (1571–1630), Newton (1642–1726) and Leibnitz (1646–1716). The latter two were the inventors of the infinitesimal calculus around 1680.

On the basis of the new scientific thinking in the late eighteenth century, Coulomb (1736–1806) showed by means of a torsion balance method that under certain conditions, using long needles, the magnetic forces follow a similar law as electric forces, i.e., the so-called Coulomb law. He succeeded in relating magnetic forces to mechanical forces which allowed a definition of magnetic charges and magnetic fields in terms of mechanical quantities (cm, g, s). Coulomb also was aware of the fact that magnetic charges cannot be separated, unlike electric charges. Nevertheless, he was a defender of the two-fluid theory supported by the ingenious mathematician Poisson (1788–1840) who introduced the concept of static scalar potentials.

After Coulomb's pioneering experiments, in the nineteenth century two further revolutionary and innovative discoveries led to the famous Maxwell equations (1831–1879) [1.7] of which Boltzmann (1844–1906) said, 'Was it God who wrote the lines?'

The first of these two discoveries was Oersted's (1777–1851) experiment where he showed that electrical currents produce magnetic fields which move a compass needle. Herewith the long-sought connection between electricity and magnetism became obvious in 1820. The quantitative description of Oersted's experiment is due to Biot and Savart. The second revolutionary discovery was due to Faraday (1791–1867). He found that all materials are magnetic, either para- or diamagnetic, and that all materials reveal magneto-optical phenomena, known as the Faraday effect or the Kerr effect. But his most innovative discovery was the law of induction in which he showed that the rate of change of the magnetic flux density, $\partial B/\partial t$, induces an electrical voltage in a closed wire. Maxwell finally succeeded in describing all electrical and magnetic phenomena, static and dynamic, by a unique set of equations. He introduced several types of magnetic fields by the equation

$$B = \mu_0 H + J = \mu_0 H + \mu_0 M = \nabla \times A, \qquad (1.1)$$

with B the magnetic flux density, H the magnetic field strength, J the magnetic polarization of the material, M its magnetization and A the vector potential. In vacuum $B = \mu_0 H$ holds with $\mu_0 = 4\pi \cdot 10^{-7}$ Vs/Am corresponding to the so-called vacuum permeability. According to eq. (1.1) the magnetic flux density, B, derives from a vector potential, A, thus B is solenoidal, fulfilling the Maxwell equation div $B = 0$. Solutions of Maxwell's equations require knowledge of the magnetic materials law

$$B = \mu_0\mu H = \mu_0 H + J \tag{1.2}$$

or

$$J = (\mu - I)\mu_0 H; \qquad M = (\mu - I) \cdot H = \chi \cdot H, \tag{1.3}$$

with μ the relative permeability tensor of rank two depending nonlinearly on the field H and I the unit tensor. Sometimes the quantity $\mu_0\mu$ is considered as the permeability, but introducing the relative permeability allows the definition of a dimensionless susceptibility $\chi = \mathrm{d}M/\mathrm{d}H$. In Maxwell's theory μ denotes a global property of a macroscopic specimen which has to be determined experimentally.

Calculation of permeabilities of para- and diamagnetic solids from first principles became possible with the discovery of the magnetic properties of the electron in the late nineteenth century and the development of quantum theory during the first three decades of the twentieth century. During this time, Pierre Curie (1859–1906) [1.8] studied the susceptibility of paramagnetic substances, Langevin (1872–1946) [1.9] explained diamagnetism, and Pierre Weiss (1865–1940) [1.10] developed molecular field theory and gave a first interpretation of the Curie–Weiss law. Weiss also presented the first ideas on magnetic domains, also called Weiss domains, in order to explain the global demagnetized state of ferromagnetic materials. Weiss's original suggestion that atomic magnetic moments should be integer multiples was finally confirmed by Pauli [1.11] on the basis of Bohr's atom model by introducing the Bohr magneton,

$$\mu_{\mathrm{B}} = 9.2740 \cdot 10^{-24} \text{ joule/tesla, } [\mathrm{Am^2}],$$

which was five times larger than the Weiss magneton. All these advances, however, did not explain the high permeabilities ranging from 1 to 10^6 and high susceptibilities measured in ferromagnetic materials. Further progress came from the development of quantum theory by Schrödinger and from Goudsmit and Uhlenbeck [1.12] who showed that the mechanical spin of the electron was $1/2$ and as a consequence the g-factor of the electron was $g = 2$. Because of their spin, $S = 1/2$, electrons behave as fermions and are subject to the Pauli and antisymmetry principles. This fact led to the explanation of ferromagnetism by Dirac [1.13] and Heisenberg [1.14] by means of the exchange energy which results from Pauli's exclusion principle and the overlap of electronic wave functions. With the introduction of the exchange energy and later on of the spin–orbit coupling energy, which led to the interpretation of the famous Hund's rules [1.15], the door was opened for a quantitative electron theoretical calculation of the intrinsic magnetic properties of spin-ordered media. It took, however, the full twentieth century to develop numerical methods to solve the many particle problem of interacting fermions [1.3].

Two main streams have been developed: the many particle theories often based on the Hubbard model [1.16] and the density functional theory [1.17] in the local-spin-density approximation and beyond [1.18]. Whereas the main lines of the quantum theoretical background for the calculation of intrinsic magnetic properties were available in the second and third decades of the twentieth century, a quantitative description of the material law, $B = \mu_0 \mu H$, and of the hysteresis loop had only just started to become a matter of microscopic theory at this time.

Maxwell's theory is purely phenomenological as global permeabilities have to be used without the possibility to derive them from first principles. On the other hand, quantum theory describes magnetic properties on an atomic level, and its application for the treatment of cooperative phenomena such as magnetization processes and hysteresis loops has not been successful so far. The need for a theory bridging the gap between Maxwell's theory and quantum theory became rather urgent after two experimental landmarks. In 1919 Barkhausen [1.19] confirmed the Weiss hypothesis of the existence of magnetic domains. He showed that the magnetization process in ferromagnets takes place discontinuously by so-called Barkhausen jumps. Later on, Sixtus and Tonks [1.20] showed that these Barkhausen jumps take place by a spontaneous displacement of a domain boundary. In 1931 Bitter [1.21] published the first pictures of magnetic domains by using the interaction between magnetic iron oxide particles and the magnetic stray fields exerted by domains and domain walls.

Theoretical landmarks towards the development of a mesoscopic theory are the first description of a domain wall by Bloch [1.22], who showed that due to the exchange energy the transition regions between the domains have a finite width, and the analysis of domain structures by Landau and Lifshitz [1.23], who introduced a continuum expression for the exchange energy and gave a first interpretation of domain patterns, which are mainly governed by the tendency to minimize predominantly the stray field energy. For the further development of a microscopic theory for a mesoscopic length scale (>1 nm) the book by Becker and Döring [1.24] became an important prerequisite. In 1939 Becker and Döring summarized the entire knowledge on ferromagnetism and gave a rigorous derivation of the magnetoelastic coupling energy for cubic crystals.

Our knowledge of microstructures in solids became important impulses during the 1930s. In 1939 Burgers [1.25] defined the dislocation as a singularity of the elastic displacement field. Hence the sources of internal stresses were discovered and in numerous articles this field became the key for understanding crystal growth and many microstructural properties such as grain structures, plasticity, epitaxial growth, radiation damage and fatigue.

William Fuller Brown Jr. [1.26, 1.27] realized the importance of the new concepts on microstructures and published in 1940 and 1941 two basic papers from which the modern theory of micromagnetism emerged. There were further papers by Kittel, Stoner, Wohlfarth, Néel, Aharoni, Shtrikman and many others who in the 1950s established the theory of micromagnetism as an efficient tool to understand the characteristic properties of magnetization processes and of the hysteresis loop. Parallel to the progress in micromagnetism, the continuum theory of dislocations and point defects was further developed and the application of transmission electron microscopy by Hirsch *et al.* [1.28] finally gave a direct proof of the existence of dislocations. Fundamental reviews on dislocations and point defects are those by Seeger [1.29], Kröner [1.30], Kosevich [1.31], Bullough and Tewary [1.32] and Leibfried and Breuer [1.33].

The present book gives a review of current knowledge of the role of microstructures with respect to the characteristic parameters of hysteresis loops. A unique theory of this topic has become possible by the interdisciplinary combination of different fields of research as shown in Fig. 1.1.

1. Material synthesis and constitution.
2. Microstructural characterization of alloys and compounds.
3. Magnetic properties – microstructure relations.

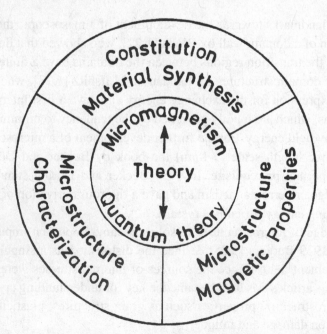

Fig. 1.1. Scheme of interdisciplinary cooperation for developing and analysing magnetic materials.

These three fields of research give the input to the theory of micromagnetism resulting in a unique description of magnetic phenomena.

In Chapters 2–5 the micromagnetic background for the description of magnetization–microstructure interactions is presented. Chapter 2 summarizes the relevant energy terms with special emphasis on the magnetostrictive terms. Chapter 3 gives a derivation of the static micromagnetic equilibrium conditions and a definition of the exchange lengths. Chapter 4 presents applications of the micromagnetic equations to domain walls taking into account the role of magnetoelastic energy terms. In Chapter 5 the interaction between domain walls and lattice defects is described for point defects, dislocations and planar defects such as grain boundaries, phase boundaries, antiphase boundaries and planar precipitations.

Chapters 6–10 deal with the application of micromagnetism to modern magnetic materials. Chapter 6 reviews the high-coercivity rare earth–transition metals intermetallic compounds which are the basis of high-quality supermagnets. Here the micromagnetic concepts from the Stoner–Wohlfarth theory to the basic theory of nucleation fields in ideal and perturbed materials are presented comprehensively. The current theories are applied for the analysis of modern systems of permanent magnets. Chapter 7 gives a review of the statistical pinning theory of domain walls with its application to dislocations, point defects and amorphous alloys. Chapter 8 deals with the law of approach to saturation with its application to point defects, dislocations and amorphous alloys. Chapter 9 reviews the influence of microstructures on domain patterns. Here the validity of micromagnetic concepts is tested for many types of domain patterns in hard and soft magnetic materials, in particular amorphous alloys. The effect of dislocations on domain patterns in crystalline materials is reviewed for nickel and iron which may be considered as characteristic examples of soft magnetic materials. Chapter 10 discusses the relaxation phenomena in amorphous alloys on the basis of two-level systems and atomic defects and gives an analysis of reversible and irreversible relaxation processes.

Chapters 11 and 12 are devoted to intrinsic properties of amorphous alloys where the microstructure plays a dominant role. The complex property of magnetostriction of amorphous alloys is treated on the basis of the elasticity theory of inhomogeneous media in Chapter 11. In Chapter 12 the magnetic materials laws in the transition regime from critical behaviour to mean field behaviour are discussed within the framework of a correlated mean field theory for second-order phase transitions.

Chapters 13 and 14 deal with the recent development of numerical micromagnetism. Chapter 13 presents results on magnetic configurations and magnetization processes in small particles and thin platelets using the method of finite elements. Critical fields and critical size parameters are determined for specimens with rectangular and spherical shapes which cannot be treated by analytical micromagnetism. Chapter 14 deals with numerical solutions of the time-dependent Landau–Lifshitz equation and presents numerical results for the relaxation times determining the reversal of magnetization in small particles.

References

[1.1] Carus, Lucretius, First century BC, translated by Th. Creech, London 1714.

[1.2] Pliny, reference by W. Gilbert: *De Magnete*, trans. by Gilbert Club, London 1900 (rev. edn. Basic Books, New York 1958) p. 8.

[1.3] Mattis, D.C., 1981, *The Theory of Magnetism I* (Springer-Verlag, Berlin–Heidelberg–New York).

[1.4] Porta, J.B., 1589, *Natural Magick (Naples)*; reprint of 1st English edition (Basic Books, New York 1957) p. 212.

[1.5] Gilbert, W., 1600, *De Magnete*, trans. by Gilbert Club, London 1900 (rev. edn. Basic Books, New York 1958).

[1.6] Descartes, R., 1644, *Principia Philosophiae*, Part IV, p. 133–183.

[1.7] Maxwell, J.C., 1873, *A Treatise on Electricity and Magnetism* (reprinted by Dover, New York 1954); 1870, *Trans. R. Soc. Edinburgh* **26**, 1.

[1.8] Curie, P., 1895, *Ann. Chim. Phys.* **5**, 289.

[1.9] Langevin, P., 1905, *Ann. Chim. Phys.* **5**, 70; 1905, *J. Phys.* **4**, 678.

[1.10] Weiss, P., 1907, *J. de Phys.* **6**, 661.

[1.11] Pauli, W., 1925, *Z. Physik* **31**, 765; in *Le Magnetisme*, 6th Solvay Conf. (Gauthiers-Villars, Paris 1932) p. 212.

[1.12] Uhlenbeck, G., and Goudsmit, S., 1925, *Naturwiss.* **13**, 593.

[1.13] Dirac, P.A.M., 1928, *Proc. R. Soc.* **117A**, 610; *ibid.* 1930, **126A**, 360.

[1.14] Heisenberg, W., 1928, *Z. Physik* **49**, 619.

[1.15] Hund, F., 1927, *Linienspektren und periodisches System der Elemente* (Springer, Berlin).

[1.16] 'Metallic Magnetism', 1980, *Topics in Current Physics* **42** (Ed. H. Capellmann, Springer-Verlag, Berlin–Heidelberg–New York).

[1.17] Dreizler, R.M., and Gross, E.K.U., 1990, *Density Functional Theory* (Springer-Verlag, Berlin).

[1.18] 'Strong Coulomb Correlations in Electronic Structure Calculations', 2000, *Advances in Condensed Matter Science*, Vol. 1 (Ed. V.I. Anisimov, Gordon and Breach Science Publishers).

[1.19] Barkhausen, H., 1919, *Phys. Z.* **20**, 401.

[1.20] Sixtus, K.J., and Tonks, L., 1931, *Phys. Rev.* **37**, 930.

[1.21] Bitter, F., 1931, *Phys. Rev.* **38**, 1903; *ibid.* 1932, **41**, 507.

[1.22] Bloch, F., 1932, *Z. Phys.* **74**, 295.

[1.23] Landau, L.D., and Lifshitz, E., 1935, *Phys. Z. Sowjetunion* **8**, 153.

[1.24] Becker, R., and Döring, W., 1939, *Ferromagnetismus* (Springer, Berlin).

[1.25] Burgers, J.M., 1939, *Proc. Kon. Nederl. Akad. Wetensch.* **42**, 293, 378.

[1.26] Brown, Jr., W.F., 1940, *Phys. Rev.* **58**, 736.

[1.27] Brown, Jr., W.F., 1941, *Phys. Rev.* **60**, 139.

[1.28] Hirsch, P.B., Partridge, P.G., and Segall, R.L., 1959, *Philos. Mag.* **4**, 721.

[1.29] Seeger, A., 1958, 'Kristallplastizität'. In *Handbuch der Physik*, Vol. 7/2 (Springer-Verlag, Berlin) p. 1; 1955, 'Theorie der Gitterfehlstellen'. In *Handbuch der Physik*, Vol. 7/1 (Springer-Verlag, Berlin) p. 383.

[1.30] Kröner, E., 1958, *Kontinuumstheorie der Versetzungen und Eigenspannungen* (Springer-Verlag, Berlin–Göttingen–Heidelberg).

[1.31] Kosevich, A.M., 1979, 'Crystal Dislocations and the Theory of Elasticity'. In *Dislocations in Solids*, Vol. 1 (Ed. F.R.N. Nabarro, North-Holland, Amsterdam) p. 33.

[1.32] Bullough, R., and Tewary, V.K., 1979, 'Lattice Theories of Dislocations'. In *Dislocations in Solids*, Vol. 2 (Ed. F.R.N. Nabarro, North-Holland, Amsterdam) p. 1.

[1.33] Leibfried, G., and Breuer, N., 1978, *Point Defects in Metals*, Vol. 1 (Springer-Verlag, Heidelberg).

Appendix

Units of magnetic properties

SI units (Système International d'Unités) have been adopted by the National Bureau of Standards with the definition of the magnetic flux density (induction)

$$B = \mu_0 H + J,$$

or

$$B = \mu_0 H + \mu_0 M,$$

with

$$J = \mu_0 M,$$

where H denotes the magnetic field strength, J the magnetic polarization, M the magnetization and μ_0 the vacuum permeability. The material law

$$B = \mu_0 \mu \cdot H$$

defines the relative permeability tensor μ as a dimensionless quantity. Similarly, the susceptibility

$$\chi = dM/dH$$

defines the dimensionless susceptibility tensor. The defining equation in Gaussian units is

$$B = H + 4\pi M,$$

with H the magnetic field strength, M the magnetization and $4\pi M$ the polarization.

Table 1.1. *Units and constants (constants taken from Cohen, E.R., and* *Giacomi, P., 1987,* Physica *146A, 1).*

Quantity	Symbol	SI unit (MKSA)	Conversion to Gaussian units (cgs unit)
magnetic charge density	ρ	A/m^2	$4\pi \cdot 10^{-5}$ Oe/cm
magnetic flux	ϕ	V s	10^8 Mx
magnetic flux density (induction)	B	T, $V s/m^2$	10^4 G
magnetic field strength	H	A/m	$4\pi \cdot 10^{-3}$ Oe
polarization	J	T	10^4 G
magnetization	M	A/m	$4\pi \cdot 10^{-3}$ Oe
magnetic potential (magnetomotive force)	U	A	$4\pi/10$ Gilbert
energy densities	ϕ_i'	J/m^3	10 erg/cm^3
specific wall energy	γ_B	J/m^2	10^3 erg/cm^2
exchange stiffness constant	A	J/m	10^5 erg/cm
spin wave stiffness constant	D	$J m^2$	$0.624\,18 \cdot 10^{42}$ meV Å^2
magnetic torque density	L	J/m^3	10 erg/cm^3
vacuum permeability	$\mu_0 = 4\pi \cdot 10^{-7}$	$V s/(A m)$	1
relative permeability	μ	1	μ
susceptibility	χ	1	χ
demagnetization factor	N	$1; \sum N_i = 1$	$1; \sum N_i = 4\pi$
Rayleigh constant	α_R	T^{-1}, $m^2/(V s)$	10^{-4} G^{-1}
Bohr magneton: $\Phi = \mu_B B$ (Ampère's definition)	$\mu_B = |e|\, \hbar/(2mc)$	$9.2740 \cdot 10^{-24}$ J/T, V A s/T, A m^2	$9.2740 \cdot 10^{-21}$ erg/G
Bohr magneton: $\Phi = \mu_B H$ (Coulomb's definition)	$\mu_B = |e|\, \hbar/(2m)$	$1.1654 \cdot 10^{-29}$ J/(A/m), V s m	$9.2740 \cdot 10^{-21}$ erg/Oe
magnetic moment	μ	V s m, T m^3	$1/(4\pi) \cdot 10^{10}$ G cm^3
gyromagnetic ratio ($\omega = -\gamma_0 H$)	$\gamma_0 = -g|e|/(2m)$	$-1.1051\, g \cdot 10^5$ m/(A s)	$\gamma_0 = -g|e|/(2mc) = -0.8794\, g \cdot 10^7$ 1/(Oe s)
Landé factor	g	2.003 193 04 electron spin	2.003 193 04
electronic charge	e	$-1.6021 \cdot 10^{-19}$ C	$-1.6021 \cdot 10^{-20}$ emu, g$^{1/2}$ cm$^{1/2}$
electronic rest mass	m	$9.1093 \cdot 10^{-31}$ kg	$9.1093 \cdot 10^{-28}$ g
velocity of light in empty space	c	$2.9979 \cdot 10^8$ m/s	$2.9979 \cdot 10^{10}$ cm/s
Planck constant	h	$6.6260 \cdot 10^{-34}$ J s	$6.6260 \cdot 10^{-27}$ erg s
	$\hbar = h/(2\pi)$	$1.0545 \cdot 10^{-34}$ J s	$1.0545 \cdot 10^{-27}$ erg s
Boltzmann constant	k_B	$1.3806 \cdot 10^{-23}$ J/K	$1.3806 \cdot 10^{-16}$ erg/K
mechanical force	P	N	10^5 dyn, 0.1 kp
elastic stress	σ	Pa, N/m^2	10 dyn/cm^2
elastic strain	ε	1	ε
elastic constant	c	Pa, N/m^2	10 dyn/cm^2, 10^{-7} kp/mm^2
Young modulus	E	Pa, N/m^2	10 dyn/cm^2, 10^{-7} kp/mm^2
shear modulus	G	Pa, N/m^2	10 dyn/cm^2, 10^{-7} kp/mm^2
Poisson ratio	ν	1	ν

V s (1 weber); T, $V s/m^2$ (1 tesla); A (1 ampere); Mx (1 maxwell); V A s, N m (1 joule); 1/m (1 ampere winding per m); G (1 gauss); Oe (1 oersted); C (1 coulomb); Pa (1 pascal) (N/m^2); g cm s^{-2} (1 dyn); $1\,eV = 1.602\,18 \cdot 10^{-19}$ joule, $1\,J = 10^7$ erg; 1 emu/cm$^3 = 10^3$ A/m; N (1 newton) (10^5 dyn); 1 kp = 10^3 p = 10^6 dyn.

2
Magnetic Gibbs free energy

2.1 Introductory remarks

In the 1940s W.F. Brown [2.1, 2.2] published his pioneering papers on the theory of micromagnetism of ferromagnetic materials and its application to the law of approach to ferromagnetic saturation. This continuum theory of ferromagnetism closed the gap between the microscopic quantum theoretical description of spin structures and Maxwell's theory of electromagnetism where the material properties are described by volume averaged permeabilities and susceptibilities. Brown's theory of micromagnetism was based on two other preceding developments which may be considered as the fundamental prerequisites of micromagnetism: the description of the exchange energy by a continuum theoretical expression derived by Landau and Lifshitz [2.3] and the derivation of the magnetoelastic coupling energy by Becker and Döring [2.4] enabling the description of magnetic and mechanical interactions. The big success of the theory of micromagnetism was due to the analysis of domain structures in small particles and thin films, the analysis of nucleation problems and magnetization processes, the calculation of the spin structures of domain walls and spin singularities, the interpretation of the characteristic parameters of the hysteresis loops, and in this context the interaction between defect structures and domain walls. The computer capabilities available nowadays make it possibility to use numerical methods to solve problems which had been unsolvable up to now.

Whereas previously the theory of micromagnetism was exclusively applied to crystalline materials based on the transition metals and garnets, the development of a tremendous number of new ferromagnetic materials, such as nanocrystalline and amorphous alloys as well as intermetallic compounds, has opened a further field of interesting applications where static fluctuations of intrinsic microscopic magnetic properties become important for an understanding of primary magnetic properties such as the Curie temperature, the temperature dependent spontaneous

11

magnetization, critical exponents, magnetic anisotropies and magnetostriction. Also secondary magnetic properties such as hysteresis loops, induced anisotropies and magnetic after-effects of these new materials may be interpreted on the basis of the theory of micromagnetism.

Different aspects of the conventional theory of micromagnetism have been presented in several review papers and books by Kittel [2.5], Brown [2.6], Kronmüller [2.7], Träuble [2.8], Aharoni [2.9], Aharoni and Shtrikman [2.10], Chikazumi [2.11], Morrish [2.12], Jiles [2.13], Hubert and Schaefer [2.14], and O'Handley [2.15]. Here we shall present the theory of micromagnetism in relation to the microstructure of crystalline alloys, intermetallic compounds and amorphous alloys. In the limit of an effective material constants theory the conventional results of the theory of micromagnetism are contained within the framework of this representation, i.e., primary material parameters in this limit are considered as space independent material properties. To take care of the real microstructure, however, in the case of nanocrystalline and amorphous materials spatially varying material parameters have to be considered.

Another new group of ferromagnetic materials are the hard magnetic, rare earth–transition metal intermetallic compounds ($SmCo_5$, Sm_2Co_{17}, $Nd_2Fe_{14}B$, $Sm_2Fe_{17}N_3$, $REFe_{12}$) characterized by extremely large magnetocrystalline anisotropies ($10^6 \, J/m^3$ – $10^7 \, J/m^3$) [2.16], where nucleation processes have become a key for the understanding of their hard magnetic properties and for further developments of high-tech permanent magnets.

Giant-magnetoresistance (GMR) thin films have become a prominent example of modern magnetism [2.17]. In these GMR systems, which are now widely used in magnetic recording as reading devices, spin structures and magnetization processes investigated by computational micromagnetism play an important role. The development during recent years clearly proves that the application of computational micromagnetism for the determination of magnetic ground states and the dynamics of magnetization processes in small particles and thin film systems has become an important tool for optimizing magnetic recording systems.

Giant-magnetostrictive alloys based on rare earth–transition metal alloys have also become interesting materials for micromechanical systems [2.18].

A further important field of research is the development of a thermodynamic theory of the temperature dependence of the spontaneous magnetization $M_s(T)$. The role of microstructure in the temperature dependence of the magnetization near the transition temperature has been determined by an expansion of the conventional Ginzburg–Landau theory to the so-called correlated molecular field theory. This latter theory gives a unique description of the second-order phase transition of disordered and amorphous alloys characterized by local fluctuations of the magnetic properties.

The basis for all these investigations is the magnetic Gibbs free energy which is composed of a series of energies to be discussed in the following.

2.2 Magnetic energy terms

As in crystalline materials, in disordered media the magnetic state is determined by the direction cosines $\gamma_i(r)$ of the spontaneous polarization, $J_s(r)$, or of the spontaneous magnetization, $M_s(r)$, defined as

$$J_s(r) = \mu_0 M_s(r) = |J_s(r)| \left(\gamma_1(r) i_1 + \gamma_2(r) i_2 + \gamma_3(r) i_3 \right), \tag{2.1}$$

with i_k the unit vectors of a cartesian coordinate system and where in general the modulus of the spontaneous polarization, $|J_s(r)|$, depends on position r. The local magnetization, $|M_s(r)|$, is related to the z-component, S_z, of spin, S, of the magnetic ion at position r,

$$M_s(r) = g\mu_B S_z(r)/\Omega(r), \tag{2.2}$$

with $g \simeq 2$ the Landé factor, μ_B = the Bohr magneton and $\Omega(r)$ the atomic volume per magnetic ion. The local magnetic moment is given by $\mu_z = g\mu_B S_z$.

Magnetic energy terms are those which depend on the polarization, $J_s(r)$. Depending on the type of problem, we have to determine either the local direction cosines, $\gamma_i(r)$, as in the case of domain walls, or the spatial averages, $\langle \gamma_i(r) \rangle_r$, or even spatial and thermal averages, $\langle \langle \gamma_i(r) \rangle_r \rangle_T$.

As a starting point for these investigations we have to consider the Gibbs free energy, ϕ_t, of a ferromagnet with the free variables, temperature T, elastic stress tensor σ, and applied magnetic field H_{ext}. In terms of energy densities we may write

$$\phi_t' = U - T \cdot S - \sigma \cdot \varepsilon - J_s \cdot H_{ext}, \tag{2.3}$$

where U, S, ε and σ denote the internal energy density, the entropy per unit volume, the strain tensor and the stress tensor, respectively. The internal energy includes exchange, anisotropy, dipolar and elastic energy terms. The last two terms correspond to elastic and magnetoelastic interaction energies. From thermodynamics it is well known that from an initial nonequilibrium state the Gibbs free energy can only decrease, and in the thermodynamic equilibrium the total Gibbs free energy,

$$\phi_t = \int \phi_t' \, dV, \tag{2.4}$$

obeys the equilibrium condition

$$\delta\phi_t = 0, \tag{2.5}$$

where for constant T, σ and H_{ext} the variation has to be performed with respect to the internal magnetic variables of the system, corresponding to the components, $J_{\text{s},i}$, or the direction cosines γ_i of the spontaneous magnetization, and to the strain tensor ε. In this context we shall not consider the effect of rotations of small volume elements due to local magnetic torques resulting from the anisotropy energy. A detailed discussion of magnetic torques as a source of local elastic stresses will be given in relation to the derivation of the magnetostriction of amorphous alloys in Chapter 12.

The internal energy density contains all intrinsic magnetic energy terms such as exchange energy, magnetocrystalline anisotropy energy, magnetic dipolar energy and the energy of deformations. The continuum theoretical expressions for the internal energy terms may be derived either from symmetry considerations or from the quantum mechanical expressions of magnetic interactions by replacing the spin variable S by the magnetization according to eq. (2.2). The quantum mechanical Hamiltonians of these energy terms for a system of localized spins S_i at positions r_i in the following are used to derive continuum theoretical expressions.

2.2.1 Exchange energy

For the derivation of a continuum theoretical expression of the exchange energy we start from the Heisenberg exchange Hamiltonian, also known as the vector model, where the exchange interaction between an ensemble of spins, S_i, is described by the following expression [2.19]:

$$H_{\text{ex}} = -2 \sum_{i \neq j} J_{ij}\left(r_{ij}\right) S_i\left(r_i\right) \cdot S_j\left(r_j\right). \tag{2.6}$$

J_{ij} denotes the exchange integral between the ions of spins S_i and S_j at positions r_i and r_j with $r_{ij} = r_j - r_i$. A continuum theoretical expression of the exchange energy is either obtained by symmetry considerations [2.3, 2.20] or by an expansion of eq. (2.6) into a Taylor series. Kittel and Herring [2.20] have chosen $(\text{grad } M_{\text{s}})^2$ to be the correct description of the excess exchange energy due to inhomogeneities of spin distributions, because the expressions div M_{s} or curl M_{s} would lead to a vanishing exchange energy in the case of zero divergence and zero curl.

2.2.1.1 Short-range exchange interactions

Equation (2.6) in principle contains all exchange interactions between the electrons. In the case of localized electrons with small variations of the spin orientation between neighbouring atoms and only nearest neighbour interactions, J_0, the exchange

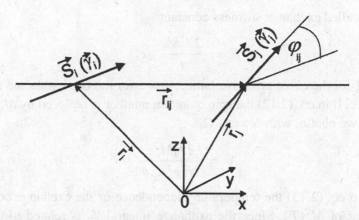

Fig. 2.1. Geometry of two neighbouring spins and their coordinates.

energy density may be written as

$$\phi'_{ex}(\boldsymbol{r}_i) = -2 S^2 J_0 \frac{1}{\Omega} \sum_{j \neq i}^{nn} \cos \varphi_{ij}, \qquad (2.7)$$

where the sum extends only over z_0 nearest neighbour interactions and φ_{ij} denotes the angle between neighbouring magnetic moments. Since $\cos \varphi_{ij}$ is given by (see Fig. 2.1)

$$\cos \varphi_{ij} = \gamma_{1,i}(\boldsymbol{r}_i)\,\gamma_{1,j}(\boldsymbol{r}_j) + \gamma_{2,i}(\boldsymbol{r}_i)\,\gamma_{2,j}(\boldsymbol{r}_j) + \gamma_{3,i}(\boldsymbol{r}_i)\,\gamma_{3,j}(\boldsymbol{r}_j), \qquad (2.8)$$

and $\gamma_{n,i}(\boldsymbol{r}_j)$ deviates only slightly from $\gamma_{n,i}(\boldsymbol{r}_i)$, we may develop $\gamma_{n,i}(\boldsymbol{r}_j)$ into a Taylor series,

$$\gamma_{n,j}(\boldsymbol{r}_j) = \gamma_{n,i}(\boldsymbol{r}_i) + (\boldsymbol{r}_{ij} \cdot \nabla)\,\gamma_{n,i}(\boldsymbol{r}_j) + \frac{1}{2}(\boldsymbol{r}_{ij} \cdot \nabla)^2\,\gamma_{n,i}(\boldsymbol{r}_i). \qquad (2.9)$$

Since in summation the linear terms and the nondiagonal terms of the quadratic terms disappear we obtain finally

$$\phi'_{ex}(\boldsymbol{r}_i) = -\frac{1}{\Omega}2S^2 J_0 \cdot z_0 + \frac{J_0 S^2}{3\Omega} \sum_{j \neq i}^{nn} \sum_{n=1}^{3} r_{ij}^2 \cdot (\nabla \gamma_{n,i})^2 + \cdots, \qquad (2.10)$$

where r_{ij} denotes the nearest neighbour distances. Here the first term corresponds to the exchange energy of a homogeneously magnetized material whereas the second term takes into account magnetic inhomogeneities. In the case of cubic lattices with edge length a of the cube and the continuum variable $\gamma_n(\boldsymbol{r})$, eq. (2.10) gives for this inhomogeneity term

$$\phi'_{ex}(\boldsymbol{r}) = A \sum_{n=1}^{3} (\nabla \gamma_n(\boldsymbol{r}))^2, \qquad (2.11)$$

with the so-called exchange stiffness constant

$$A = \frac{2J_0 S^2}{a} \cdot c, \tag{2.12}$$

where $c = 1$ for the cubic primitive lattice, $c = 2$ for the bcc lattice and $c = 4$ for the fcc lattice. If in eq. (2.12) the spin quantum number is replaced by M_s as given by eq. (2.2) we obtain, with $N = 1/\Omega$,

$$A = \frac{2J_0 M_s^2 c}{N^2 g^2 \mu_B^2 a}. \tag{2.13}$$

According to eq. (2.13) the temperature dependence of the exchange constant is given by that of $M_s^2(T)$. Since the exchange integral J_0 is related to the Curie temperature, T_c, within the framework of molecular field theory (see e.g., [2.12]) eq. (2.13) may also be written as [2.7]

$$A = \frac{3k_B T_c S \cdot c'}{2a(S+1)}, \tag{2.14}$$

with $c' = 1/6, 1/4, 1/3$ for sc, bcc and fcc lattices. Equation (2.11) holds for cubic crystals. In the case of uniaxial crystals such as hexagonal, tetragonal or trigonal lattices, the exchange energy density may be written as

$$\phi'_{ex} = A_\perp \sum_{i=1,2} (\nabla \gamma_i)^2 + A_\parallel (\nabla \gamma_3)^2, \tag{2.15}$$

where $i = 1, 2$ refer to the coordinate axes perpendicular to the preferred symmetry axis taken parallel to the z-axis. The exchange constants $A_\perp, A_\parallel$ in the case of hexagonal close-packed lattices are related to the corresponding exchange integrals and lattice constants as follows (a = lattice constant in the basal plane, c = lattice constant along the c-axis):

$$A_\perp = \frac{2J_\perp S^2}{c} \frac{8 \cdot \sqrt{3}}{3} \left(\frac{1}{3} + \frac{c^2}{4a^2} \right), \qquad A_\parallel = \frac{2J_\parallel S^2}{c} \frac{8 \cdot \sqrt{3}}{3}. \tag{2.16}$$

For ideal close packed lattices with $c/a = (2/3)\sqrt{6}$, $A_\perp = A_\parallel = \frac{J_0 S^2}{a} 4 \cdot \sqrt{2}$ holds.

In the case of an orthorhombic lattice, ϕ'_{ex} is given by

$$\phi'_{ex} = A_1 (\nabla \gamma_1) + A_2 (\nabla \gamma_3)^2 + A_3 (\nabla \gamma_3)^2, \tag{2.17}$$

with the exchange constants for a simple orthorhombic lattice,

$$A_1 = \frac{2J_a S^2}{bc} a, \qquad A_2 = \frac{2J_b S^2}{ac} b, \qquad A_3 = \frac{2J_c S^2}{ab} c, \tag{2.18}$$

where a, b, c denote the edge lengths of the orthorhombic unit cell and J_a, J_b and J_c the exchange integrals between nearest neighbours in the direction of

Table 2.1. *Intrinsic magnetic material parameters of transition metals and intermetallic compounds. The exchange stiffness constant, A, is obtained either from the spin wave stiffness constant (eq. (8.93)), or from the specific wall energy (eq. (4.13)).*

Magnet	J_s [T]	T_c [K]	D_{sp} [meV Å^2] $^*\gamma_B$ [mJ/m^2]	A [pJ/m]
α-Fe	2.185	1043	281 [2.21]	20.7
			310 [2.22]	22.8
			266 [2.22]	
Co	1.79	1403	490 [2.23]	30.2
			510 [2.24]	31.4
Ni	0.62	627	340 [2.25]	7.2
			400 [2.26]	8.5
Ni$_3$Fe	1.1	873	300 [2.27]	7.1
Nd$_2$Fe$_{14}$B	1.61	588	133 [2.28]	7.3
			*24 [2.29]	8.4
Pr$_2$Fe$_{14}$B	1.56	565	*33 [2.30]	12
SmCo$_5$	1.05	993	*57 [2.31]	12
Sm$_2$Co$_{17}$	1.29	1070	*31 [2.32]	14
Sm$_2$Fe$_{17}$N$_3$	1.56	749	*42 [2.33]	12
BaFe$_{12}$O$_{19}$	0.48	723	*5.0 [2.33]	6.3

the corresponding unit cell edges. Exchange stiffness constants as determined by eq. (2.14) in general lead to a lower bound for A because the molecular field theory omits the low energy spin wave excitations. Therefore, values of A derived from the spin wave stiffness constant determined by inelastic neutron scattering or from Bloch's $T^{3/2}$ law (see Section 8.4.8.5) are more realistic for temperatures $T < T_c$. Also values of A determined from the specific domain wall energy are reliable numbers (see Section 4.2). Table 2.1 presents exchange stiffness constants of ferromagnetic materials which were obtained by the above-mentioned methods.

2.2.1.2 Long-range exchange interactions

In the preceding section the continuum expression for the exchange energy has been derived under the assumption of a Heisenberg-type exchange interaction between nearest neighbour atoms. As is well known, especially in the metallic ferromagnets, the relevant exchange interactions are the intra-atomic and the indirect exchange mechanisms. The latter exchange mechanism is important in the case of rare earth metals and their alloys and also plays some role in transition metals [2.34–2.36]. Since the indirect exchange interaction includes long-range interactions, an expression of the type eq. (2.11) does not give an adequate micromagnetic description. The theory of indirect exchange interactions was developed originally by Zener [2.34], Ruderman and Kittel [2.37], Kasuya [2.35] and Yosida [2.36] (RKKY-interaction). These authors considered the exchange interaction between localized magnetic moments, either nuclear or electronic, and the sea of delocalized conduction electrons. Kasuya showed in his fundamental paper that localized spin moments, e.g., due to d- or f-electrons, produce an oscillating spin polarization of the s-electrons. Another localized spin moment at r_j will interact with the induced spin density. Thus, distant localized moments become magnetically exchange coupled via the spin polarization of the s-electrons. Assuming the s–f interaction to correspond to a δ-function the effective exchange integral between the localized spin moments with distance $|r_{ij}| = |r_i - r_j| = r$ may be written as

$$J(r) = \frac{\text{const. } J_{\text{sf}}^2}{(2k_{\text{F}}r)^4}(2k_{\text{F}}r \, \cos 2k_{\text{F}}r \; - \; \sin 2k_{\text{F}}r), \qquad (2.19)$$

where k_{F} denotes the modulus of the Fermi wave vector, and J_{sf} the exchange integral between s- and f-electrons. Replacing the sum of eq. (2.6) by an integral, the exchange energy density may be written as [2.38]

$$\phi_{\text{ex}}'(r) = -\frac{2}{\Omega^2} \int J(|r - r'|) \; S(r) \cdot S(r') \, \mathrm{d}^3 r' \qquad (2.20)$$

$$= -\frac{2}{(g\mu_{\text{B}})^2} \int J(|r - r'|) \; M(r) \cdot M(r') \, \mathrm{d}^3 r',$$

where the magnetization operators obey the commutation relations

$$[M(r) \times M(r')] = \frac{1}{2}ig\mu_{\text{B}} \, M(r) \, \delta(r - r'), \qquad (2.21)$$

$$[M^+(r) \times M^-(r')] = 2g\mu_{\text{B}} \, M_z(r) \, \delta(r - r'),$$

where $M^{\pm}(r) = M_x(r) \pm iM_y(r)$. With the Fourier transforms

$$J(k) = \int J(r) \, e^{ik \cdot r} \, \mathrm{d}^3 r, \qquad (2.22)$$

$$M(k) = \int M(r) \, e^{ik \cdot r} \, \mathrm{d}^3 r,$$

the exchange energy can be expressed by using the convolution theorem [2.38],

$$\phi_{\text{ex}}' = -\frac{2}{(2\pi)^3 (g\mu_{\text{B}})^2 \Omega} \int M(k) \cdot M(-k) \, J(k) \, \mathrm{d}^3 k, \qquad (2.23)$$

where the Fourier transform of $J(k)$ follows from eq. (2.19) and eq. (2.22) according to Ruderman and Kittel [2.37]:

$$J(k) = \frac{3\Omega z J_{sd}^2}{16 E_F} \left\{ 1 - \frac{4k_F^2 - k^2}{4k_F k} \ln \frac{2k_F + k}{2k_F - k} \right\}. \tag{2.24}$$

(z = number of s-electrons per atom, E_F = Fermi energy). In the Ruderman–Kittel approximation of a point-like localized spin, $J(r)$ becomes infinite for $r \to 0$. Actually, the s–d or s–f interaction has some finite range. According to Yosida [2.36] this may be taken into account by using instead of a constant J_{sd} in eq. (2.24) a k-dependent $J_{sd}(k)$ decreasing with increasing k and taking into account that the s–d interaction extends over a range of the ion diameter. A suitable ansatz for $J(k)$ has been found to be

$$J(k) = \text{const.} \qquad \text{for} \quad k \leq 2k_F, \tag{2.25}$$
$$J(k) = 0 \qquad \qquad \text{for} \quad k > 2k_F.$$

Figure 2.2 shows the k-dependent $J(k)$ for the Ruderman–Kittel and the Yosida approximations.

In our representation so far we have introduced Fourier-integrals as appropriate for a continuum theory. Actually, however, these Fourier-integrals should be represented by Fourier-series as a consequence of the discrete structure of the spin lattice. $J(\mathbf{k})$ is then given by

$$J(\mathbf{k}) = \Omega \sum_{i \neq j} J_{ij}\big(|\mathbf{R}_j - \mathbf{R}_i|\big) e^{i\mathbf{k} \cdot (\mathbf{R}_i - \mathbf{R}_j)}, \tag{2.26}$$

where the summation extends over all lattice sites. Equation (2.26) may be used to derive simple expressions for one-dimensional spin structures as in strongly uniaxial materials such as the rare earth metals or the rare earth–transition metal intermetallic compounds. In uniaxial materials, the ground state of the spin structure is composed of planes with nearest distance, d, and each plane may be treated as

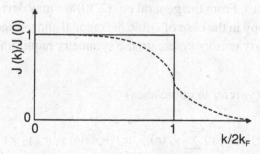

Fig. 2.2. The Fourier transform $J(k)$ of the indirect exchange integral for the Ruderman–Kittel approximation [2.37] and the model proposed by Yosida [2.36].

a unit interacting with the spin of the nth planes of distance d. Then $J(k)$ may be written as

$$J(k) = \sum_{n=0}^{\infty} A_n \cos(ndk_z). \tag{2.27}$$

$J(k)$ depends only on k_z because of the one-dimensionality of the problem. The exchange parameters A_n are obtained from

$$A_n = \Omega \sum_j J_{ij}\left(|R_j^{(n)}|\right), \tag{2.28}$$

where we have put $R_i \equiv 0$ and the summation extends over all lattice sites $R_j^{(n)}$ of the nth plane. Equation (2.27) has been successfully applied to describe the spiral and conical spin structures of rare earth metals [2.38–2.42].

2.2.2 *Magnetocrystalline anisotropy energy*

The magnetocrystalline anisotropy energy in disordered or amorphous media in general obeys triclinic symmetry and is described by an even function in M_i because the energy expression must be invariant with respect to rotations of M by 180°. This leads to the general expression [2.43, 2.44]

$$\phi_K'(r) = k_0(r) + \sum_{i \neq j} k_{ij}(r)\, \gamma_i(r)\, \gamma_j(r) + \sum_{i \neq j,h,l} k_{ijkl}(r)\, \gamma_i(r)\, \gamma_j(r)\, \gamma_k(r)\, \gamma_l(r). \tag{2.29}$$

If the material property tensors k_{ij} etc. are diagonalized $\phi_K'(r)$ becomes

$$\phi_K'(r) = k_0(r) + \sum_i k_{1i}(r)\, \gamma_i^2(r) + \sum_i k_{2i}(r)\, \gamma_i^4(r) + \cdots \tag{2.30}$$

$$+ \sum_{i \neq j} k_{3ij}(r)\, \gamma_i^2(r) \cdot \gamma_j^2(r) + \cdots.$$

Equation (2.30) corresponds to an orthorhombic symmetry where the γ_i refer to the three two-fold axes. From the general eq. (2.30) we may derive the expressions for the local anisotropy in the case of cubic, hexagonal and tetragonal symmetry by specifying the property tensors k_{1i} etc. to the symmetry requirements of the special point groups:

1. Cubic symmetry (γ_i refer to cubic axes)

$$\phi_K'(r) = K_0(r) + K_1(r) \sum_{i \neq j} \gamma_i^2(r)\, \gamma_j^2(r) + K_2(r)\, \gamma_1^2(r)\, \gamma_2^2(r)\, \gamma_3^2(r). \tag{2.31}$$

K_1 and K_2 are denoted as the first and second anisotropy constants.

2. Hexagonal symmetry

   Here we introduce $\gamma_3(r) = \cos\varphi(r)$, $\gamma_1(r) = \sin\varphi\cos\psi(r)$, $\gamma_2 = \sin\varphi(r)\,\psi(r)$, where φ corresponds to $\angle(M_s, z\text{-axis})$ and ψ the azimuthal angle of the projection of M_s on the hexagonal plane with respect to the x-axis. In this case eq. (2.29) can be rearranged as

$$\phi'_K(r) = K_0(r) + K_1(r)\sin^2\varphi(r) + K_2(r)\sin^4\varphi(r) \qquad (2.32)$$
$$+ K_3(r)\sin^6\varphi(r) + K_4(r)\sin^6\varphi(r)\cos^6\psi(r).$$

3. Tetragonal symmetry

   With a similar definition of the angles φ and ψ as in the hexagonal case we find from eq. (2.29)

$$\phi_K(r) = K_0(r) + K_1(r)\sin^2\varphi(r) + K_2(r)\sin^4\varphi(r) \qquad (2.33)$$
$$+ K_3(r)\sin^4\varphi(r)\cos^4\psi(r) + \cdots.$$

Table 2.2 gives a review of the anisotropy constants of some relevant magnetic materials.

Within the magnetic domains of a macroscopic sample the spontaneous magnetization is oriented parallel to the so-called easy directions which are determined by the minima of the magnetocrystalline energy, i.e., where the conditions $\partial\phi'_K/\partial\gamma_i = 0$ and $\partial^2\phi'_K/\partial\gamma_i^2 > 0$ hold. In the case of a cubic crystal the following three types of easy directions exist, which are represented in Fig. 2.3 and in the phase diagram of Fig. 2.4:

1. $\langle 100\rangle$-directions: $K_1 > 0$, $K_2 > 0$; $K_1 > -\frac{1}{9}K_2$, $K_2 < 0$.
2. $\langle 111\rangle$-directions: $K_1 < -\frac{1}{9}K_2$, $K_2 < 0$; $K_1 < -\frac{4}{9}K_2$, $K_2 > 0$.
3. $\langle 110\rangle$-directions: $-\frac{4}{9}K_2 < K_1 < 0$, $K_2 > 0$.

At room temperature case 1 holds for α-Fe and case 2 for Ni.

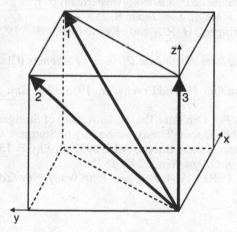

Fig. 2.3. Possible easy directions in cubic crystals. The numbered arrows indicate the sequence of easy directions for $K_2 > 0$ and K_1 changing from negative to positive values.

Table 2.2. *Crystal anisotropy constants K_1 and K_2 of transition metals, oxides and intermetallic compounds at room temperature and low temperatures.*

Material	Room temperature		4.2 K (*20 K)		Ref.
	$K_1 \, [\text{J/m}^3]$	$K_2 \, [\text{J/m}^3]$	$K_1 \, [\text{J/m}^3]$	$K_2 \, [\text{J/m}^3]$	
α-Fe (bcc)	4.8×10^4	-1.0×10^4	5.2×10^4	-1.8×10^4	[1,2]
Co (hcp)	45.3×10^4	14.5×10^4	7×10^5	1.8×10^5	[3,4]
Ni (fcc)	-4.5×10^3	-2.5×10^3	-12×10^4	3.0×10^4	[5,6]
Fe_3O_4	-11×10^3	-3×10^3	-28×10^3		[7]
$Nd_2Fe_{14}B$	4.3×10^6	0.65×10^6	$^*-18 \times 10^6$	$^*48 \times 10^6$	[8,9]
$Pr_2Fe_{14}B$	5.6×10^6	≈ 0	24×10^6	-7×10^6	[9,10]
$SmCo_5$	1.7×10^7		2.6×10^7		[11,12]
Sm_2Co_{17}	4.2×10^6		6.5×10^6		[13,14]
$Sm_2Fe_{17}N_3$	8.6×10^6	1.9×10^6	12×10^6	3×10^6	[12,15]
$BaFe_{12}O_{19}$	3.2×10^5	$< 0.1 \times 10^6$	4.5×10^5	$< 0.3 \times 10^6$	[16,17]

References
 [1] Gengnagel, H., and Hofmann, U., 1968, *Phys. Stat. Sol.* **29**, 91.
 [2] O'Handley, R.C., 2000, *Modern Magnetic Materials, Principles and Applications* (J. Wiley & Sons, Inc., New York) p. 189.
 [3] Pauthenet, R., Barnier, Y., and Rimet, E., 1962, *J. Phys. Soc. Japan* **17**, Suppl., 309.
 [4] Barnier, Y., Pauthenet, R., and Rimet, G., 1961, *Compt. Rend.* **252**, 2839.
 [5] Pussei, I.M., 1957, *Isv. Akad. Nauk. S.S.S.R., Ser. Physics* **21**, 1088.
 [6] Franse, J.J.M., and de Vries, G., 1968, *Physica* **39**, 477.
 [7] Abe, K., Miyamoto, Y., and Chikazumi, S., 1976, *J. Phys. Soc. Japan* **41**, 1894.
 [8] Hock, St., and Kronmüller, H., 1987, *Proc. 5th Int. Symp. on Magnetic Anisotropy and Coercivity in RE-Transition Metal Alloys* (Eds. C. Herget, H. Kronmüller and R. Poerschke, DPG mbH, Bad Honnef) p. 275.
 [9] Buschow, K.H.J., 1988, 'Permanent Magnet Materials Based on 3d-Rich Ternary Compounds'. In *Ferromagnetic Materials*, Vol. 4 (Eds. E.P. Wohlfarth and K.H.J. Buschow, Elsevier Science Publishers, Amsterdam) p. 1.
[10] Sagawa, M., 1985, *J. Magn. Soc. Japan* **9**, 25.
[11] Kütterer, R., Hilzinger, H.-R., and Kronmüller, H., 1977, *J. Magn. Magn. Mater.* **4**, 1.
[12] Coey, J.M.D., 1996, *Rare-Earth Iron Permanent Magnets* (Clarendon Press, Oxford) p. 45.
[13] Durst, K.-D., Kronmüller, H., and Ervens, W., 1988, *Phys. Stat. Sol. (A)* **108**, 403; **108**, 705.
[14] Zeppelin, F., von, 1999, Diploma Thesis, University of Stuttgart, Germany.
[15] Kleinschroth, I., 1995, Doctor Thesis, University of Stuttgart, Germany.
[16] Rathenau, G.W., Smit, J., and Stuyts, A.L., 1952, *Z. Physik* **133**, 250; Pauthenet, R., and Rimet, G., 1959, *Compt. Rend.* **249**, 1875.
[17] Smit, J., and Wijin, H.P.J., 1954, *Ferrites* (John Wiley, New York) p. 204.

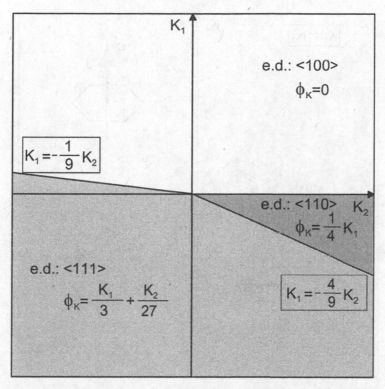

Fig. 2.4. Phase diagram of easy directions in cubic crystals as a function of K_1 and K_2.

In the case of uniaxial hexagonal crystals the easy directions are described by the angle φ between the spontaneous magnetization and the c-axis. Again three types of easy directions have to be distinguished if we take into account only K_1 and K_2 (see Fig. 2.5):

1. $\langle 0001 \rangle$-directions ($\pm\,c$-axis): $K_1 > 0$, $K_2 > 0$; $K_1 > -K_2$, $K_2 < 0$.
2. All directions within the basal plane: $K_1 < -K_2$, $K_2 < 0$; $K_1 < -2K_2$, $K_2 > 0$.
3. All directions on the surface of a cone with angle $\sin\varphi = \sqrt{-K_1/2K_2}$: $-2K_2 < K_1 < 0$, $K_2 > 0$.

Case 1 holds for Co for $T < 520$ K, case 2 holds for $T > 605$ K and case 3 holds for 520 K $< T <$ 605 K.

The origin of the magnetocrystalline anisotropy is based on the coupling between spin moments and the electronic orbital moments (L–S-coupling) and their coupling to the anisotropic crystal field acting on an atom. In transition metals with their nearly quenched orbital moments of the 3d-electron, the coupling of the orbital moments to the crystal field remains small, leading to moderate anisotropies [2.45] (see Table 2.2). In contrast, in the case of rare earth metals with strong L–S-coupling

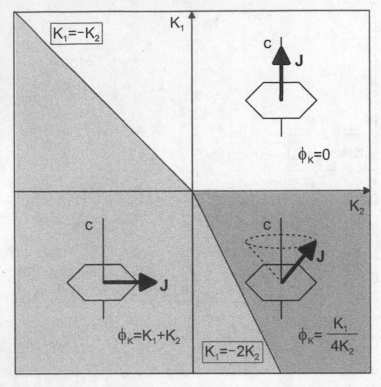

Fig. 2.5. Phase diagram of easy directions in hexagonal crystals as a function of K_1 and K_2.

of the 4f-electrons, the coupling of the anisotropic 4f-charge cloud (see Fig. 2.6) to the anisotropic crystal field leads to large anisotropies [2.46, 2.47].

The anisotropy constants, K_1, in general are strongly temperature dependent [2.8, 2.15]. By thermally activated spin waves the directions of local spins deviate from the easy directions, which leads to an increase of the local anisotropy energy, but to a decrease of the effective anisotropy constants, K_n. If the index n corresponds to the order of the spherical harmonics describing the angular dependence of the atomic anisotropy, the temperature dependence according to Zener [2.48] and Callen and Callen [2.49] is given by

$$\frac{K_n(T)}{K_n(0)} = \left(\frac{M_s(T)}{M_s(0)} \right)^{n(n+1)/2}. \qquad (2.34)$$

In the case of uniaxial crystals with $n = 2$ this gives $K_n(T) \propto (M_s(T))^3$ and for cubic crystals with $n = 4$, $K_n(t) \propto (M_s(T))^{10}$ is obtained. It should be noted, however, that higher order spherical harmonics may modify the temperature dependence of $K_n(T)$, in particular at higher temperatures.

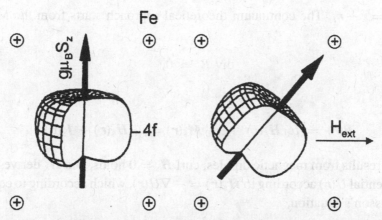

Fig. 2.6. Single-ion model of magnetocrystalline anisotropy in rare earth intermetallic compounds. Left side: ground state at $H = 0$. Right side: rotation of anisotropic 4f charge cloud under the action of an applied field in the crystal field of neighbouring ions.

2.2.3 Magnetostatic energies

Two types of magnetostatic energies have to be considered: the magnetostatic energy of the external field and the dipolar energy also denoted as stray field energy. The magnetostatic energy, also known as the Zeeman energy, is written

$$\phi_H = -g\mu_B \sum_i S_i(r_i) \cdot B_{ext}(r_i) = -g\mu_0\mu_B \sum_i S_i(r_i) \cdot H_{ext}(r_i). \qquad (2.35)$$

Introducing continuum variables $M_s(r)$ or $J_s(r)$ according to eq. (2.2), the magnetostatic energy density is written

$$\phi'_H(r) = -\mu_0 H_{ext}(r) \cdot M_s(r) = -H_{ext}(r) \cdot J_s(r). \qquad (2.36)$$

The importance of the magnetostatic stray fields for the formation of domain patterns has been described previously by Landau and Lifshitz [2.3] as well as by Kittel and Galt [2.5]. The role of stray fields in micromagnetic problems has been stressed by Néel [2.50, 2.51] and Brown has given an extensive description of magnetostatic principles [2.52].

A very characteristic property of ferromagnetic materials is their magnetostatic stray field which may be determined by summation over all dipole fields of the individual magnetic moments $\mu_i(r) = g\mu_B S_i(r)$

$$H_s(r) = \frac{1}{4\pi} \sum_i \left(\frac{\mu_i(r_i)}{R^3} - \frac{3(\mu_i(r_i) \cdot R) \cdot R}{R^5} \right), \qquad (2.37)$$

where $R = r - r_i$. The continuum theoretical approach starts from the Maxwell relation

$$\text{div } B = 0, \tag{2.38}$$

with

$$B(r) = \mu_0 H_s(r) + \mu_0 M_s(r) = \mu_0 H_s(r) + J_s(r). \tag{2.39}$$

Since H_s results from magnetic dipoles, curl $H_s = 0$ holds, and H_s derives from a scalar potential $U(r)$ according to $H_s(r) = -\nabla U(r)$, which according to eq. (2.38) obeys Poisson's equation,

$$\Delta U = \text{div } M_s(r) = \frac{1}{\mu_0} \text{div } J_s(r) = -\rho(r), \tag{2.40}$$

with $\rho(r)$ the magnetic charge density. H_s is denoted as the magnetic stray field because it only appears if $M_s(r)$ is spatially inhomogeneous either in the modulus $|M_s|$ or in orientation. The general solution of eq. (2.40) is given by

$$U(r) = -\frac{1}{4\pi} \int_V \frac{\rho(r')}{|r - r'|} \, d^3r' + \frac{1}{4\pi} \int_S \frac{\sigma(r') \cdot df'}{|r - r'|}, \tag{2.41}$$

where the first term in eq. (2.41) takes care of the volume charges, $\rho(r) = -\text{div } M_s(r)$, within the volume V, and the second integral extends over the material surfaces and takes into account the effect of the surface charges, $\sigma(r) = M_s(r) \cdot n$, as presented in Fig. 2.7.

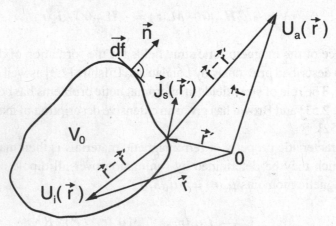

Fig. 2.7. Integration variables of the Poisson equation for the internal potential $U_i(r)$ and the external potential $U_a(r)$ for surface charges $\sigma(r') = M_s(r') \cdot n$.

The magnetostatic stray field is related to a magnetostatic self-energy density, also denoted as dipolar energy given by

$$\phi_s'(r) = \frac{\mu_0}{2} \cdot H_s^2(r).$$ (2.42)

If we consider an individual magnetic moment, $\mu_i(r_i)$, its total interaction energy with all the other dipoles may be written as

$$\phi_s^{(i)}(r_i) = -\mu_0 \mu_i(r_i) \cdot \sum_{j \neq i} H_s^{(j)}(r_i),$$ (2.43)

where the sum extends over the field contribution of all other dipoles, j, at r_i. Accordingly, the total stray field energy is given by

$$\phi_s = -\frac{1}{2} \sum_i \phi_s^{(i)} = -\frac{1}{2} \mu_0 \sum_{j \neq i} \mu_i(r_i) \cdot H_s^{(j)}(r_i),$$ (2.44)

where the factor $1/2$ takes into account that $\mu_i(r_i) \cdot H_s^{(j)}(r_i) = H_s^{(i)}(r_j) \cdot \mu_i(r_j)$ holds and the interaction energy between two dipoles may be counted only once. Introducing the local magnetization $M_s(r)$ by

$$\mu_i(r_j) = M_s(r) \, dV,$$ (2.45)

eq. (2.44) transforms into an integral,

$$\phi_s = -\frac{1}{2} \int_{V_0} H_s(r) \cdot J_s(r) \, d^3r = \frac{1}{2} \mu_0 \int_V H_s^2(r) \, d^3r,$$ (2.46)

where the first integral extends over the volume of the magnet and the second one over the total space. Applying Green's theorem, eq. (2.46) may be transformed into a surface and a volume integral,

$$\phi_s = -\frac{1}{2} \int_S U(r) \cdot J_s(r) \cdot df + \frac{1}{2} \int_V U(r) \, \text{div} \, J_s(r) \, d^3r.$$ (2.47)

Here the first term corresponds to the contribution of the surface charges and the second term to that of the volume charges as visualized in Fig. 2.7.

In terms of surface charges σ and volume charges ρ, eq. (2.48) is written

$$\phi_s = -\frac{1}{2} \mu_0 \int_S \sigma(r) \, U(r) \, df - \frac{1}{2} \mu_0 \int_{V_0} \rho(r) \, U(r) \, d^3r,$$ (2.48)

where the first term extends over the surface, S, and the second one over the volume, V_0, of the ferromagnet.

2.2.4 Elastic potential of a ferromagnet

2.2.4.1 Strain tensor in ferromagnetic materials

The interaction of elastic stresses with the spontaneous magnetization is of particular significance for ferromagnetic materials because these interactions are responsible for the arrangement of domains, pinning of domain walls and deviations from ferromagnetic saturation. Accordingly, these interactions are responsible for the influence of the microstructure on microscopic as well as macroscopic magnetic properties. A first treatment of magnetoelastic interactions in ferromagnetic materials was given by Becker [2.53] and Becker and Döring [2.4]. Later on their ideas were used by Brown [2.1, 2.2] to develop the modern theory of micromagnetism and to introduce modern aspects of lattice defect theory into the field of magnetism. More recent fundamental self-consistent theories of the magnetoelastic interactions have been published by Brown [2.54, 2.55], Rieder [2.56] and Kronmüller [2.7]. The basic problem of elastic interactions in ferromagnetic materials in general is of high complexity because not only the strains but also the rotations of volume elements may be responsible for magnetoelastic interactions. Rotations of volume elements corresponding to the asymmetric part of the distortion tensor are due to the magnetic torques being related to anisotropic energies treated in the preceding sections. It will be shown in the magnetostrictive Chapter 11 that magnetic torques for the usual ferromagnetic materials in fact do not result in significant magnetoelastic interactions. Therefore, we shall concentrate in this section exclusively on the symmetric strain tensor within the framework of linear elasticity theory. Also this approximation has to be considered as a rather complex problem because the total strain tensor according to

$$\varepsilon^{\mathrm{T}} = \varepsilon^{\mathrm{ext}} + \varepsilon^{\mathrm{def}} + \varepsilon^{\mathrm{Q}} + \varepsilon^{\mathrm{el}}, \tag{2.49}$$

in general is composed of four contributions which have the following origins:

1. The external strain tensor, $\varepsilon^{\mathrm{ext}}$, results from external volume (f^{ext}) and surface forces (F^{ext}).
2. $\varepsilon^{\mathrm{def}}$ describes the so-called internal elastic strain tensor due to defect structures such as point defects or dislocations.
3. ε^{Q} denotes the quasiplastic, spontaneous magnetostrictive strain tensor due to spin ordering.
4. $\varepsilon^{\mathrm{el}}$ describes the elastic strain tensor related to inhomogeneous spontaneous magnetostrictive deformation ε^{Q}.
5. $\varepsilon^{\mathrm{m}} = \varepsilon^{\mathrm{Q}} + \varepsilon^{\mathrm{el}}$ corresponds to the total strain tensor due to magnetostriction.

2.2.4.2 Determination of strain tensors

The different contributions to the total strain tensor ε^{T} have to be determined by quite different procedures which, however, are all based on the linear elastic continuum theory assuming that between elastic strains, ε, and elastic stresses, σ, Hooke's law holds:

$$\sigma = c \cdot\cdot\, \varepsilon, \tag{2.50}$$

where c denotes the fourth rank tensor of elastic constants. In the following we assume ε^{ext}, ε^{def} and the corresponding elastic stresses σ^{ext} and σ^{def} to be determined by conventional elasticity theory based on the mechanical equilibrium and surface conditions ($n =$ surface normal):

$$\begin{aligned} \text{Div } \sigma^{\text{ext}} + f^{\text{ext}} &= 0 \\ \text{Div } \sigma^{\text{def}} + f^{\text{def}} &= 0 \\ n \cdot \sigma^{\text{ext}}|_{\text{surface}} &= F^{\text{ext}}, \end{aligned} \tag{2.51}$$

where f^{ext} denotes the external volume forces, f^{def} the internal forces producing the defect structure, and F^{ext} corresponds to the surface forces.

The magnetostrictive terms ε^{Q} and ε^{el} are responsible for the interaction between external and internal elastic stresses and the magnetic state.

Magnetostriction is due to the fact that in a distorted solid not only does the elastic energy change, but also all the magnetic energy terms such as exchange, anisotropy and dipolar energy may be influenced by deformations. It is common to distinguish between four different types of magnetostrictive effects according to their different origins:

1. Volume magnetostrictions are those deformations which are due to the volume dependence of the intrinsic material parameters such as spontaneous magnetization, exchange integrals, anisotropy constants.
2. Forced magnetostrictions are due to the volume dependence of the magnetostatic energy term $- \int J_{\text{s}} \cdot H_{\text{ext}} \, d^{3}r$.
3. The so-called form effect results from the dependence of the dipolar energy (stray field energy) on the shape of the specimen.
4. The shape magnetostriction, also called spontaneous magnetostriction, takes into account the anisotropy of spontaneous deformations due to anisotropic energy terms as the anisotropy energy which results mainly from the spin–orbit coupling energy.

A discussion of these different types of magnetostriction may be found in the books of Kneller [2.57], Brown [2.55] and Landau and Lifshitz [2.58]. Since in usual ferromagnetic materials the first three effects are small in comparison to the shape magnetostriction we shall concentrate on this latter effect, which predominantly determines the interaction of the defects with the spontaneous magnetization. The spontaneous magnetostriction according to Becker and Döring [2.4] may be written as

$$\varepsilon_{ij}^{Q} = \lambda_{ijkl} \gamma_{k} \gamma_{l}, \tag{2.52}$$

where Einstein's sum convention holds (summation from 1 to 3 has to be performed if a suffix appears twice in the same term). The fourth rank so-called magnetostrictive tensor possesses the symmetry of the atomic arrangement, i.e., of crystal symmetry or short-range order in otherwise disordered materials. In the case of a cubic crystal ε^Q is given by the following symmetrical tensor:

$$\varepsilon^Q = \begin{pmatrix} \frac{3}{2}\lambda_{100}\left(\gamma_1^2 - \frac{1}{3}\right) & \frac{3}{2}\lambda_{111}\gamma_1\gamma_2 & \frac{3}{2}\lambda_{111}\gamma_1\gamma_3 \\ 0 & \frac{3}{2}\lambda_{100}\left(\gamma_2^2 - \frac{1}{3}\right) & \frac{3}{2}\lambda_{111}\gamma_2\gamma_3 \\ 0 & 0 & \frac{3}{2}\lambda_{100}\left(\gamma_3^2 - \frac{1}{3}\right) \end{pmatrix}. \tag{2.53}$$

λ_{100} and λ_{111} are called magnetostriction constants. λ_{100} corresponds to the fractional change in length upon saturation in the $\langle 100 \rangle$-direction and λ_{111} has the same meaning for saturation in the $\langle 111 \rangle$-direction. For hexagonal crystals ε^Q may be written as

$$\varepsilon^Q = \begin{pmatrix} \lambda_{11}\gamma_1^2 + \lambda_{12}\gamma_2^2 & (\lambda_{11} - \lambda_{12})\gamma_1\gamma_2 & \frac{1}{2}\lambda_{44}\gamma_1\gamma_3 \\ 0 & \lambda_{12}\gamma_1^2 + \lambda_{11}\gamma_2^2 & \frac{1}{2}\lambda_{44}\gamma_2\gamma_3 \\ 0 & 0 & \lambda_{33}\gamma_3^2 \end{pmatrix}. \tag{2.54}$$

In the case of an isotropic material we obtain for the diagonal strain components:

$$\varepsilon_{ii}^Q = \frac{3}{2}\lambda_s\left(\gamma_i^2 - \frac{1}{3}\right), \qquad i = 1, 2, 3, \tag{2.55}$$

and the off-diagonal strain components:

$$\varepsilon_{ij}^Q = \frac{3}{2}\lambda_s\gamma_i\gamma_j, \qquad i \neq j, \tag{2.56}$$

where λ_s denotes the isotropic magnetostriction constant.

Once the spontaneous magnetostrictive strains ε^Q are available the corresponding elastic magnetostrictive strains ε^{el} may be determined from the theory of elasticity by using the condition that the total magnetostrictive strain field,

$$\varepsilon^m = \varepsilon^Q + \varepsilon^{el}, \tag{2.57}$$

can be derived from a displacement field s^m according to

$$\varepsilon^m = \text{Def } s^m, \tag{2.58}$$

where the definition of the operator Def follows from

$$\varepsilon_{ij}^m = \frac{1}{2}\left(\frac{\partial s_i^m}{\partial x_j} + \frac{\partial s_j^m}{\partial x_i}\right), \tag{2.59}$$

and the elastic stress due to magnetostriction,

$$\sigma^m = c \cdot\cdot \varepsilon^{el}, \tag{2.60}$$

obeys the mechanical equilibrium condition

$$\text{Div } \sigma^m = 0. \tag{2.61}$$

Since σ^m may be expressed as $\sigma^m = c \cdot \cdot (\varepsilon^m - \varepsilon^Q)$, eq. (2.61) can be written as

$$\text{Div}(c \cdot \cdot \varepsilon^{el}) = \text{Div}(c \cdot \cdot \text{Def } s^m) - \text{Div}(c \cdot \cdot \varepsilon^Q) = 0. \tag{2.62}$$

Equation (2.62) represents three sets of differential equations which in components ε_{ij}^m and ε_{ij}^Q are given by

$$\partial_i \, c_{ijkl} \, \varepsilon_{kl}^m = \partial_i \, c_{ijkl} \, \varepsilon_{kl}^Q, \tag{2.63}$$

where ∂_i denotes the operator $\partial / \partial x_i$. The three sets of equations (2.62) or (2.63) are not sufficient to determine all six components of ε^m if we do not use the second term of eq. (2.62). However, since ε^m derives from a displacement field s^m by $\varepsilon^m = \text{Def } s^m$, additional relations hold for ε^m, usually written in the comprehensive form [2.59]

$$\text{Ink } \varepsilon^m = \nabla \times \varepsilon^m \times \nabla = 0, \tag{2.64}$$

with the operator $\text{Ink } \varepsilon^m = \nabla \times \varepsilon^m \times \nabla$. Equation (2.64) takes into account that the total magnetostrictive strain tensor ε^m should describe a so-called compatible deformation without any abrupt changes of the displacement field. In the case of elastic isotropy the combination of eq. (2.63) and eq. (2.64) leads to the so-called Beltrami equations [2.59], where the elastic stresses σ^m are related to their sources ε^Q:

$$\Delta \sigma_{ij}^m + \frac{\nu}{1+\nu} \left(\nabla_i \nabla_j \sigma_{kk}^m - \Delta \sigma_{kk}^m \delta_{ij} \right) = 2G \eta_{ij}^Q, \tag{2.65}$$

($\nu = $ Poisson ratio, $G = $ shear modulus, $\delta_{ij} = 0$ for $i \neq j$, and $\delta_{ij} = 1$ for $i = j$, $\sigma_{kk}^m = \sum_{i=1}^3 \sigma_{ii}^m$).

In eq. (2.65) we have introduced the so-called incompatibility tensor [2.59] of the spontaneous magnetostrictions:

$$\eta^Q = (-) \text{Ink } \varepsilon^Q = \text{Ink } \varepsilon^{el}. \tag{2.66}$$

Equations (2.65) and (2.66) clearly show that the sources of the elastic magnetostrictive stresses are inhomogeneous spontaneous strains. In the case of a homogeneous tensor ε^Q, σ^m as well as ε^{el} vanish.

2.2.4.3 Derivation of the magnetoelastic potential

In order to determine the internal elastic energy density we consider as a reference state a hypothetical nonmagnetic material. This hypothetical material may contain internal strains, $\varepsilon^{\mathrm{def}}$, and stresses, σ^{def}, due to defects, and in addition may be exposed to external forces producing $\varepsilon^{\mathrm{ext}}$ and σ^{ext}. If the magnetic state is now steadily switched on, the probe is also exposed to the spontaneous strains ε^{Q}.

Here we have to consider the case of a homogeneously magnetized material with a homogeneous magnetostrictive strain tensor ε^{Q} and a vanishing strain tensor $\varepsilon^{\mathrm{el}}$ as well as an inhomogeneously magnetized material where both ε^{Q} and $\varepsilon^{\mathrm{el}}$ are spatially varying quantities. The difference between the elastic behaviour of these two cases is demonstrated in Fig. 2.8 by cutting the probe into individual volume elements. In the case of a homogeneously magnetized probe, after cutting, all volume elements remain in their homogeneously strained state without any change in

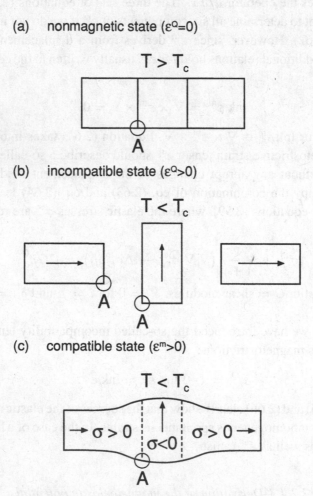

Fig. 2.8. Schematic representation of the origin of elastic stresses σ^{m} due to inhomogeneous magnetostrictive strains ε^{Q}. In order to keep compatibility the sites A have to be connected leading to the compatible state with elastic stresses σ.

volume or shape. These volume elements therefore can be fitted together without any additional strain. In the case of the inhomogeneously magnetized probe, each volume element is characterized by a different spontaneous strain tensor $\varepsilon^Q(\gamma_i)$. Accordingly, after cutting, the volume elements deform into different shapes which no longer fit together. Therefore, this state is also called the incompatible state of the probe. In order to fit the individual volume elements together again we have to submit each volume element to an elastic strain tensor ε^{el} in order to produce a compatible state without holes and cracks. This consideration gives a direct illustration of the stress-producing role of the incompatibility tensor η^Q introduced in the preceding section. It also becomes obvious that elastic magnetostrictive strains and stresses are a consequence of inhomogeneous spontaneous magnetostrictions (quasiplastic deformations), i.e., of inhomogeneous magnetic states.

In the following we assume the elastic problems of the magnetic systems to be solved for a given strain tensor ε^Q. Then we may derive a Gibbs free energy expression only containing the direction cosines γ_i as free variables. Writing for the Gibbs free energy,

$$\phi_{el} = \int \left(\phi_0 + \frac{1}{2} \varepsilon^T \cdot\cdot \, c \cdot\cdot \, \varepsilon^T - \sigma \cdot\cdot \, \varepsilon^T \right) d^3r, \qquad (2.67)$$

where ϕ_0 includes all energy terms independent of deformations (ϕ'_{ex}, ϕ'_K and ϕ'_s), the minimization with respect to ε^T leads to Hooke's law as the equilibrium condition,

$$\sigma = c \cdot\cdot \, \varepsilon^T. \qquad (2.68)$$

Inserting eq. (2.68) into eq. (2.67) then gives for the Gibbs free energy,

$$\phi_{el} = \int \left(\phi_0 - \frac{1}{2} \varepsilon^T \cdot\cdot \, c \cdot\cdot \, \varepsilon^T \right) d^3r. \qquad (2.69)$$

From eq. (2.69) we may now derive the different types of magnetoelastic energy terms and interaction energies which are important to describe the interaction between microstructures and magnetic states. Inserting into eq. (2.69) our expression for ε^T as given by eq. (2.49), we find for the elastic part of the Gibbs free energy,

$$\phi_{el} = -\frac{1}{2} \int \varepsilon^T \cdot\cdot \, c \cdot\cdot \, \varepsilon^T \, d^3r$$

$$= -\frac{1}{2} \int \left(\varepsilon^{ext} \cdot\cdot \, c \cdot\cdot \, \varepsilon^{ext} + \varepsilon^{def} \cdot\cdot \, c \cdot\cdot \, \varepsilon^{def} + \varepsilon^Q \cdot\cdot \, c \cdot\cdot \, \varepsilon^Q + \varepsilon^{el} \cdot\cdot \, c \cdot\cdot \, \varepsilon^{el} \right.$$

$$\left. + 2 \, \varepsilon^Q \cdot\cdot \, c \cdot\cdot \, \varepsilon^{el} \right) d^3r$$

$$- \int \left(\varepsilon^{def} + \varepsilon^{ext} \right) \cdot\cdot \, c \cdot\cdot \, \left(\varepsilon^Q + \varepsilon^{el} \right) d^3r - \int \varepsilon^{ext} \cdot\cdot \, c \cdot\cdot \, \varepsilon^{def} \, d^3r. \qquad (2.70)$$

Here the first terms on the right side of eq. (2.70) correspond to the self energies of the different types of strains. The term with the integral $\varepsilon^Q \cdots c \cdots \varepsilon^{el}$ is transformed by means of eq. (2.57) into $\varepsilon^Q \cdots c \cdots \varepsilon^{el} = -\varepsilon^{el} \cdots c \cdots \varepsilon^{el} + \varepsilon^m \cdots c \cdots \varepsilon^{el}$. Since Ink $\varepsilon^m = 0$ holds it may be shown that the integral of the latter terms vanishes. Since in the following we have to consider only elastic terms of magnetic origin we finally obtain for the elastic potential,

$$\phi_{el} = -\frac{1}{2} \int \left\{ \varepsilon^Q \cdots c \cdots \varepsilon^Q - \varepsilon^{el} \cdots c \cdots \varepsilon^{el} \right\} \, d^3r \qquad (2.71)$$
$$- \int \left(\varepsilon^{def} + \varepsilon^{ext} \right) \cdots c \cdots \left(\varepsilon^Q + \varepsilon^{el} \right) \, d^3r.$$

This elastic potential contains all the elastic self energies and interaction energies required for the treatment of microstructural effects. To increase the transparency of this expression we consider three different situations of practical importance.

1. Homogeneously magnetized ferromagnet without defect structures (ideal ferromagnet)

   In this case no elastic stresses are present (ε^{ext}, ε^{def}, $\varepsilon^{el} = 0$). Only spontaneous magnetostrictive strains ε^Q have to be considered and consequently ϕ_{el} is written

   $$\phi_{el} = -\frac{1}{2} \int \varepsilon^Q \cdots c \cdots \varepsilon^Q \, d^3r. \qquad (2.72)$$

This energy term corresponds to the well-known magnetostrictive contribution to the crystal anisotropy of a ferromagnet. From an energetic point of view the energy term corresponds to the negative elastic free energy of the spontaneous magnetostrictive strains. This energy becomes free if the probe becomes nonmagnetic. Inserting the ε^Q from eq. (2.53) into eq. (2.72) gives, for the cubic crystal, an elastic energy density of

$$\phi'_{el} = \left\{ \frac{9}{4}(c_{11} - c_{12})\lambda_{100}^2 - \frac{9}{4}c_{44}\lambda_{111}^2 \right\} \sum_{i \neq j} \gamma_i^2 \gamma_j^2, \qquad (2.73)$$

where c_{ij} corresponds to elastic constants in Voigt's notation. Equation (2.73) has the same symmetry as the anisotropy energy and therefore adds to ϕ_K.

2. Inhomogeneous magnetization

   Inhomogeneous arrangements of the direction of the spontaneous magnetization in an ideal ferromagnet occur in domain walls, near the surface of the ferromagnet and in excited spin waves. If no defects and external forces are present the only strains are ε^Q and ε^{el}, and ϕ_{el} is given by

   $$\phi_{el} = -\frac{1}{2} \int \left\{ \varepsilon^Q \cdots c \cdots \varepsilon^Q - \varepsilon^{el} \cdots c \cdots \varepsilon^{el} \right\} \, d^3r. \qquad (2.74)$$

As compared with the homogeneously magnetized ferromagnet in eq. (2.74) the self energy of the elastic magnetostrictive strains appears. This energy term is of particular interest for some domain walls in cubic crystals which otherwise would split up into partial domain walls of infinite distance (e.g. (100) – 180°-walls in α-Fe, or the (100) – 109°-walls in Ni) [2.7, 2.56, 2.60].

3. Homogeneous magnetization and internal stresses

In the case of a homogeneously magnetized ferromagnet, i.e., $\varepsilon^{el} = 0$, containing internal stresses, σ^{def}, the magnetoelastic potential is given by

$$\phi_{el} = -\frac{1}{2} \int \varepsilon^Q \cdots c \cdots \varepsilon^Q \, d^3r - \int \sigma^{def} \cdots \varepsilon^Q \, d^3r. \tag{2.75}$$

Here the first term corresponds to the magnetostrictive contribution to the anisotropy energy. The second term is known as the magnetoelastic coupling energy, ϕ_σ, originally derived by Becker and Döring [2.4]. A similar expression holds for external stresses σ^{ext}. Inserting our results for the spontaneous magnetostriction, ε^Q, we then obtain the following expression for the magnetoelastic coupling energy density of cubic crystals ($\sigma^{def}, \sigma^{ext} \equiv \sigma$):

$$\phi'_\sigma = -\frac{3}{2}\lambda_{100} \sum_{i=1}^{3} \sigma_{ii}\gamma_i^2 - \frac{3}{2}\lambda_{111} \sum_{i \neq j} \sigma_{ij}\gamma_i\gamma_j. \tag{2.76}$$

where σ and the γ_i refer to the cubic coordinate system. In the case of isotropic magnetostriction eq. (2.76) gives for any arbitrary coordinate system

$$\phi'_\sigma = -\frac{3}{2}\lambda_s \left\{ \sum_{i=1}^{3} \sigma_{ii}\gamma_i^2 + \sum_{i \neq j} \sigma_{ij}\gamma_i\gamma_j \right\}. \tag{2.77}$$

In eq. (2.76) and eq. (2.77) the stress tensor σ may correspond either to internal stresses resulting from defect structures or to externally applied stresses. Table 2.3 summarizes the magnetostriction constants of transition metals and of some alloys and intermetallic compounds.

2.3 Summary

In the preceding sections five energy terms have been discussed leading to a total magnetic Gibbs free energy density, which for $T = 0$ is given by

$$\phi'_t = \phi'_{ex} + \phi'_K + \phi'_s + \phi'_{el} + \phi'_H. \tag{2.78}$$

In many cases not all these energy terms need be taken into account in their general form. For example, concerning ϕ'_{el} two cases are of importance. For a treatment of internal stresses due to defects ϕ'_{el} is replaced by ϕ'_σ. For a crystal of cubic

Table 2.3. *Magnetostriction constants and elastic moduli at 4.2 K and at room temperature for transition metals and intermetallic compounds. E, G denote the Youngs modulus and the shear modulus of the isotropic materials.*

Material	$\lambda \cdot 10^6$ Room temperature	4.2 K (*73 K)	c_{ij} [10^{11} Pa] Room temperature	Ref.
α-Fe	$\lambda_{100} = 22$ $\lambda_{111} = -21$	$\lambda_{100} = 26$ $\lambda_{111} = -30$	$c_{11} = 2.41$ $c_{12} = 1.46, c_{44} = 1.12$	[1]
Co	$\lambda_{11} = -45$ $\lambda_{12} = -45$ $\lambda_{33} = 110$ $\lambda_{44} = -260$	$*\lambda_{11} = -66$ $\lambda_{12} = -123$ $\lambda_{33} = 126$ $\lambda_{44} = -323$	$c_{11} = 3.07, c_{33} = 3.58$ $c_{12} = 1.65, c_{44} = 0.76$ $c_{13} = 1.03$	[2]
Ni	$\lambda_{100} = -55$ $\lambda_{111} = -23$	$\lambda_{100} = 57$ $\lambda_{111} = 24$	$c_{11} = 2.51$ $c_{44} = 1.24$ $c_{12} = 1.50$	[3] [4]
Fe_3O_4	$\lambda_{100} = -20$ $\lambda_{111} = 78$	$\lambda_{100} \approx 0$ $\lambda_{111} = 50$		[5] [6]
Ni_3Fe	$\lambda_{100} = 18$ $\lambda_{111} = 5$		$c_{11} = 2.46, c_{44} = 1.24$ $c_{12} = 1.48$	[7]
Fe_3Al	$\lambda_{100} = 42$ $\lambda_{111} = 2.8$		$c_{11} = 1.71, c_{44} = 1.31$ $c_{12} = 1.31$	[8]
$SmCo_5$	$\lambda_{11} = -762$ $\lambda_{12} = -181$ $\lambda_{33} = -47$ $\lambda_{44} = 243$ $\lambda_{66} = -291$		$c_{11} = 1.97, c_{33} = 2.40$ $c_{12} = 1.03, c_{44} = 0.48$ $c_{13} = 1.05, c_{66} = 0.47$	[9]
$(Tb_{0.3}Dy_{0.7})Fe_2$	$\lambda_{100} = 90$ $\lambda_{111} = 1600$		$E = 0.65$ $G = 0.43$	[10]
$TbFe_2$	$\lambda_{111} = 2600$			[11]
$BaFe_{12}O_{19}$	$\lambda_{11} = -15$ $\lambda_{12} = 16$ $\lambda_{33} = -11$ $\lambda_{44} = -48$		$E = 1.52$ $G = 1.48$	[12]

References
 [1] Lee, E.W., 1955, *Rep. Prog. Phys.* **18**, 184.
 [2] McSkimin, H.J., 1955, *J. Appl. Phys.* **26**, 406.
 [3] Corner, W.D., and Hunt, G.H., 1955, *Proc. Phys. Soc. (London)* A **68**, 138.
 [4] Corner, W.D., and Hutchinson, F., 1958, *Proc. Phys. Soc. (London)* A **72**, 1049.
 [5] Miyata, N., and Funatogama, 1962, *J. Phys. Soc. Japan* **17**, 279.
 [6] Blickford, Jr., L.R., Puppis, J., and Stull, J.L., 1955, *Phys. Rev.* **99**, 1210.
 [7] Lichtenberger, F., 1932, *Ann. Physik* **10**, 45; Turdre, P., Plique, F., and Calvayrac, Y., 1975, *Scr. Mat.* **9**, 797.
 [8] Hall, R.C., 1957, *J. Appl. Phys.* **28**, 707; Leany, H.J., Gibson, E.D., and Kagro, F.X., 1967, *Acta Met.* **15**, 1827.
 [9] Doane, D.A., 1977, *J. Appl. Phys.* **48**, 2062; 1977, **48**, 2591.
[10] Jiles, D.C., 1998, *Introduction to Magnetism and Magnetic Materials*, Vol. 1, (Chapman & Hall, New York).
[11] Clark, A.E., 1980, 'Magnetostriction of RE-iron Compounds'. In *Ferromagnetic Materials*, Vol. 1, Ed. E.P. Wohlfarth (North-Holland, Amsterdam) p. 531.
[12] Mason, W.P., 1954, *Phys. Rev.* **96**, 302.

symmetry ϕ' is then written,

$$\phi'_t = A \sum_i (\nabla \gamma_i)^2 + K_1 \sum_{i>j} \gamma_i^2 \gamma_j^2 + K_2 \gamma_1^2 \gamma_2^2 \gamma_3^2$$

$$- \frac{3}{2} \lambda_{100} \sum_i \sigma_{ii} \gamma_i^2 - \frac{3}{2} \lambda_{111} \sum_{i \neq j} \sigma_{ij} \gamma_i \gamma_j$$

$$- \frac{1}{2} \boldsymbol{H}_s \cdot \boldsymbol{J}_s - \boldsymbol{H}_{\text{ext}} \cdot \boldsymbol{J}_s . \tag{2.79}$$

Here it should be noted that γ_i and σ_{ij} refer to the cubic coordinate system.

Another important example is 180°-walls in Ni and Fe and the 109°-wall in Ni which have finite width only if the elastic self energy, $\varepsilon^{\text{el}} \cdot \cdot c \cdot \cdot \varepsilon^{\text{el}}$, is taken into account.

Each of the energy terms in eq. (2.78) has a special influence on the direction of the spontaneous magnetization:

1. ϕ'_{ex} smooths inhomogeneous distributions of $\boldsymbol{J}_s$.
2. ϕ'_K aligns the magnetization into preferred directions, so-called 'easy directions'.
3. The magnetoelastic coupling energy ϕ'_σ similar to ϕ'_K tries to align $\boldsymbol{J}_s$ parallel or perpendicular to the stress axes depending on the sign of the stresses and magnetostriction constants.
4. The stray field energy ϕ'_s tries to avoid magnetic volume and surface charges which leads to the formation of domain patterns in macroscopic samples and tries to align $\boldsymbol{J}_s$ parallel to $\boldsymbol{H}_s$.
5. The magnetostatic energy ϕ'_H tries to align $\boldsymbol{J}_s$ parallel to $\boldsymbol{H}_{\text{ext}}$.

It is of interest to note that the continuum theoretical expression for ϕ'_t has a quantum theoretical analogue in a Hamiltonian given by

$$H = - 2 \sum_{i \neq j} J_{ij} \, \boldsymbol{S}_i \cdot \boldsymbol{S}_j + \lambda \sum_i \boldsymbol{L}_i \cdot \boldsymbol{S}_i$$

$$- \frac{\mu_0 (g \mu_B)^2}{4\pi} \sum_{i \neq j} \left(\frac{\boldsymbol{S}_i \cdot \boldsymbol{S}_j}{R_{ij}^3} - \frac{3(\boldsymbol{S}_i \cdot \boldsymbol{R}_{ij})(\boldsymbol{S}_j \cdot \boldsymbol{R}_{ij})}{R_{ij}^5} \right) - g \mu_0 \mu_B \sum_i \boldsymbol{S}_i \cdot \boldsymbol{H}_{\text{ext}}, \tag{2.80}$$

where we have omitted the elastic energy terms, and the magnetocrystalline energy has been related to the spin–orbit coupling with $\boldsymbol{L}_i$ denoting the orbital magnetic moment operator.

References

[2.1] Brown, W.F., 1940, *Phys. Rev.* **58**, 736.
[2.2] Brown, W.F., 1941, *Phys. Rev.* **60**, 132.
[2.3] Landau, L., and Lifshitz, E., 1935, *Phys. Z. Sowjetunion* **8**, 153.
[2.4] Becker, R., and Döring, W., 1939, *Ferromagnetismus* (Springer, Berlin).

[2.5] Kittel, C., and Galt, J.K., 1956, *Solid State Phys.* **3**, 437.

[2.6] Brown, W.F., Jr., 1963, *Micromagnetics* (Wiley Interscience, New York, London).

[2.7] Kronmüller, H., 1966, 'Magnetisierungskurve der Ferromagnetika'. In *Moderne Probleme der Metallphysik*, Vol. 2 (Ed. A. Seeger, Springer-Verlag, Berlin) p. 24.

[2.8] Träuble, H., 1966, 'Magnetisierungskurve und Hysterese ferromagnetischer Einkristalle'. In *Moderne Probleme der Metallphysik*, Vol. 2 (Ed. A. Seeger, Springer-Verlag, Berlin) p. 157.

[2.9] Aharoni, A., 1996, *Introduction to the Theory of Ferromagnetism* (Clarendon Press, Oxford).

[2.10] Aharoni, A., and Shtrikman, S., 1958, *Phys. Rev.* **109**, 1522.

[2.11] Chikazumi, S., 1997, *Physics of Ferromagnetism* (Clarendon Press, Oxford).

[2.12] Morrish, A.H., 1965, *The Physical Principles of Magnetism* (J. Wiley, New York).

[2.13] Jiles, D., 1990, *Introduction to Magnetism and Magnetic Materials* (Chapman and Hall, London–New York–Tokyo–Melbourne–Madras).

[2.14] Hubert, A., and Schaefer, R., 1998, *Magnetic Domains* (Springer, Berlin-Heidelberg-New York).

[2.15] O'Handley, R.C., 2000, *Modern Magnetic Materials, Principles and Applications* (John Wiley & Sons Inc., New York).

[2.16] Buschow, K.H.J., 1988, 'Permanent Magnet Materials Based on 3d-Rich Ternary Compounds'. In *Ferromagnetic Materials*, Vol. 4 (Eds. E.P. Wohlfarth and K.H.J. Buschow, North-Holland, Amsterdam).

[2.17] Fert, A., and Grünberg, P., 1995, *J. Magn. Magn. Mater.* **140–144**, 1.

[2.18] Clark, A.E., and Abbundi, R., 1977, *IEEE Trans. Magn.* **13**, 1519.

[2.19] Heisenberg, W., 1926, *Z. Physik* **38**, 441.

[2.20] Kittel, C., and Herring, C., 1951, *Phys. Rev.* **81**, 869.

[2.21] Collins, M.F., Minkiewicz, V.J., Nathans, R., Russell, L., and Shirane, G., 1969, *Phys. Rev.* **179**, 417.

[2.22] Loony, C.-K., Carpenter, J.M., Lyme, J.W., Robinson, R.A., and Mook, A., 1984, *J. Appl. Phys.* **55**, 1895.

[2.23] Alperin, H., Steinsvoll, O., Shirane, G., and Nathans, R., 1956, *J. Appl. Phys.* **37**, 1052.

[2.24] Shirane, G., Minkiewicz, V.J., and Nathans, R., 1968, *J. Appl. Phys.* **39**, 383.

[2.25] Riste, T., Shirane, G., Alperin, H.A., and Pickart, S.J., 1965, *J. Appl. Phys.* **36**, 1076.

[2.26] Minkiewicz, V.J., Collins, M.F., Nathans, R., and Shirane, G., 1969, *Phys. Rev.* **182**, 624.

[2.27] Hermion, M., Hermion, B., Castets, A., and Tochetti, D., 1975, *Solid State Commun.* **17**, 899.

[2.28] Mayer, H.M., Steiner, M., Stüßer, N., Weinfurter, H., Dorner, B., Lindgard, P.A., Clausen, K.N., Hock, S., and Verhoef, R., 1992, *J. Magn. Magn. Mater.* **104–107**, 1295.

[2.29] Durst, K.-D., and Kronmüller, H., 1986, *J. Magn. Magn. Mater.* **59**, 86.

[2.30] Szymczak, R., Burzo, E., and Wallace, W.E., 1985, *J. Physique* **9**, C6-309.

[2.31] Kütterer, R., Hilzinger, H.R., and Kronmüller, H., 1977, *J. Magn. Magn. Mater.* **4**, 1.

[2.32] Durst, K.-D., Kronmüller, H., and Ervens, W., 1988, *Phys. Stat. Sol. (A)* **108**, 403; **108**, 705.

[2.33] Coey, J.M.D., 1946, *Rare-Earth Iron Permanent Magnets* (Clarendon Press, Oxford) p. 45; Skomski, R., and Coey, J.M.D., 1999, *Permanent Magnetism* (Institute of Physics Publishing, Bristol).

[2.34] Zener, C., 1951, *Phys. Rev.* **81**, 440; *ibid.* **82**, 403; **83**, 299.

[2.35] Kasuya, T., 1956, *Progr. Theor. Phys. (Kyoto)* **16**, 45; *ibid.* 58.

[2.36] Yosida, K., 1957, *Phys. Rev.* **106**; 893.

[2.37] Ruderman, M.A., and Kittel, C., 1954, *Phys. Rev.* **96**, 99.

[2.38] Kronmüller, H., 1972, *Int. J. Magn.* **3**, 211.

[2.39] Elliot, R.J., 1961, *Phys. Rev.* **124**, 346.

[2.40] Enz, V., 1961, *J. Appl. Phys.* **32**, 225.

[2.41] Kronmüller, H., and Schmid, W., 1975, *Physica B* **80**, 330.

[2.42] Herz, R., and Kronmüller, H., 1978, *J. Magn. Magn. Mater.* **9**, 273; 1978, *Phys. Stat. Sol. (A)* **47**, 451.

[2.43] Birss, P.R., 1964, *Symmetry and Magnetism* (Ed. E.P. Wohlfarth, North-Holland, Amsterdam).

[2.44] Akulov, N.S., 1928, *Z. Physik* **52**, 389; *ibid.* 1929, **54**, 582; 1929, **57**, 249; 1930, **59**, 254; 1931, **69**, 78.

[2.45] Coehoorn, R., and Dalderop, G.H.O., 1992, *J. Magn. Magn. Mater.* **104–107**, 1081.

[2.46] Cadogan, J.M., and Coey, J.M.D., 1984, *Phys. Rev.* **B30**, 7326.

[2.47] Coehoorn, R., 1991, 'Electron Structure Calculations for Rare Earth Transition Metal Compounds'. In *Supermagnets, Hard Magnetic Materials*, Vol. 331, NATO ASI Series C (Eds. G.J. Long and F. Grandjean, Kluwer, Dordrecht) p. 133.

[2.48] Zener, C., 1954, *Phys. Rev.* **96**, 1335.

[2.49] Callen, E.R., and Callen, H.B., 1960, *Z. Phys. Chem. Solids*, **16**, 310.

[2.50] Néel, L., 1945, *Compt. Rend.* **220**, 814.

[2.51] Néel, L, 1948, *J. Phys. Rad.* **9**, 184; *ibid.* **9**, 193.

[2.52] Brown, W.F., Jr., 1962, *Magnetostatic Principles in Ferromagnetism* (North-Holland, Amsterdam).

[2.53] Becker, R., 1930, *Z. Physik* **62**, 253; *ibid.* 1932, **33**, 905; 1934, **87**, 547.

[2.54] Brown, W.F., Jr., 1966, *Magnetoelastic Interactions*, Springer Tracts in Natural Philosophy, Vol. 9 (Ed. C. Truesdell, Springer-Verlag, Berlin–Heidelberg–New York).

[2.55] Brown, W.F., Jr., 1953, *Rev. Mod. Phys.* **25**, 131.

[2.56] Rieder, G., 1954, *Abh. Braunschw. Wiss. Ges.* **11**, 20.

[2.57] Kneller, E., 1962, *Ferromagnetismus* (Springer-Verlag, Berlin–Göttingen–Heidelberg).

[2.58] Landau, L.D., and Lifshitz, E.M., 1960, *Electrodynamics of Continuum Media* (Pergamon Press, Oxford–London–New York–Paris) p. 155.

[2.59] Kröner, E., 1958, *Kontinuumstheorie der Versetzungen und Eigenspannungen* (Springer Verlag, Berlin–Göttingen–Heidelberg).

[2.60] Lilley, B.A., 1950, *Philos. Mag.* **41**, 792.

3

Basic micromagnetic equilibrium conditions

3.1 Static micromagnetic equations

The static micromagnetic equilibrium conditions are usually formulated as a torque equation [2.2, 2.6, 2.9, 2.12, 3.1],

$$L = [J_s \times H_{eff}] = 0, \tag{3.1}$$

where H_{eff} corresponds to an effective field composed of the external field and contributions of exchange, anisotropy, dipolar and magnetoelastic energies. The condition of a vanishing torque may also be interpreted as a local material law,

$$J_s = \mu_0(\mu - 1)H_{eff}, \qquad M_s = \chi H_{eff}, \tag{3.2}$$

with a local space and orientation dependent effective permeability or susceptibility,

$$\mu - 1 = \frac{J_s}{\mu_0 |H_{eff}|}, \qquad \chi = \frac{M_s}{|H_{eff}|}. \tag{3.3}$$

From these considerations it becomes obvious that the main purpose of micromagnetism is the determination of the effective field H_{eff}, i.e., of the direction of the spontaneous magnetization. This point of view may be taken if the modulus of the spontaneous magnetization is assumed to be a constant with respect to magnetic fields. As is well known, this approximation is valid for $T < T_c$; however, it fails for temperatures $T \simeq T_c$ where the magnetic phase transition takes place. This case will be treated in Chapter 12. In the derivation of the micromagnetic equations, Maxwell's equations only play a role in the determination of the stray field using the equation of continuity div $B = 0$. Furthermore, it should be noted that the material laws, eq. (3.2), open the possibility of introducing a space-dependent permeability in Maxwell's equations.

In order to derive the micromagnetic equilibrium conditions (eq. (3.1)) and the effective field, the Gibbs free energy under the constraint,

$$\sum_{i=1}^{3} \gamma_i^2 = 1 \quad \text{or} \quad \sum_{i=1}^{3} J_{s,i}^2 = J_s^2, \tag{3.4}$$

has to be minimized with respect to the direction cosines γ_i. This leads to the variational problem,

$$\delta_{\gamma_i} \phi_t = \delta_{\gamma_i} \int \left\{ \phi_{ex}' + \phi_K' + \phi_{el}' + \phi_s' + \phi_H' + \lambda \left(\sum_{i=1}^{3} \gamma_i^2 - 1 \right) \right\} d^3 r = 0, \tag{3.5}$$

where the last term in eq. (3.5) takes care of the constraint (3.4) by a Lagrange parameter λ. In eq. (3.5) in general not all energy terms are explicitly known as functions of the direction cosines γ_i. The exchange energy contains $\nabla \gamma_i$, the magnetoelastic energy term, the elastic strains ε^{el} and the dipolar energy of the stray field H_s. Both quantities, H_s and ε^{el}, are determined by differential equations (eq. (2.41), eq. (2.62)) the solutions of which contain the unknown direction cosines γ_i or the components $M_{s,i}$ of the spontaneous magnetization, respectively. H_s as well as ε^{el} in principle can be represented by integrals and consequently the variational problem (eq. (3.5)) leads to integro-differential equations or to a set of coupled differential equations. For solving the variational problem we rearrange eq. (3.5) as follows:

$$\delta \phi_t = \int \left\{ 2A \sum_{i=1}^{3} \nabla \gamma_i \, \delta(\nabla \gamma_i) - \frac{1}{2} (H_s \cdot \delta J_s + J_s \cdot \delta H_s) \right. \tag{3.6}$$

$$\left. + \sum_{i=1}^{3} (\partial \phi_K' / \partial \gamma_i + \partial \phi_{el}' / \partial \gamma_i - J_s H_{ext,i} + 2 \lambda \gamma_i) \delta \gamma_i \right\} d^3 r.$$

The exchange term contains the variation of $\delta(\nabla \gamma_i)$ and must be transformed into $\delta \gamma_i$ by partial integration. Using the operator $\partial_i = \partial / \partial x_i$, the exchange term may be rewritten as (using Einstein's sum convention and the interchange of ∂_i and δ)

$$\partial_j \gamma_i \cdot \delta(\partial_j \gamma_i) = \partial_j \{ (\partial_j \gamma_i) \delta \gamma_i \} - \partial_j^2 \gamma_i \, \delta \gamma_i. \tag{3.7}$$

Since the first term corresponds to a divergence it can be transformed by means of Gauss's theorem into a surface integral, thus giving for the exchange term

$$\int 2A \sum_{i=1}^{3} \nabla \gamma_i \, \delta \nabla \gamma_i \, d^3 r = 2A \int_S \nabla \gamma_i \, \delta \gamma_i \, df - 2A \int_V \Delta \gamma_i \, \delta \gamma_i \, d^3 r. \tag{3.8}$$

In the case of the dipolar energy we deal with variation of $\delta J_s(\delta\gamma_i)$ as well as δH_s because both fields are coupled with each other via the continuity condition div $\boldsymbol{B} = 0$. From vector analysis it follows that

$$\int A_1 \cdot A_2 \, d^3r = 0, \qquad (3.9)$$

if div $A_1 = 0$ and curl $A_2 = 0$ holds. Applying this relation to $\boldsymbol{J}_s$ and $\boldsymbol{H}_s$ it may be shown that

$$\int \boldsymbol{H}_s \cdot \delta \boldsymbol{J}_s \, d^3r = \int \boldsymbol{J}_s \cdot \delta \boldsymbol{H}_s \, d^3r, \qquad (3.10)$$

$$\frac{1}{2} \int (\boldsymbol{H}_s \cdot \delta \boldsymbol{J}_s + \boldsymbol{J}_s \cdot \delta \boldsymbol{H}_s) \, d^3r = \int \boldsymbol{H}_s \cdot \delta \boldsymbol{J}_s d^3r. \qquad (3.11)$$

Inserting our results from eq. (3.8) to eq. (3.11) into eq. (3.6) gives for the variation with respect to γ_i,

$$\delta_{\gamma_i} \phi_t = \int_V \left\{ -2A \, \Delta\gamma_i + \left(\partial\phi'_K/\partial\gamma_i + \partial\phi'_{el}/\partial\gamma_i \right) - H_{s,i} J_s - H_{ext,i} J_s \right.$$

$$\left. + 2\lambda\gamma_i \right\} \delta\gamma_i \, d^3r + 2A \int_S \nabla\gamma_i \, \delta\gamma_i \, df = 0, \qquad i = 1, 2, 3.$$

$$(3.12)$$

The variations $\delta_{\gamma_i}\phi$ vanish if the integrands of the volume and surface integrals are zero for all variations $\delta\gamma_i$. The differential equations and boundary conditions for the magnetic equilibrium state therefore are written,

$$2A \, \Delta\gamma_i - \partial\phi'_K/\partial\gamma_i - \partial\phi'_{el}/\partial\gamma_i + \left(H_{s,i} M_s + H_{ext,i} \right) J_s - 2\lambda\gamma_i = 0,$$

$$i = 1, 2, 3, \qquad (3.13)$$

in the volume, and

$$\boldsymbol{n} \cdot \nabla_n\gamma_i = 0, \qquad (3.14)$$

on the surface, where $\boldsymbol{n}$ denotes the surface normal, and $\nabla_n\gamma_i$ the gradient of γ_i parallel to $\boldsymbol{n}$.

Equation (3.1) still contains the Lagrange parameter λ, which in principle can be determined together with the γ_i from eq. (3.1) and the constraint of eq. (3.4).

The equilibrium conditions in the form of a torque equation may be derived from eq. (3.1) by eliminating the term $-2\lambda\gamma_i$. This is easily done by multiplying the differential equation for γ_i by γ_j and vice versa and subtracting the corresponding expressions. By this procedure we obtain three expressions of the type,

$J_{s,j} H_{\text{eff},i} - J_{s,i} H_{\text{eff},j} = 0$, which may be considered as the components of the vector product of $\boldsymbol{J}_s$ and an effective field, $\boldsymbol{H}_{\text{eff}}$, thus leading to the micromagnetic equilibrium condition in the volume

$$[\boldsymbol{J}_s \times \boldsymbol{H}_{\text{eff}}] = 0, \tag{3.15}$$

and

$$[\boldsymbol{J}_s \times \nabla_n \boldsymbol{J}_s] = 0,$$

on the surface with the components of the effective field

$$H_{\text{eff},i} = -\partial\phi'_t/\partial\boldsymbol{J}_s = (2A/J_s)\,\Delta\gamma_i + H_{K,i} + H_{\sigma,i} + H_{s,i} + H_{\text{ext},i}, \tag{3.16}$$

where we have introduced effective fields $\boldsymbol{H}_K$ and $\boldsymbol{H}_\sigma$ of the anisotropy energy and elastic energy defined as

$$H_{K,i} = -(1/J_s)\partial\phi'_K/\partial\gamma_i, \tag{3.17}$$
$$H_{\sigma,i} = -(1/J_s)\partial\phi'_{\text{el}}/\partial\gamma_i.$$

In addition to the equilibrium conditions (eqs. (3.15)), Poisson's equation,

$$\Delta U^{(i)} = \text{div}\,\boldsymbol{M}_s(\boldsymbol{r}), \tag{3.18}$$
$$\boldsymbol{H}_s^{(i)} = -\nabla U^{(i)},$$

inside the volume and the Laplace equations,

$$\Delta U^{(o)} = 0, \tag{3.19}$$
$$\boldsymbol{H}_s^{(o)} = -\nabla U^{(o)},$$

outside the volume have to be taken into account. For $U^{(i)}$ and $U^{(o)}$ the following boundary conditions,

$$U^{(i)}(\boldsymbol{r}) = U^{(o)}(\boldsymbol{r}), \tag{3.20}$$

hold on the surface, and from the continuity equation it follows for the normal components of $\boldsymbol{H}_s^{(i,o)}$ and $\boldsymbol{M}_s$:

$$H_{s,n}^{(i)} + M_{s,n} = H_{s,n}^{(o)}. \tag{3.21}$$

The full problem is finally formulated by taking into account also the elasticity equilibrium equation:

$$\text{Div}\,(c \cdot\cdot\, \varepsilon^{\text{el}}) = -\,\text{Div}\,(c \cdot\cdot\, \varepsilon^Q). \tag{3.22}$$

The micromagnetic magnetostatic and elasticity differential equations given by eqs. (3.15) to (3.22) correspond to a system of coupled differential equations for the unknown variables γ_i. If $\boldsymbol{H}_s$ is represented as an integral using eq. (2.41) and the elasticity differential equation (3.22) is integrated [2.7], it becomes obvious that the system of coupled equations represent a rather complicated nonlinear integro-differential equation. Solutions of these equations have been found mainly for three fundamental micromagnetic problems: domain walls in bulk materials and thin films [2.7, 2.56, 2.58, 2.60, 3.1, 3.2], nucleation problems [2.6, 2.9, 2.10], and the magnetization process in the approach to ferromagnetic saturation [2.7, 3.3–3.5].

3.2 Micromagnetic equations in polar coordinates

Sometimes it is desirable to have equilibrium equations which are free from the Lagrange parameter λ. One way to achieve this is to introduce spherical angular coordinates φ and θ, where φ denotes the azimuthal and θ the polar angle, thus leading to

$$\gamma_1 = \sin\theta\,\cos\varphi, \qquad \gamma_2 = \sin\theta\,\sin\varphi, \qquad \gamma_3 = \cos\theta. \tag{3.23}$$

Then the fundamental equations of micromagnetism become:

$$2A\,\Delta\theta \;-\; A\,\sin 2\theta(\nabla\varphi)^2 \;-\; \frac{\partial}{\partial\theta}\big(\phi'_K + \phi'_{el} + \phi'_s - \boldsymbol{J}_s\cdot\boldsymbol{H}_{ext}\big) = 0$$

$$2A\big\{\sin^2\theta\,\Delta\varphi \;+\; \sin 2\theta\,(\nabla\varphi)\cdot(\nabla\theta)\big\} \;-\; \frac{\partial}{\partial\varphi}\big(\phi'_K + \phi'_{el} + \phi'_s - \boldsymbol{J}_s\cdot\boldsymbol{H}_{ext}\big) = 0.$$
$$\tag{3.24}$$

Equations (3.24) are especially suitable to treat Bloch and Néel walls. Naturally, in addition to eqs. (3.24), Poisson's equation and the micromagnetic and magnetostatic boundary conditions also have to be taken into account.

3.3 Micromagnetic equations in terms of swirls and magnetic charges

The solution of the torque equation (3.1) according to eq. (3.16) may be written as the following differential equation:

$$\frac{2A}{J_s^2}\,\Delta\boldsymbol{J}_s \;+\; \boldsymbol{H}_K \;+\; \boldsymbol{H}_\sigma \;+\; \boldsymbol{H}_s \;+\; \boldsymbol{H}_{ext} = \lambda\,\boldsymbol{M}_s, \tag{3.25}$$

where $\lambda = 1/\chi = |\boldsymbol{H}_{eff}|/M_s$ corresponds to a Lagrange parameter with the meaning of a local, spin-dependent reciprocal susceptibility. The vector field of $\boldsymbol{M}_s$ can be derived from a scalar potential $U(\boldsymbol{r})$ and a vector potential $\boldsymbol{A}(\boldsymbol{r})$ according to

$$\boldsymbol{M}_s(\boldsymbol{r}) = \operatorname{grad} U(\boldsymbol{r}) + \operatorname{curl} \boldsymbol{A}(\boldsymbol{r}), \tag{3.26}$$

where

$$U(r) = -\frac{1}{4\pi} \int \frac{\operatorname{div} M_s(r')}{|r - r'|} d^3 r', \tag{3.27}$$

$$A(r) = \frac{1}{4\pi} \int \frac{\operatorname{curl} M_s(r')}{|r - r'|} d^3 r'.$$

In order to satisfy the magnetostatic boundary condition $U(r)$ has to include a solution U_0 of the Laplace equation $\Delta U_0 = 0$ with the boundary condition $\nabla_n U_0 = M_n$ on the surface.

According to eq. (3.26) and eqs. (3.27), the vector field $M_s(r)$ is fully determined by $\operatorname{div} M_s$ and $\operatorname{curl} M_s$. Introducing these new scalar and vector fields as new variables,

$$\psi_1 = \operatorname{div} M_s(r), \qquad \psi_2(r) = \operatorname{curl} M_s(r), \tag{3.28}$$

the differential equations for ψ_1 and ψ_2 are found from eq. (3.25) by applying the operators div and curl. This leads to the following set of differential equations (div $H_{\text{ext}} \equiv 0$) [3.6]:

$$\frac{2A}{\mu_0 M_s^2} \Delta \psi_1 - \{1 + \lambda\}\psi_1 = -\operatorname{div}(H_K + H_\sigma) + M_s \cdot \nabla \lambda,$$

$$\frac{2A}{\mu_0 M_s^2} \Delta \psi_2 - \lambda \psi_2 = -\operatorname{curl}(H_K + H_\sigma + H_{\text{ext}}) + [\nabla \lambda \times M_s].$$

$$\tag{3.29}$$

In contrast to the conventional micromagnetic equations, eqs. (3.29) contain only local quantities because the stray field terms are replaced by the volume charges ψ_1. On the other hand the parameter λ, related to the modulus of the effective field by $\lambda = |H_{\text{eff}}|/M_s$, complicates the solution of these equations. Simplified equations are obtained in the following cases:

1. In the approach to ferromagnetic saturation λ may be considered as a constant, $|H_{\text{ext}}|/M_s$, or in the case of large magnetocrystalline anisotropy it is given by $2K_1/\mu_0 M_s^2$.
2. In the case of stray field free configurations $\psi_1 \equiv 0$ holds.

Equations (3.29) have been used to investigate stripe and ripple domains [3.6, 3.7] leading to different results as obtained by Hoffmann [3.8]. Using recursion techniques eqs. (3.29) can be solved iteratively by a suitable zero-order ansatz for λ.

3.4 Linearized micromagnetic equations

In many cases the spontaneous magnetization deviates only slightly from a preferred direction determined either by a large applied magnetic field or by a strong uniaxial anisotropy. In such a case the magnetic torques producing the magnetic inhomogeneity, e.g., defect structures, may be treated as a perturbation. If the angular deviations of the inhomogeneities in the direction of M_s are smaller than $30°$ the micromagnetic equations may be linearized, leading in general to inhomogeneous differential equations of second order with constant coefficients. In the following we choose the y-coordinate as the preferred direction, whereas the x- and z-coordinates are either chosen arbitrarily or according to some symmetry requirements if these are present. Within this coordinate system [2.2, 3.3, 3.5], known as Brown's coordinate system, the direction cosines obey the following relations:

$$\gamma_1 \ll 1, \qquad \gamma_2 \simeq 1, \qquad \gamma_3 \ll 1, \tag{3.30}$$

$$\gamma_2 = \sqrt{1 - \gamma_1^2 - \gamma_3^2} \simeq 1 - \frac{1}{2}(\gamma_1^2 + \gamma_3^2).$$

By putting $\gamma_2 \simeq 1$ we have fulfilled approximately our constraint $\sum_{i=1}^{3} \gamma_i^2 = 1$ and thus may omit the Lagrange term in our variational problem (eq. (3.5)). The variation $\delta_{\gamma_i} \phi_t = 0$ then has to be determined with respect to γ_i referring to Brown's coordinate system. Whereas exchange energy, stray field energy and magnetostatic energy may be easily expressed by Brown's γ_i, the anisotropic energy terms, ϕ_k' or ϕ_{el}', are presented by γ_i'. If the direction cosines between Brown's coordinates (x, y, z) and the symmetry axes (x', y', z') are given by β_{ki}, we have the following transformation:

$$\gamma_i' = \sum_k \beta_{ki} \gamma_k. \tag{3.31}$$

By means of eq. (3.31), ϕ_k' as well as ϕ_{el}' can be easily expressed by the γ_k. Within the framework of a linearized theory all energy terms up to the second order $\gamma_i \gamma_j$, with $\gamma_2 = 1 - 1/2(\gamma_1^2 + \gamma_3^2)$, are retained, thus giving

$$\phi_k'(\gamma_i) = g_0^k + g_i^k \gamma_i + \frac{1}{2} g_{ij}^k \gamma_i \gamma_j, \qquad i, j \neq 2, \tag{3.32}$$

$$\phi_{el}'(\gamma_i) = g_0^{el} + g_i^{el} \gamma_i + \frac{1}{2} g_{ij}^{el} \gamma_i \gamma_j, \qquad i, j \neq 2.$$

The tensor components g_i and g_{ij} follow from the transformation by eq. (3.31) if this is inserted into the original energy terms. The Gibbs free energy density submitted to variation with respect to γ_1 and γ_3 now may be written as

$$\phi_t' = A\{(\nabla \gamma_1)^2 + (\nabla \gamma_3)^2\} + \sum_{i=1,3} (g_i^k + g_i^{el}) \gamma_i \tag{3.33}$$

$$+ \frac{1}{2} \sum_{i=1,3} (g_{ij}^k + g_{ij}^{el}) \gamma_i \gamma_j - \frac{1}{2} H_s \cdot J_s - H_{ext} J_s \gamma_2,$$

with γ_2 given by eq. (3.30). Here the variation of the stray field energy requires special consideration. According to eqs. (3.10) and (3.11) we have

$$\delta \int \phi_s \, d^3r \; = \; - \int H_s \cdot \delta J_s \, d^3r . \tag{3.34}$$

In components and in terms of γ_i the integrand of eq. (3.34) may be decomposed into

$$\delta \phi_s' \; = \; -J_s H_{s,1} \delta \gamma_1 \; - \; J_s H_{s,2} \delta \gamma_2 \; - \; J_s H_{s,3} \delta \gamma_3 . \tag{3.35}$$

Replacing in eq. (3.35) $\delta \gamma_2 = - \gamma_1 \delta \gamma_1 - \gamma_3 \delta \gamma_3$, which follows from eq. (3.30), we find

$$\delta \phi_s' \; = \; -J_s \big(H_{s,1} - H_{s,2} \gamma_1 \big) \delta \gamma_1 \; - \; J_s \big(H_{s,3} - H_{s,2} \gamma_3 \big) \delta \gamma_3 . \tag{3.36}$$

If the variation of the total Gibbs free energy is performed with the energy density expression eq. (3.33) by taking into account eq. (3.36), the following micromagnetic equilibrium equations for the volume are obtained:

$$2A \, \Delta \gamma_i - J_s \big(H_{\mathrm{ext}} + H_{s,2} \big) \gamma_i - \big(g_{ii}^k + g_{ii}^{\mathrm{el}} \big) \gamma_i$$
$$- \big(g_{ij}^k + g_{ij}^{\mathrm{el}} \big) \gamma_j + J_s H_{s,i} = g_i^k + g_i^{\mathrm{el}} , \qquad i, j = 1, 3, \qquad i \neq j . \tag{3.37}$$

On the surface the boundary conditions

$$\nabla_n \gamma_i \; = \; 0, \qquad i = 1, 3, \tag{3.38}$$

hold. Poisson's equation for the potential U within the framework of the linearized equations is given by

$$\Delta U \; = \; M_s \left(\frac{d\gamma_1}{dx} + \frac{d\gamma_3}{dz} \right) . \tag{3.39}$$

The solution of these linearized coupled, second-order, inhomogeneous differential equations is most effectively performed by introducing Fourier transforms [3.4]:

$$\tilde{\gamma}_i(k) \; = \; \frac{1}{(2\pi)^{3/2}} \int_V \gamma(r) \, e^{-ik \cdot r} d^3r ,$$

$$\gamma_i(r) \; = \; \frac{1}{(2\pi)^{3/2}} \int_{V_k} \tilde{\gamma}(k) \, e^{ik \cdot r} d^3k ,$$

$$\tilde{U}(k) \; = \; \frac{1}{(2\pi)^{3/2}} \int_V U(r) \, e^{-ik \cdot r} d^3r ,$$

$$U(r) \; = \; \frac{1}{(2\pi)^{3/2}} \int_{V_k} \tilde{U}(k) \, e^{ik \cdot r} d^3k ,$$

$$\tilde{g}_i(\mathbf{k}) = \frac{1}{(2\pi)^{3/2}} \int_V g_i(\mathbf{r}) \, e^{-i\mathbf{k}\cdot\mathbf{r}} \mathrm{d}^3\mathbf{r},$$

$$g_i(\mathbf{r}) = \frac{1}{(2\pi)^{3/2}} \int_{V_k} \tilde{g}_i(\mathbf{k}) \, e^{i\mathbf{k}\cdot\mathbf{r}} \mathrm{d}^3\mathbf{k}. \tag{3.40}$$

Introducing eqs. (3.40) into eqs. (3.37) to (3.39) gives the following solution for $\tilde{\gamma}_i$ and $\tilde{U}$, where we neglect the second-order terms $g_{ij}^{k,\mathrm{el}}$ and $H_{\mathrm{s},2}$:

$$\tilde{\gamma}_i(\mathbf{k}) = -\frac{1}{2A}\frac{\tilde{g}_i}{k^2+\kappa_H^2} + \frac{\kappa_s^2}{2A} \times \frac{k_i}{k^2+\kappa_H^2} \times \frac{\tilde{g}_x k_x + \tilde{g}_z k_z}{k^4 + (\kappa_H^2+\kappa_s^2)k^2 - \kappa_s^2 k_y^2}, \tag{3.41a}$$

where $\tilde{g}_i$ means $\tilde{g}_i^k$ or $\tilde{g}_i^{\mathrm{el}}$ or the sum of both, and

$$\tilde{U}(\mathbf{k}) = -\frac{M_s}{k^2}(ik_x\tilde{\gamma}_1 + ik_z\tilde{\gamma}_3),$$

$$\tilde{H}_s(\mathbf{k}) = -i\mathbf{k} \times \tilde{U}(\mathbf{k})/(2\pi)^{3/2}. \tag{3.41b}$$

Taking into account the RKKY long-range exchange interactions in eq. (3.41a), the term Ak^2 has to be replaced by $(J(0) - J(\mathbf{k})) \cdot M_s^2/(g\mu_B)^2$. Equations (3.41) include the so-called exchange lengths [2.7, 3.7] of the external field and of the stray field:

$$l_H = \kappa_H^{-1} = \sqrt{\frac{2A}{J_s H_{\mathrm{ext}}}}, \qquad l_s = \kappa_s^{-1} = \sqrt{\frac{2A}{J_s M_s}}. \tag{3.42}$$

If in the derivation of eq. (3.41a) the magnetocrystalline energy ϕ_K' or the magnetoelastic coupling energy ϕ_σ' are taken into account, two additional exchange lengths play a role [3.7]:

$$l_K = \kappa_K^{-1} = \sqrt{\frac{A}{K_1}}, \qquad l_\sigma = \kappa_\sigma^{-1} = \sqrt{\frac{2A}{3\lambda_{100}\sigma_{11}}}. \tag{3.43}$$

Here we have considered uniaxial crystals and a uniaxial stress state along the $\langle 100\rangle$-direction. In Chapter 4, l_K also is denoted as the Bloch wall parameter δ_0. The meaning of these exchange lengths can easily be obtained by considering planar δ-function like perturbations, $g_i(\mathbf{r}) = g_0 \delta(y - y_0)$, which for one-dimensional problems leads to exponential solutions,

$$\gamma_i(y) = \frac{\pi g_0}{2A\kappa^2} \exp[-|y|/l], \tag{3.44}$$

Table 3.1. *Exchange lengths of transition metals and intermetallic compounds using the material parameters of Tables 2.1–2.3. All lengths in units of nm = 10 Å (H_{ext} = 1000 Oe = $10^6/(4\pi)$ A/m; σ_{11} = 10 MN/m²).*

Material	l_K	l_s	l_H	l_σ
α-Fe	21	3.3	15.6	80
Co	8.3	4.9	20.4	56
Ni	42	8.7	16.1	31
Ni_3Fe	57	3.7	12.7	51
$Nd_2Fe_{14}B$	1.4	2.8	11.2	
$Pr_2Fe_{14}B$	1.4	3.1	14.0	
$SmCo_5$	0.84	5.3	17.1	
Sm_2Co_{17}	1.83	4.6	16.6	
$Sm_2Fe_{17}N_3$	1.18	4.4	14.0	
$BaFe_{12}O_{19}$	4.45	8.3	18.2	

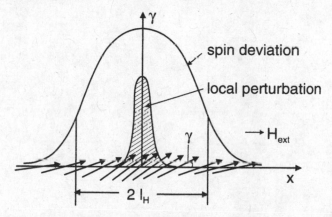

Fig. 3.1. Distribution of magnetization around a planar perturbation of the orientation of magnetization under the action of an applied field. Depending on the energy terms which exert a torque on J_s the extension of the spin perturbation is governed by one of the exchange lengths.

where for κ and l the dominant effective fields have to be inserted. Due to the assumption of a δ-function type perturbation the boundary condition $d\gamma_i/dy = 0$ at $y = 0$ is violated. Actually, the distribution of γ_i' is smoothed out, as shown in Fig. 3.1. The exchange lengths defined by eqs. (3.42) and (3.43) show large variations depending on the different types of materials, as presented in Table 3.1. In general, all types of exchange lengths play some role; however, in the case of inhomogeneous spin states the exchange length resulting in the lowest energy state determines the extension of the spin inhomogeneity. For example, in Bloch walls l_K is dominant, and in Néel walls l_s (see Sections 4.2 and 4.6).

References

[3.1] Kronmüller, H., 1995, 'Micromagnetism in Modern Magnetic Materials'. In *Aspects of Modern Magnetism* (Eds. F.C. Pu, Y.Z. Wang, and C.H. Shang, World Scientific, Singapore) p. 33.

[3.2] Hubert, A., 1974, *Theorie der Domänenwände in geordneten Medien* (Springer Verlag, Berlin–Heidelberg–New York).

[3.3] Brown, Jr., W.F., 1941, *Phys. Rev.* **60**, 132.

[3.4] Seeger, A., and Kronmüller, H., 1960, *J. Phys. Chem. Solids* **12**, 298.

[3.5] Kronmüller, H., and Seeger, A., 1961, *J. Phys. Chem. Solids* **18**, 93.

[3.6] Kronmüller, H., 1971, *Z. Angew. Physik* **32**, 49.

[3.7] Brown, Jr., W.F., 1970, *IEEE Trans. Magn.* **6**, 121.

[3.8] Hoffmann, H., 1964, *Phys. Kondens. Materie* **2**, 22.

4

Domain walls in crystalline and amorphous solids

4.1 General remarks

Domain walls are regions separating neighbouring magnetic domains of different orientations of the spontaneous magnetization. The driving energy for the splitting up of a macroscopic ferromagnet into domains is due to the stray field energy ϕ_s. The reduction of ϕ_s is accompanied by an increase of the total domain wall energy. The equilibrium state therefore is characterized by the minimum of the stray field energy and the total domain wall energy. According to eq. (3.1) the general equilibrium condition, valid for all magnetic states – for zero and applied fields – predicts that the spontaneous magnetization aligns either parallel to the effective field or that the effective field vanishes. A general analytical solution of the micromagnetic equilibrium conditions for macroscopic three-dimensional ferromagnets which are split into domains still does not exist. For a quantitative treatment, however, it turns out that an approximate solution of the equilibrium equation is obtained by determining the total domain wall energy and the energy connected with the domains for a given domain arrangement separately [2.3–2.5]. In this approximation the inhomogeneous regions of magnetization are concentrated to the domain walls, whereas within the domains only homogeneous magnetic states are considered. The domain arrangement with the lowest energy then has to be considered as the macroscopic magnetic ground state. As a first step for a quantitative understanding of domain patterns, therefore, the spin arrangement and the magnetic parameters of domain walls have to be determined.

4.2 Bloch walls

Bloch walls nowadays are defined as those walls which turn their magnetization vector without creating magnetic stray fields. In the case of a planar domain wall, where the rotation angle φ of the magnetization depends only on the z-coordinate,

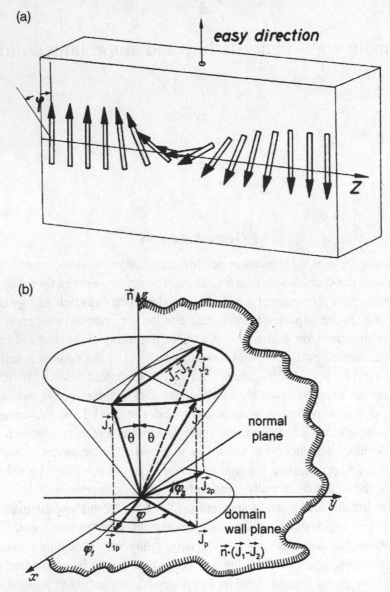

Fig. 4.1. (a) Distribution of magnetization in a 180°-domain wall. (b) Definition of Bloch wall parameters. Bloch wall plane is defined by $n \cdot (J_1 - J_2) = 0$. Bloch wall angle $\Theta = \measuredangle(J_1, n) = \text{const.}$ Azimuthal angle $\varphi = \measuredangle(x, J_p)$.

which is taken perpendicular to the domain wall plane, this condition requires constancy of the angle Θ between J_s and the wall normal n throughout the wall, as determined by Fig. 4.1:

$$J_s \cdot n = J_s \cos \Theta. \tag{4.1}$$

In terms of the spontaneous polarization J_s^I and J_s^{II} within the neighbouring domains this condition is written

$$\left(J_s^I - J_s^{II}\right) \cdot n = 0. \tag{4.2}$$

Since planar Bloch walls are fully described by the space dependent azimuthal rotation angle, $\varphi(z)$, and the constant polar angle, Θ, the micromagnetic equation describing the domain wall is easily derived from eqs. (3.24), giving

$$2A \sin^2 \Theta \frac{d^2\varphi}{dz^2} - \frac{\partial}{\partial\varphi}(\phi_B') = 0, \tag{4.3}$$

where $\phi_B' = \phi_K' + \phi_{el}'$ includes the anisotropy energy and the elastic energies related to magnetostrictive deformations. In the following we assume that ϕ_B' is given as a function of φ. In this case eq. (4.3) may be easily integrated, which leads to

$$A \sin^2 \Theta \left(\frac{d\varphi}{dz}\right)^2 - \left(\phi_B'(\varphi) - \phi_B'(\varphi_I)\right) = 0, \tag{4.4}$$

or

$$z = \sqrt{A} \sin \Theta \int_{\varphi_I}^{\varphi} \frac{d\varphi}{\left(\phi_B'(\varphi) - \phi_B'(\varphi_I)\right)^{1/2}}, \tag{4.5}$$

where $\phi_B'(\varphi_I)$ denotes the value of ϕ_B' in domain I, i.e., for $z \to -\infty$. For a first calculation we neglect the elastic energy terms and consider only $\phi_K' = K_1 \sin^2 \varphi$ for a uniaxial crystal or an amorphous material with an induced anisotropy constant $K_1 = K_u$. Integration of eq. (4.5) gives us, for a 180°-wall ($\Theta = \pi/2$, rotation of M_s from $\varphi_I = 0$ to $\varphi_{II} = \pi$),

$$\operatorname{tg} \frac{\varphi}{2} = e^{z/\delta_0}, \tag{4.6}$$

or

$$\sin \varphi = \frac{1}{\operatorname{ch}(z/\delta_0)}, \qquad \cos \varphi = \operatorname{th}(z/\delta_0), \tag{4.7}$$

with

$$\delta_0 = \kappa_K^{-1} = \sqrt{A/K_1}, \tag{4.8}$$

denoting the so-called Bloch wall parameter. It is obvious from eq. (4.6) and eq. (4.7) that the rotation of J_s for an isolated domain wall extends from $z = -\infty$ to $z = +\infty$. Nevertheless, it is useful to define a wall width because the main rotation of J_s is restricted to a region of width δ_0. According to Lilley [2.60] the wall width is defined by the largest slope of the tangent at the $\varphi(z)$ curve and its intersection

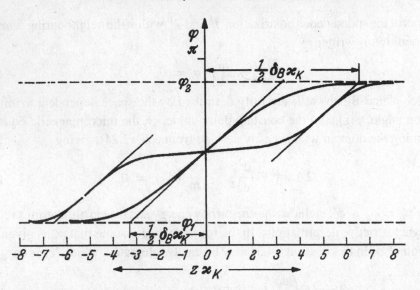

Fig. 4.2. Determination of the wall width by the tangents of inflection points for type I and type II walls. The case of three inflection points is discussed in Section 4.5.

points for φ_I and φ_{II} (see Fig. 4.2):

$$\delta_B = (\varphi_{II} - \varphi_I)/(\mathrm{d}\varphi/\mathrm{d}z)_{\max}. \tag{4.9}$$

In the case of the above 180°-wall we find

$$\delta_B = \pi\sqrt{A/K_1} = \pi \cdot l_K, \tag{4.10}$$

or

$$\delta_B = \pi\sqrt{A/(K_1 + K_2)}$$

if the second anisotropy constant is taken into account. The wall energy γ_B per unit area of the domain wall is obtained from

$$\gamma_B = \int_{-\infty}^{+\infty} (\phi'_{ex} + \phi'_B(\varphi) - \phi'_B(\varphi_I))\,\mathrm{d}z. \tag{4.11}$$

According to eq. (4.4), within the wall we have $\phi'_{ex} = \phi'_B(\varphi) - \phi'_B(\varphi_I)$ and $\mathrm{d}z = \sqrt{A}\sin\Theta\,(\phi'_B(\varphi) - \phi'_B(\varphi_I))^{-1/2}\,\mathrm{d}\varphi$, giving finally

$$\gamma_B = 2\sqrt{A}\sin\Theta\int_{\varphi_I}^{\varphi_{II}} (\phi'_B(\varphi) - \phi'_B(\varphi_I))^{1/2}\,\mathrm{d}\varphi. \tag{4.12}$$

For the 180°-wall we then obtain for $\varphi_I = 0$, $\varphi_{II} = \pi$, $\Theta = \pi/2$,

$$\gamma_B = 4 \cdot \sqrt{AK_1} = 4K_1 \cdot \delta_0, \tag{4.13}$$

Table 4.1. *Wall width and specific wall energy of some uniaxial crystals ($J/m^2 = 10^3$ erg/cm^2).*
δ_B has been determined according to
$$\delta_B = \pi(A/(K_1 + K_2))^{1/2}.$$

Ferromagnet	δ_B [nm]	γ_B [10^{-2} J/m^2]
Co	22.3	1.49
SmCo$_5$	2.64	5.71
Sm$_2$Co$_{17}$	5.74	3.07
Nd$_2$Fe$_{14}$B	3.82	2.24
Sm$_2$Fe$_{17}$N$_3$	3.36	4.06
BaFe$_{12}$O$_{19}$	13.94	0.57

or

$$\gamma_B = 2\sqrt{AK_1}\left\{1 + [(1 + \kappa)/\kappa^{1/2}]\text{ arc sin}\left[\frac{1}{(1 + \kappa)^{1/2}}\right]\right\},$$

with $\kappa = K_1/K_2$ if the second anisotropy constant K_2 is taken into account. Some values of δ_B and γ_B are summarized in Table 4.1.

4.3 Effect of magnetostrictive deformations

In the preceding section we have neglected the influence of elastic energies on the domain wall configuration. In fact, a calculation of these energies requires a solution of the elastic equilibrium equations as discussed in Section 2.2.4. As has been shown in fundamental papers of Rieder [2.56] and Lilley [2.60], in the case of planar domain walls explicit results can be derived because of the one-dimensionality of the problem. Assuming that there are no external stresses or internal stresses resulting from defect structures we have only to consider the magnetostrictive elastic stresses giving rise to an elastic energy density

$$\phi'_{el} = (1/2)\varepsilon^{el} \cdot\cdot c \cdot\cdot \varepsilon^{el}. \tag{4.14}$$

The strain tensor ε^{el} has to be determined from the equilibrium conditions (eqs. (2.61) to (2.63)) where the sources of the elastic strains are the spontaneous magnetostrictive strains ε^Q. In the case of planar domain walls the elastic strains are most simply derived from Albenga's theorem [2.59]. This theorem says that the volume average of each elastic stress component vanishes. In addition, all stress components, σ_{3i}^m, have to be zero because Div $\sigma^m = 0$ and all components depend only on z. Accordingly, we obtain

$$\langle \sigma^m \rangle = c \cdot\cdot \langle \varepsilon^m - \varepsilon^Q \rangle = 0, \tag{4.15}$$

and with the condition Ink $\varepsilon^{\mathrm{m}} = 0$,

$$\varepsilon^{\mathrm{m}} = \langle \varepsilon^{\mathrm{m}} \rangle = \langle \varepsilon^{\mathrm{Q}} \rangle \tag{4.16}$$

holds, which gives, with $\varepsilon^{\mathrm{el}} = \varepsilon^{\mathrm{m}} - \varepsilon^{\mathrm{Q}}$,

$$\varepsilon_{ij}^{\mathrm{el}} = \langle \varepsilon_{ij}^{\mathrm{Q}} \rangle - \varepsilon_{ij}^{\mathrm{Q}}, \quad i, j = 1, 2, \tag{4.17}$$

$$\varepsilon_{3i}^{\mathrm{el}} = \varepsilon_{3i}^{\mathrm{m}} - \varepsilon_{3i}^{\mathrm{Q}}, \quad i = 1, 2, 3. \tag{4.18}$$

Since the stress components, σ_{3i}^{m}, are zero, only the planar strain components given by eq. (4.17) have to be considered. The elastic strains $\varepsilon^{\mathrm{el}}$ and stresses σ^{m} have been calculated by Rieder [2.56, 4.1, 4.2] for the most important domain walls of cubic and hexagonal lattices. For our present purpose we consider the case of an isotropic amorphous medium with respect to magnetostriction and elasticity, i.e., we introduce an isotropic magnetostriction constant, λ_{s}, and elastic constants G (shear modulus) and ν (Poisson ratio). The relevant planar elastic strains and stresses are then given by

$$\varepsilon_{11}^{\mathrm{el}} = \frac{3}{2}\lambda_{\mathrm{s}} \sin^2 \varphi; \quad \varepsilon_{22}^{\mathrm{el}} = -\frac{3}{2}\lambda_{\mathrm{s}} \sin^2 \varphi; \quad \varepsilon_{12}^{\mathrm{el}} = -\frac{3}{4}\lambda_{\mathrm{s}} \sin 2\varphi;$$

$$\sigma_{11}^{\mathrm{m}} = 3G\lambda_{\mathrm{s}} \sin^2 \varphi; \quad \sigma_{22}^{\mathrm{m}} = -3G\lambda_{\mathrm{s}} \sin^2 \varphi; \quad \sigma_{12}^{\mathrm{m}} = -\frac{3}{2}G\lambda_{\mathrm{s}} \sin 2\varphi. \tag{4.19}$$

The elastic energy density is found to be

$$\phi_{\mathrm{el}}' = \frac{G}{\nu - 1}\left\{ \nu \left[(\varepsilon_{11}^{\mathrm{el}})^2 + (\varepsilon_{22}^{\mathrm{el}})^2 \right] + 2\varepsilon_{11}^{\mathrm{el}} \cdot \varepsilon_{22}^{\mathrm{el}} \right\} + G(\varepsilon_{12}^{\mathrm{el}})^2 = \frac{9}{2}G\lambda_{\mathrm{s}}^2 \cdot \sin^2 \varphi. \tag{4.20}$$

According to this result, ϕ_{el}' corresponds to a uniaxial anisotropy to be added to the anisotropy ϕ_K'. The wall width and the wall energy are now given by

$$\delta_{\mathrm{B}} = \pi \sqrt{\frac{A}{K_1 + \frac{9}{2}G\lambda_{\mathrm{s}}^2}},$$

$$\gamma_{\mathrm{B}} = 4 \cdot \sqrt{A\left(K_1 + \frac{9}{2}G\lambda_{\mathrm{s}}^2 \right)}. \tag{4.21}$$

In magnetostrictive materials with $\lambda_{\mathrm{s}} = 30 \cdot 10^{-6}$ and a shear modulus of $G = 45.5$ GPa we find $(9/2)G\lambda_{\mathrm{s}}^2 = 184$ J/m^3, whereas for a low magnetostrictive material, e.g., $\lambda_{\mathrm{s}} = 3 \cdot 10^{-6}$, a value of only 1.84 J/m^3 is obtained. It is obvious from these results that the magnetostrictive effects are only of importance in soft magnetic materials with $K_1 < 10^3$ J/m^3. In particular, in amorphous alloys with small induced anisotropies, $K_1 \equiv K_u < 100 \; \mathrm{J/m^3}$, the magnetostrictive contributions to the wall parameter become appreciable.

4.4 Effect of internal stresses

In Sections 4.2 and 4.3 we have considered the effect of anisotropy and magnetostrictive self energies on a 180°-wall. If, in addition to the magnetostrictive stresses, internal or external stresses are present, we have to take into account the magnetoelastic coupling energy, ϕ_σ, defined by eqs. (2.76) and (2.77). Assuming a tensile or compressive stress component σ_{11} only, ϕ_σ' is written

$$\psi_\sigma' = -\frac{3}{2}\lambda_s\sigma_{11} \; \cos^2\varphi. \tag{4.22}$$

With this energy contribution we obtain for the energy function, $\phi_B'(\varphi) - \phi_B'(\infty)$, of the domain wall

$$\phi_B'(\varphi) - \phi_B'(\infty) = K_1 \sin^2\varphi + \frac{9}{2}G\lambda_s^2 \sin^2\varphi + \frac{3}{2}\lambda_s\sigma_{11} \sin^2\varphi, \tag{4.23}$$

and the domain wall parameters are given by

$$\delta_B = \pi \sqrt{\frac{A}{K_1 + \frac{9}{2}G\lambda_s^2 + \frac{3}{2}\lambda_s\sigma_{11}}},$$

$$\gamma_B = 4 \cdot \sqrt{A\left(K_1 + \frac{9}{2}G\lambda_s^2 + \frac{3}{2}\lambda_s\sigma_{11}\right)}. \tag{4.24}$$

4.5 Bloch walls in cubic crystals

Bloch walls in cubic crystals are denoted as walls of type I or type II according to their magnetostrictive elastic stresses. Type I Bloch walls are those for which the elastic stresses within the domains are finite, and type II Bloch walls are those with elastic stresses only within the domain walls and vanishing elastic stress within the domains. Well-known examples of type I domain walls are the $(001) - 90°$-wall in α-Fe and the $(110) - 70°$ and $109°$-walls in Ni.

In the presence of other stress sources, e.g., dislocations, type I domain walls interact with all stress sources within the domains and therefore become strongly pinned, showing only a small mobility. Domain walls of type II only interact with stress sources lying within the domain walls themselves. These domain walls therefore are easily mobile and determine the magnetization process. In particular, 180°-walls belong to the type II walls. In Table 4.2 the elastic strains ε^{el} of some

Table 4.2. The magnetostrictive elastic tensors and the trace ε_I of some dws [2.8, 2.58]. β is the transformation tensor of the domain wall coordinate system with respect to the cubic axes. The strain components in the cubic coordinate system are

$$\varepsilon_{ij}^{el} = \sum_{l,k} \beta_{il}\beta_{jk}\varepsilon_{lk}^{el}.$$

The last column gives the direction cosines, α_i, of the spontaneous magnetization with respect to the cubic axes.

Wall type	$\varepsilon_{11}^M,\ \varepsilon_{22}^M,\ \varepsilon_{12}^M$	β	$\alpha_1,\ \alpha_2,\ \alpha_3$
Ni (001) – 109°	$\varepsilon_{11}^{el} = -\varepsilon_{22}^{el} = \lambda_{100}\sin\varphi\cos\varphi$ $\varepsilon_{12}^{el} = \lambda_{111}\sin^2\varphi$ $\varepsilon_I^{el} = 0$	$\begin{pmatrix} 1 & 0 & 0 \\ 0 & 1 & 0 \\ 0 & 0 & 1 \end{pmatrix}$	$\alpha_1 = \left(\dfrac{\sqrt{2}}{2}\sin\varphi + \dfrac{\sqrt{2}}{2}\cos\varphi\right)\dfrac{\sqrt{6}}{3}$ $\alpha_2 = \left(-\dfrac{\sqrt{2}}{2}\sin\varphi + \dfrac{\sqrt{2}}{2}\cos\varphi\right)\dfrac{\sqrt{6}}{3}$ $\alpha_3 = \dfrac{\sqrt{3}}{3}$
Ni (110) – 180°	$\varepsilon_{11}^{el} = \sin\varphi\left\{\lambda_{100}\left(-\dfrac{1}{4}\sin\varphi + \dfrac{\sqrt{2}}{2}\cos\varphi\right)\right.$ $\left. - \lambda_{111}\left(\dfrac{3}{4}\sin\varphi + \dfrac{\sqrt{2}}{2}\cos\varphi\right)\right\}$ $\varepsilon_{22}^{el} = \dfrac{3}{2}\lambda_{111}\sin^2\varphi$ $\varepsilon_{12}^{el} = \dfrac{1}{2}\sin\varphi\left\{\lambda_{100}\left(-\dfrac{\sqrt{2}}{2}\sin\varphi + 2\cos\varphi\right)\right.$ $\left. + \lambda_{111}\left(\dfrac{\sqrt{2}}{2}\sin\varphi + \cos\varphi\right)\right\}$ $\varepsilon_I^{el} = \dfrac{1}{4}(\lambda_{111} - \lambda_{100})\{\sin^2\varphi - \sqrt{2}\sin 2\varphi\}$	$\begin{pmatrix} \dfrac{\sqrt{6}}{6} & -\dfrac{\sqrt{3}}{3} & \dfrac{\sqrt{2}}{2} \\ -\dfrac{\sqrt{6}}{6} & \dfrac{\sqrt{3}}{3} & \dfrac{\sqrt{2}}{2} \\ -\dfrac{\sqrt{6}}{3} & -\dfrac{\sqrt{3}}{3} & 0 \end{pmatrix}$	$\alpha_1 = \dfrac{\sqrt{6}}{6}\sin\varphi + \dfrac{\sqrt{3}}{3}\cos\varphi$ $\alpha_2 = -\dfrac{\sqrt{6}}{6}\sin\varphi + \dfrac{\sqrt{3}}{3}\cos\varphi$ $\alpha_3 = -\dfrac{\sqrt{6}}{3}\sin\varphi - \dfrac{\sqrt{3}}{3}\cos\varphi$
Ni ($\bar{1}\bar{1}2$) – 180°	$\varepsilon_{11}^{el} = \left(-\dfrac{1}{4}\lambda_{100} - \dfrac{5}{4}\lambda_{111}\right)\sin^2\varphi$ $\varepsilon_{22}^{el} = \dfrac{3}{2}\lambda_{111}\sin^2\varphi$ $\varepsilon_{12}^{el} = \dfrac{1}{2}(2\lambda_{100} + \lambda_{111})\sin\varphi\cos\varphi$ $\varepsilon_I^{el} = -\dfrac{1}{4}(\lambda_{100} - \lambda_{111})\sin^2\varphi$	$\begin{pmatrix} \dfrac{\sqrt{2}}{2} & \dfrac{\sqrt{3}}{3} & -\dfrac{\sqrt{6}}{6} \\ -\dfrac{\sqrt{2}}{2} & \dfrac{\sqrt{3}}{3} & -\dfrac{\sqrt{6}}{6} \\ 0 & \dfrac{\sqrt{3}}{3} & \dfrac{\sqrt{6}}{3} \end{pmatrix}$	$\alpha_1 = \dfrac{\sqrt{2}}{2}\sin\varphi + \dfrac{\sqrt{3}}{3}\cos\varphi$ $\alpha_2 = -\dfrac{\sqrt{2}}{2}\sin\varphi + \dfrac{\sqrt{3}}{3}\cos\varphi$ $\alpha_3 = \dfrac{\sqrt{3}}{3}\cos\varphi$

Ni (001) – 71°

$$\varepsilon_{11}^{el} = -\varepsilon_{22}^{el} = -\frac{\lambda_{100}}{2}\sin 2\varphi$$
$$\varepsilon_{12}^{el} = -\frac{\lambda_{111}}{2}\cos 2\varphi$$
$$\varepsilon_I^{el} = 0$$

$$\begin{pmatrix} 1 & 0 & 0 \\ 0 & 1 & 0 \\ 0 & 0 & 1 \end{pmatrix}$$

$$\alpha_1 = -\frac{\sqrt{6}}{3}\sin\left(\varphi + \frac{\pi}{4}\right)$$
$$\alpha_2 = \frac{\sqrt{6}}{3}\cos\left(\varphi + \frac{\pi}{4}\right)$$
$$\alpha_3 = \frac{\sqrt{3}}{3}$$

Fe (001) – 180°

$$\varepsilon_{11}^{el} = \frac{3}{2}\lambda_{100}\sin^2\varphi$$
$$\varepsilon_{22}^{el} = -\frac{3}{2}\lambda_{100}\sin^2\varphi$$
$$\varepsilon_{12}^{el} = -\frac{3}{2}\lambda_{111}\sin\varphi\cos\varphi$$
$$\varepsilon_I^{el} = 0$$

$$\begin{pmatrix} 0 & -1 & 0 \\ 1 & 0 & 0 \\ 0 & 0 & 1 \end{pmatrix}$$

$$\alpha_1 = -\sin\varphi$$
$$\alpha_2 = \cos\varphi$$
$$\alpha_3 = 0$$

Fe (001) – 90°

$$\varepsilon_{11}^{el} = -\frac{3}{4}\lambda_{100}\sin 2\varphi$$
$$\varepsilon_{22}^{el} = +\frac{3}{4}\lambda_{100}\sin 2\varphi$$
$$\varepsilon_{12}^{el} = \frac{3}{4}\lambda_{111}\cos 2\varphi$$
$$\varepsilon_I^{el} = 0$$

$$\begin{pmatrix} 0 & -1 & 0 \\ 1 & 0 & 0 \\ 0 & 0 & 1 \end{pmatrix}$$

$$\alpha_1 = -\sin\left(\varphi + \frac{\pi}{4}\right)$$
$$\alpha_2 = \cos\left(\varphi + \frac{\pi}{4}\right)$$
$$\alpha_3 = 0$$

Co

$(klm\,0) – 180°$

$$\varepsilon_{11}^{el} = \lambda_{33}\sin^2\varphi$$
$$\varepsilon_{22}^{el} = -\lambda_{11}\sin^2\varphi,$$
$$\varepsilon_{12}^{el} = -\frac{\lambda_{44}}{2}\sin\varphi\cos\varphi,$$
$$\varepsilon_I^{el} = (\lambda_{33} - \lambda_{11})\sin^2\varphi$$

$x\|c$ axis (hexagonal axis)

$\varphi\measuredangle(J_s, c\text{-axis})$

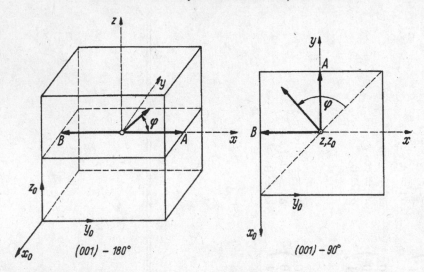

Fig. 4.3. Definition of domain wall coordinate systems in bcc α-Fe.

representative type II walls are presented. Figures 4.3 and 4.4 show the coordinate systems used in general for the description of the spin arrangement of domain walls in cubic crystals.

In the preceding sections we have treated 180°-walls in uniaxial crystalline and amorphous materials leading to the well-known classical results for δ_B and γ_B. In multiaxial materials such as α-Fe with $\langle 100 \rangle$-easy directions or Ni with $\langle 111 \rangle$-easy directions it turns out that some of the domain walls only have a well-defined configuration if the magnetoelastic energy ϕ'_{el} is taken into account. Otherwise these walls split into subwalls leading to an undefined wall width. As an example we consider the $(001) - 180°$-wall in α-Fe where the spontaneous magnetization rotates within the (x, y)-plane from $(+)x$ to $(-)x$ where the (x, y, z)-coordinates are identical with the cubic axes [100], [010] and [001], respectively. The direction cosines are $\gamma_1 = \cos\varphi$, $\gamma_2 = \sin\varphi$, $\gamma_3 = 0$. The magnetocrystalline energy density within the wall is now written

$$\phi'_K = \frac{1}{4}K_1 \sin^2 2\varphi, \tag{4.25}$$

and the contribution of K_2 vanishes because $\gamma_3 = 0$. Inserting eq. (4.25) into eq. (4.5) gives for the spin distribution

$$z = (A/K_1)^{1/2} \ln|\tan\varphi|, \tag{4.26a}$$

and for the wall energy

$$\gamma_B = 2\sqrt{AK_1}. \tag{4.26b}$$

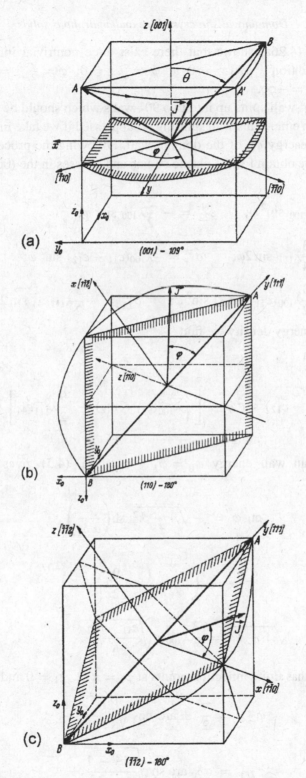

Fig. 4.4. Definition of domain wall coordinate systems in Ni. (a) (001) − 109°, (b) (110) −
180°, (c) ($\bar{1}\bar{1}2$) −180°.

Analysis of eq. (4.26a) shows that there exist three nontrivial inflection points obeying the condition $d^2\varphi/dz^2 = 0$ at $\varphi = \frac{\pi}{4}, z = 0$; $\varphi = \frac{\pi}{2}, z = \infty$; $\varphi = \frac{3\pi}{4}$, $z = 0$; $\varphi = \pi, z = -\infty$.

Thus, the 180°-wall splits up into two 90°-walls which should be infinitely separated from each other. This wall splitting is suppressed if we take into account the magnetoelastic energy ϕ_{el} of the domain wall. Following the procedure outlined in Section 4.3 we obtain for the elastic strains and stresses in the (001)–180°-wall [2.7, 2.56]

$$\varepsilon_{11}^{el} = \frac{3}{2}\lambda_{100}\sin^2\varphi; \qquad \varepsilon_{22}^{el} = -\frac{3}{2}\lambda_{100}\sin^2\varphi;$$

$$\varepsilon_{12}^{el} = -\frac{3}{4}\lambda_{111}\sin 2\varphi; \qquad \sigma_{11}^{m} = \frac{3}{2}\lambda_{100}(c_{11} - c_{12})\sin^2\varphi;$$

$$\sigma_{22}^{m} = -\frac{3}{2}\lambda_{100}(c_{11} - c_{12})\sin^2\varphi; \qquad \sigma_{12}^{m} = -\frac{3}{2}\lambda_{111}c_{44}\sin 2\varphi. \qquad (4.27)$$

For the elastic energy density we find

$$\phi_{el}' = \frac{1}{2}\sigma_{ij}^{m} \cdot \varepsilon_{ij}^{el}$$

$$= \frac{9}{4}\lambda_{100}^2(c_{11} - c_{12})\sin^2\varphi - \left[\frac{9}{4}\lambda_{100}^2(c_{11} - c_{12}) - \frac{9}{2}\lambda_{111}^2 c_{44}\right]\sin^2\varphi \cos^2\varphi. \qquad (4.28)$$

With the domain wall energy $\phi_B' = \phi_K' + \phi_{el}'$, eq. (4.5) gives for the spin arrangement

$$\operatorname{ctg}\varphi = -\sqrt{1 - k^2}\ \operatorname{sh}\left(\frac{z}{\alpha\,\delta_0}\right), \qquad (4.29)$$

with

$$\alpha = \left(1 + \frac{1}{K_1} \cdot \frac{9}{2}\lambda_{111}^2 c_{44}\right)^{-1/2}$$

$$\sqrt{1 - k^2} = \frac{3}{2}\lambda_{100}\alpha \sqrt{\frac{c_{11} - c_{12}}{K_1}}. \qquad (4.30)$$

Equation (4.29) has finite inflection points at $\varphi_1 = \pi/2$, $z_1 = 0$ and at

$$\varphi_{2,3} = \frac{\pi}{2} \pm \arccos\sqrt{\frac{2k^2 - 1}{2k^2}},$$

$$z_{2,3} = \alpha\delta_0 \operatorname{arc\,sh}\sqrt{\frac{2k^2 - 1}{1 - k^2}}. \qquad (4.31)$$

The wall width is then given by

$$\delta_B = 2z_2 + 2\varphi_2 \frac{dz}{d\varphi}\bigg|_{\varphi_2} \tag{4.32}$$

$$= 2\alpha\delta_0 \left\{ \text{arc sh} \sqrt{\frac{2k^2 - 1}{1 - k^2}} + 2k\left(\frac{\pi}{2} - \text{arc cos} \sqrt{\frac{2k^2 - 1}{2k^2}}\right)\right\}. \tag{4.33}$$

For α-Fe we have $\alpha = 1 + 2.2 \cdot 10^{-4}$ and $(1 - k^2)^{1/2} = 2.94 \cdot 10^{-2}$. With these parameters eq. (4.33) can be approximated by

$$\delta_B = \alpha\delta_0 \left\{ \frac{\pi}{2}(1 + k^2) - \ln(1 - k^2) + 2\ln 2 \right\}, \tag{4.34}$$

giving for iron $\delta_B = 10.5 \, \delta_0$. This result clearly shows that the 180°-walls in cubic crystals in general are much wider than the classical result $\pi \, \delta_0$. For the specific wall energy integration of eq. (4.12) gives

$$\gamma_B = \frac{2}{\alpha}\sqrt{A K_1} \left(1 - \frac{1 - k'^2}{2k'}\ln\frac{1 - k'}{1 + k'}\right), \tag{4.35}$$

with $k'^2 = 1 - \frac{9}{4}\lambda_{100}^2 \frac{c_{11} - c_{12}}{K_1}$. From these results it becomes obvious that the magnetostrictive energy terms become relevant for soft magnetic materials with $K_1 < 10^3 \, J/m^3$.

The domain walls in Ni with the $\langle 111 \rangle$-easy direction obey a similar equation as eq. (4.29),

$$\text{ctg}\,\varphi = -\sqrt{1 - k^2} \, \text{sh}\frac{z}{\alpha\delta_0} + \rho. \tag{4.36}$$

The parameters at room temperature are summarized in Table 4.4.

Table 4.4. *Parameters of domain walls in Ni and α-Fe at room temperature.*

Metal	Domain wall	$\sqrt{1 - k^2}$	α	ρ	$\gamma_B(\sqrt{A|K_1|})$
Ni	$(001) - 109°$	0.248	1.17	0	$\approx 2(2/3)^{3/2}$
	$(001) - 71°$	0.248	1.22	0	$\approx (2/3)^{3/2}$
	$(112) - 180°$	0.453	1.24	0	≈ 2
	$(1\bar{1}0) - 180°$	0.310	1.17	-0.32	≈ 2
α-Fe	$(110) - 180°$	0.50	1	0	2.76
	$(100) - 180°$	$2.94 \cdot 10^{-2}$	≈ 1	0	≈ 2

4.6 Néel walls in bulk materials and thin films

4.6.1 General remarks

As shown in the preceding sections, Bloch walls are characterized by the condition Θ = const., which results in vanishing magnetic volume charges in the case of planar walls. There exists another extreme case of domain walls, the so-called Néel walls for which $\varphi = 0$ holds and Θ is considered as the angular variable. Changes of the direction of J_s in this case are connected with magnetic stray fields because magnetic volume charges $\rho = -\text{div} \, M_s$ are produced. The different behaviour of the two types of walls with respect to their stray fields is demonstrated in Fig. 4.5. Whereas in Néel walls the stray field extends throughout the entire wall, the stray field of Bloch walls is restricted to the region where the Bloch wall enters the surface, thus producing surface charges. In the case of Néel walls surface charges are suppressed at the expense of volume charges. Since in thin films the stray field energy of surface charges becomes larger than that of the volume charges, the Néel walls become stable in thin films. This transition from Bloch walls to Néel walls as a function of the film thickness has been investigated by several authors [3.2, 4.3–4.6]. It should be noted, however, that explicit analytical solutions for Bloch and Néel walls in thin films are not yet available. Numerical and approximate solutions have been published by Néel [4.3], Dietze and Thomas [4.4], Brown and La Bonte [4.5], Kirchner and Döring [4.6], Holz and Hubert [4.7], Hubert [3.2, 4.10], Riedel and Seeger [4.8], La Bonte [4.9], and Aharoni [4.11–4.14].

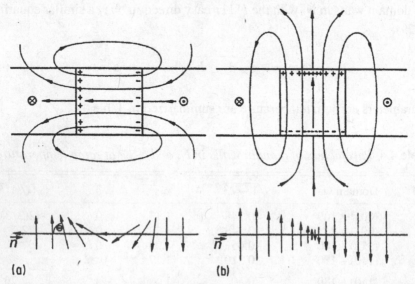

Fig. 4.5. Distribution of stray fields and spontaneous magnetization in Néel and Bloch walls. (a) Néel wall, (b) Bloch wall.

4.6.2 Néel walls in bulk crystals

Explicit solutions can be given for infinitely extended planar Néel walls. Assuming the wall is lying in the $(x–y)$-plane and variation of the angle $\Theta(z)$ along the z-axis, corresponding to the wall normal, we obtain from $\mathrm{div}\,\boldsymbol{B} = 0$

$$\frac{\mathrm{d}H_{s,z}}{\mathrm{d}z} = -M_s \frac{\mathrm{d}(\cos\Theta)}{\mathrm{d}z}, \tag{4.37}$$

and by integration

$$H_{s,z} = -M_s(\cos\Theta - \cos\Theta_0). \tag{4.38}$$

Here $2\Theta_0$ denotes the total change of the orientation of $\boldsymbol{J}_s$ along the z-axis. With the stray field energy

$$\phi'_s = -\frac{1}{2}H_{s,z} \cdot J_s \cos\Theta. \tag{4.39}$$

The differential equation of the Néel wall is obtained from eq. (3.24) giving

$$A\left(\frac{\mathrm{d}\Theta}{\mathrm{d}z}\right)^2 = (\phi'_K + \phi'_s). \tag{4.40}$$

Equation (4.40) can be solved explicitly, giving for the angle $\Theta(z)$ and the characteristic parameters of a $180°$-wall ($\Theta_0 = \frac{\pi}{2}$) in a uniaxial material:

$$\mathrm{ctg}\,\frac{\Theta}{2} = \exp\left[-z\sqrt{\kappa_K^2 + \kappa_s^2}\right],$$

$$\delta_{\text{Né}} = \frac{\pi}{\sqrt{\kappa_K^2 + \kappa_s^2}}, \tag{4.41}$$

$$\gamma_{\text{Né}} = 4 \cdot \sqrt{A\left(K_1 + \frac{1}{2}M_s J_s\right)}.$$

For easy directions $\Theta_0 < \frac{\pi}{2}$ no explicit solutions of eq. (4.40) are achievable. Just the wall width $\delta_{\text{Né}}$ can be determined by eq. (4.9):

$$\delta_{\text{Né}}^{2\Theta_0} = \frac{\sqrt{2}\Theta_0}{\sin(\Theta_0/2)}\frac{1}{\kappa_s}\frac{1}{[1 + 2Q\cos^2(\Theta_0/2)]^{1/2}}, \tag{4.42}$$

with the parameter $Q = (2A/M_s J_s)$. The specific wall energy of the Néel wall as a function of Θ_0 is presented in Fig. 4.6 for $K_1 = 0$. The results shown in Fig. 4.6 reveal a strong dependence of $\gamma_{\text{Né}}$ on Θ_0 which can be represented approximately by $\sin^3\Theta_0$. It is of interest to note that a $90°$ Néel wall possesses only 32% of the wall energy of the $180°$-wall. In soft magnetic materials where $\kappa_K^2 \le \kappa_s^2$ holds (see Table 4.1), the properties of the Néel wall are essentially determined by the stray

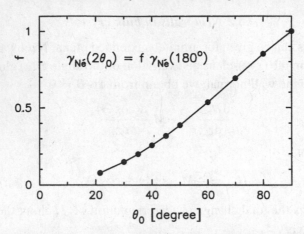

Fig. 4.6. Relative wall energy $f = \gamma_{N\acute{e}}(2\Theta_0)/\gamma_{N\acute{e}}(180°)$ of Néel walls as a function of the wall angle $2\Theta_0$ obtained by computational micromagnetism (courtesy of D. Goll).

field energy, giving for the 180°-wall

$$\delta_{N\acute{e}} = \pi\kappa_s^{-1} = \pi\sqrt{\frac{2A}{\mu_0 J_s^2}} = \pi l_s,$$

$$\gamma_{N\acute{e}} = 4\sqrt{A \cdot M_s J_s/2}. \tag{4.43}$$

A comparison of the two wall energies, γ_B and $\gamma_{N\acute{e}}$, of 180°-walls clearly shows that in thick layers the Bloch wall always has the lower specific wall energy and therefore Bloch walls are also realized in bulk materials, whereas the Néel walls are found only under particular boundary conditions such as in thin films or on the surface regions of hard magnetic materials.

4.6.3 Néel walls in thin films

In thin films Bloch and Néel walls are characterized by quantitatively different stray fields. In the case of Bloch walls the stray fields are of the order of M_s, whereas the volume charges of Néel walls generate somewhat smaller stray fields. In thin films it is therefore energetically more favourable to avoid surface charges and to rotate the magnetization within the film plane by producing volume charges. This type of spin configuration corresponds to a linearly extended finite magnetic dipole with a complex stray field, which at larger distances from the wall centre decreases as $1/r^2$. Due to the special behaviour of the stray field of the volume charges in films with finite thickness the rotation angle Θ within the wall shows a nonexponential behaviour. Figure 4.7 represents the direction cosine $\cos\Theta$ for different parameters, $Q = 2K_1/(M_s J_s)$, in a logarithmic scale. In Fig. 4.7 three

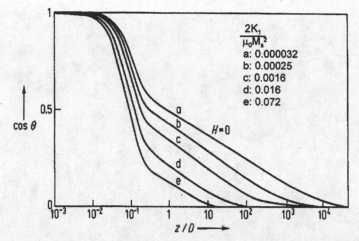

Fig. 4.7. Rotation angle Θ of a one-dimensional symmetric Néel wall for different parameters $Q = 2K_1/(M_s J_s)$ in reduced length scale $z' = z/D$ (D = film thickness) [4.7].

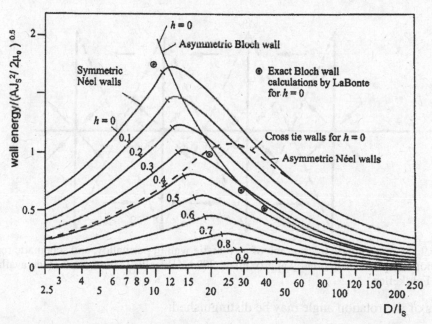

Fig. 4.8. Wall energies of Néel and Bloch type walls as a function of film thickness for different applied fields in units of $h = H_{ext} J_s/2K_u$ and $Q = 0.00025$. All wall energies are upper bounds and were obtained by the Ritz method by Hubert [4.10]. The circles with centred points correspond to results by La Bonte [4.9]. The transition to asymmetric Néel walls is indicated by markings. The energy of cross-tie walls is given by the dashed line.

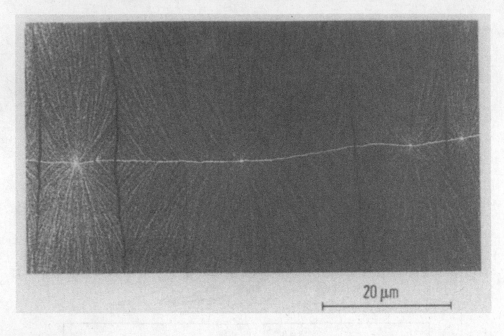

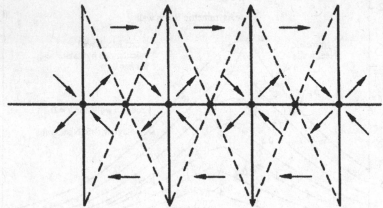

Fig. 4.9. Magneto-optic micrograph of a cross-tie wall in permalloy, and schematic representation of spin distribution by approximate 90° Néel walls. (Micrograph made available
by E. Feldkeller.)

stages of the rotation angle may be distinguished:

1. The kernel region with $\cos \Theta(z) \propto \Delta_N^2/(\Delta_N^2 + z^2)$.
2. The logarithmic region with $\cos \Theta(z) \propto \ln(z/D)$.
3. The dipolar tail with $\cos \Theta(z) \propto 1/z^2$.

Whereas in stage 1 the crystal anisotropy governs the rotation angle, in stages 2
and 3 the stray field is the dominant energy term. In stage 2 the local short-range
stray field determines $\cos \Theta$, whereas in stage 3 it is the dipolar long-range field.

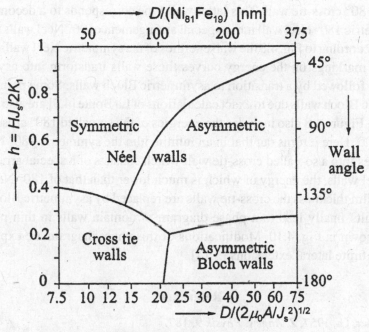

Fig. 4.10. Phase diagram of the different wall types in permalloy with $Q = 0.00025$ as derived from Fig. 4.8 according to Hubert and Schaefer [2.14].

It is of interest to compare wall widths and specific wall energies of Bloch and Néel walls in very thin films with $D < \kappa_s^{-1}$. In this case the Bloch wall width becomes that of the Néel wall in bulk materials and vice versa. The same exchange takes place for the specific wall energies. The wall parameter Δ_N describing the kernel of the Néel wall for thin films is given by

$$\Delta_N \simeq 2\sqrt{\left(\sqrt{2}-1\right)A/K_1 - \frac{J_s^2 D^2}{96\mu_0 K_1}}, \qquad (4.44)$$

and the specific wall energy is written

$$\gamma_{Né}^{180°} \simeq 2\pi\sqrt{\sqrt{2}-1} \cdot \sqrt{AK_1} + \frac{\pi}{16}J_s M_s D, \qquad (4.45)$$

where we have used the approximate solutions given by Aharoni [2.9].

4.6.4 Phase diagrams of Néel and Bloch walls in thin films

In thin films a large variety of wall types may exist dependent on film thickness, applied magnetic field and lateral extensions of the film. Wall energies of Néel and Bloch walls as a function of film thickness and applied magnetic field are shown for permalloy, $Q = 0.00025$, in Fig. 4.8 [4.10], together with results obtained for the so-called 180° cross-tie wall. This latter wall type corresponds to a decomposition of a symmetric 180° Néel wall into a special arrangement of 90° Néel walls shown in Fig. 4.9. According to Fig. 4.8 the wall energies of the symmetric Néel walls increase and at the markings on the energy curves these walls transform into asymmetric Néel walls followed by a transition to asymmetric Bloch walls. Some points for the asymmetric Bloch walls due to exact calculations of La Bonte [4.9] are also included in Fig. 4.8. Figure 4.8 also includes the energies of the so-called 180° cross-tie wall for $H_{ext} = 0$. Here it turns out that in an infinite film the symmetric 180° Néel wall decomposes into a so-called cross-tie wall which consists of a special arrangement of 90° Néel walls, the energy of which is much lower than that of 180° Néel walls. At larger film thickness the cross-tie walls are replaced by asymmetric Bloch walls. These results finally lead to a phase diagram of domain walls in thin permalloy films as shown in Fig. 4.10. Modifications of this phase diagram are expected for films with finite lateral extensions [2.14].

References

[4.1] Rieder, G., 1957, *Z. Angew. Physik* **9**, 187.
[4.2] Rieder, G., 1958, *Z. Angew. Physik* **10**, 140.
[4.3] Néel, L., 1955, *C.R., Acad. Sci. (Paris)* **241**, 533.
[4.4] Dietze, H.D., and Thomas, H., 1961, *Z. Physik* **163**, 523.
[4.5] Brown, W.F., and La Bonte, A.E., 1965, *J. Appl. Phys.* **36**, 1380.
[4.6] Kirchner, R., and Döring, W., 1968, *J. Appl. Phys.* **39**, 855.
[4.7] Holz, A., and Hubert, A., 1969, *Z. Angew. Physik* **26**, 145.
[4.8] Riedel, H., and Seeger, A., 1971, *Phys. Stat. Sol. (B)* **46**, 377.
[4.9] La Bonte, A.E., 1969, *J. Appl. Phys.* **40**, 2450.
[4.10] Hubert, A., 1970, *Phys. Stat. Sol.* **38**, 699.
[4.11] Aharoni, A., 1973, *Phys. Stat. Sol. A* **18**, 661.
[4.12] Aharoni, A., 1975, *J. Appl. Phys.* **46**, 908.
[4.13] Aharoni, A., 1976, *J. Appl. Phys.* **47**, 3329.
[4.14] Aharoni, A., and Jakubovics, J.P., 1990, *IEEE Trans. Magn.* **26**, 2810.

5

Interaction of domain walls with defects

5.1 Introductory remarks

In real materials crystal lattice defects or the microstructure of amorphous materials influence the magnetic ground state as well as magnetization processes under an applied external field. Magnetic stray fields and internal stresses produce inhomogeneous distributions of the magnetization. Consequences of these inhomogeneities of the ground state are extended field ranges of the law of approach to magnetic saturation, and modifications of the so-called nucleation fields governing the coercive field of single domain particles. In soft magnetic materials the interactions of lattice defects with domain walls sensitively influence the reversible as well as irreversible magnetization processes due to domain wall displacements. The reversible rotation processes may also be influenced significantly by magnetic inhomogeneities. In the following we describe the interaction between a domain wall (dw) and individual defects and then in Chapter 7 we describe the interaction of a large number of defects with dws on the basis of the statistical potential theory. The results obtained herewith for the characteristics of the statistical potential will be used to determine the relevant parameters of the hysteresis loop, such as initial susceptibility, coercive field and the Rayleigh constant. In particular, the coercive field, H_c, is related to the total interaction energy, e, of a 180° domain wall with lattice defects according to

$$H_c = \frac{1}{2J_s} \left| \frac{de}{dz} \right|_{max} = \frac{1}{2J_s} \left| P_3^{max} \right|,$$

where $\left| P_3^{max} \right|$ denotes the maximum pinning force acting on a domain wall parallel to the domain wall normal.

5.2 Interaction energy of domain walls with point defects

Atomic defects in crystalline materials are vacancies, self-interstitials, impurity interstitials, substitutional impurity atoms or atomic disorder in ordered alloys. As

71

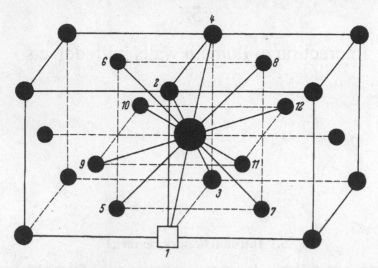

Fig. 5.1. Biatomic complex of a vacancy with an impurity atom of orthorhombic symmetry. Positions 5, 7, 11, 9 correspond to nearest neighbour sites with respect to the vacancy. All other sites require several jumps of the vacancy.

an example, Fig. 5.1 shows a biatomic defect composed of an impurity atom and a vacancy. According to

$$e = e_{ex} + e_K + e_{el}, \qquad (5.1)$$

the dominant magnetic energy terms e_{ex}, e_K, e_{el} are perturbations of the exchange energy, ϕ_{ex}, of the magnetocrystalline energy, ϕ_K, and of the magnetoelastic coupling energy, ϕ_{el}. The first two terms are of short-range type whereas the magnetoelastic term, e_{el}, is of long-range type due to the elastic stresses decreasing according to a $1/r^3$-law. The short-range terms in principle can be attributed to the local modifications of the exchange integrals, J_{ij}, or of the spin–orbit coupling constant λ_i. Calculations of these perturbations require electron theoretical ab-initio calculations so far not performed. In order to determine the pinning of dws by point defects within the framework of micromagnetism, the dependence of the interaction energy, e, on the direction cosines, γ_i, of the spontaneous magnetization, M_s, has to be derived. In a crystal any point defect destroys the translational symmetry and the original point group of the perfect crystal is replaced by that of the point group of the point defect. Accordingly, the angular dependence of the perturbation energy, e, is equivalent to the expressions derived for the exchange and the magnetocrystalline energies of the different crystal systems, but with different material constants. If we denote the magnetic energy densities of the unperturbed material by an upper index 0, the terms e_{ex} and e_K are given by

$$e_{ex} = \int_{\Omega_{n.n.}} \left(\phi_{ex}(\nabla \gamma_i) - \phi_{ex}^{(0)}(\nabla \gamma_i) \right) d^3 r,$$

$$e_K = \int_{\Omega_{n.n.}} \left(\phi_K(\gamma_i) - \phi_K^{(0)}(\gamma_i) \right) d^3 r, \tag{5.2}$$

where the integrals extend over the volume of the nearest neighbour atoms with respect to the atomic defect. ϕ_{ex} and ϕ_K take care of the symmetry and of the space dependent material parameters $A_i(r)$ and $K_i(r)$.

The magnetoelastic coupling energy e_{el} is described as the interaction between elastic strains, ε^{el}, of the dw and the stresses resulting from the point defect.

For a quantitative description of this energy term we introduce the double force dipole tensor, P, [2.59] giving

$$e_{el} = -P \cdot \cdot \varepsilon^{el}, \tag{5.3}$$

or alternatively

$$e_{el} = -Q \cdot \cdot \sigma^m, \tag{5.4}$$

where P and the so-called displacement tensor Q obey the relation

$$P = c \cdot \cdot Q, \tag{5.5}$$

and c denotes the fourth rank tensor of elastic constants. The trace $Q_I = \sum Q_{ii}$ has the simple meaning of the volume dilatation introduced by the point defect. For defects of orthorhombic and lower symmetry P has the form

$$P = \begin{pmatrix} P_{11} & & \\ & P_{22} & \\ & & P_{33} \end{pmatrix}. \tag{5.6}$$

In the case of uniaxial defects of hexagonal, tetragonal or trigonal symmetry, the condition $P_{11} = P_{22} \neq P_{33}$ holds, and for defects of isotropic or cubic symmetry we have $P_{11} = P_{22} = P_{33} = P$. In general, the coordinate systems appropriate to describe the defects and the spontaneous magnetization within the dw do not coincide. Therefore, in most cases transformations either of M_s to the coordinate system of the defects or vice versa are necessary. In the following, we consider the interaction of point defects with 180°-dws in amorphous and crystalline materials.

5.3 180°-wall in amorphous alloys with uniaxial anisotropy

As described in Section 4.2, the spontaneous magnetization in the case of a 180°-dw rotates from the $(+)x$-axis into the $(-)x$-axis. The dw normal is taken as the z-axis. The rotation angle φ of M_s in the case of an amorphous alloy with induced uniaxial anisotropy, K_u, obeys the following differential equation (see Section 4.3):

$$A\left(\frac{d\varphi}{dz}\right)^2 = \phi_B^{180°}(\varphi), \qquad (5.7)$$

with

$$\phi_B^{180°}(\varphi) = K_u \sin^2 \varphi + \frac{9}{2}G\lambda_s^2 \sin^2 \varphi. \qquad (5.8)$$

For a calculation of the three energy contributions of eq. (5.1) we have to introduce the direction cosines, $\gamma_1 = \cos\varphi$, $\gamma_2 = \sin\varphi$, $\gamma_3 = 0$, into the energy expressions of eq. (5.2) and eq. (5.3). For ϕ_{ex} and ϕ_K we have to use the expressions corresponding to the point group of the defect as derived in Section 2.2.

For ε^{el} we have to insert into eq. (5.3) the elastic strains derived for the 180°-walls in Section 4.3:

1. Isotropic defects

$$e = a_0\phi_B^{180°}(\varphi) + k_0 \sin^2 \varphi. \qquad (5.9)$$

2. Tetragonal defects with the tetragonal axes parallel to the x (1)-, y (2)- or z (3)-axis

$$e^{(1)} = (a_{33} \cos^2 \varphi + a_{11} \sin^2 \varphi)\phi_B^{180°}(\varphi) + (k_1 - k_3) \sin^2 \varphi - \frac{3}{2}(P_{33} - P_{11})\lambda_s \sin^2 \varphi,$$

$$e^{(2)} = (a_{33} \sin^2 \varphi + a_{11} \cos^2 \varphi)\phi_B^{180°}(\varphi) + (k_3 - k_1) \sin^2 \varphi + \frac{3}{2}(P_{33} - P_{11})\lambda_s \sin^2 \varphi,$$

$$e^{(3)} = a_{33}\phi_B^{180°}(\varphi) + k_1. \qquad (5.10)$$

The interaction constants a_0, a_{ii} and k_i describe the modifications of the exchange constants A and of the magnetocrystalline constants K_i. From eq. (5.2) it follows that

$$a_{ii} = \int_{\Omega_{n.n.}} \left(A_{ii}(r) - A^{(0)}\right) d^3r,$$

$$k_i = \int_{\Omega_{n.n.}} \left(K_i(r) - K_i^{(0)}\right) d^3r. \qquad (5.11)$$

5.4 180°-wall in α-Fe

The domain wall energy density of a (001)–180°-wall in α-Fe has been derived in Section 4.4 and is given by

$$\phi_B^{180°}(\varphi) = \frac{1}{4}\left\{ K_1 - \frac{9}{4}\lambda_{100}^2(c_{11} - c_{12}) + \frac{9}{2}\lambda_{111}^2 c_{44} \right\} \cdot \sin^2 2\varphi$$

$$+ \frac{9}{4}(c_{11} - c_{12})\lambda_{100}^2 \sin^2 \varphi. \tag{5.12}$$

The interaction energy expressions for defects of cubic (vacancies, substitutional impurity atoms) and tetragonal (impurity interstitials of H, N, C) symmetry are equivalent to eqs. (5.9) and (5.10) derived for the 180°-wall in amorphous alloys with K_u to be replaced by K_1 in the case of cubic defects and $K_u = 0$ for tetragonal defects. λ_s has to be substituted by λ_{100}.

5.5 Interaction forces of domain walls with point defects

The force exerted by a point defect at position z on a dw always acts parallel or antiparallel to the z-axis and is given by

$$P_z = -\frac{de}{dz} = -\frac{de}{d\varphi} \cdot \frac{d\varphi}{dz} = -\frac{1}{\sqrt{A}}\frac{de}{d\varphi} \left(\phi_B^{180°}(\varphi) \right)^{1/2}. \tag{5.13}$$

5.6 Interaction of Bloch walls with dislocations

In crystalline materials the main types of dislocations are the edge (sign $\perp$) and screw (sign $\odot$) dislocations. Schematic models of these dislocations are shown in Fig. 5.2 and Fig. 5.3. Edge dislocations correspond to an additional net plane which ends within the bulk of the crystal. The inner edge of this half-plane describes the line direction l and the distance between the net planes corresponds to the Burgers vector b. In the case of fcc crystals b and l correspond to (a/2) $\langle 110 \rangle$ and $\langle 112 \rangle$, respectively. For edge dislocations b and l are perpendicular to each other. Screw dislocations are characterized by a shear deformation, where part of the crystal is displaced by a nearest neighbour distance with respect to the other part. Burgers vector and line direction are parallel to each other and correspond to the $\langle 110 \rangle$ directions in fcc crystals. There exist also mixed dislocations with Burgers vector oblique to the dislocation line. A common property of all dislocations is their long-range stress field which decreases according to a $1/r$-law. The basic properties of dislocations which exist in as-grown crystals as well as in plastically deformed materials are described in a series of books [2.59, 5.1–5.4].

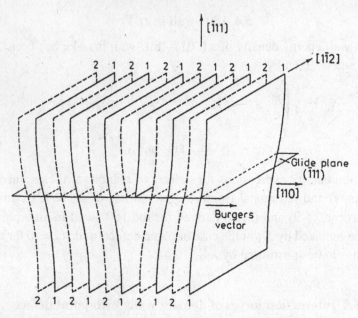

Fig. 5.2. Schematic model of an edge dislocation ($\perp$) in fcc lattices with Burgers vector $(a/2)$ [110], line direction [1$\bar{1}$2] and glide plane normal [$\bar{1}$11] [5.2].

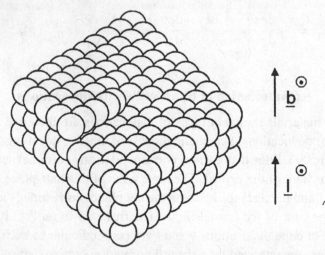

Fig. 5.3. Schematic model of a screw dislocation ($\odot$ in fcc lattices with Burgers vector [110], line direction [110] and arbitrary glide plane normal).

The interaction between dislocations and dws was originally investigated by Vicena [5.5] by means of the magnetoelastic coupling energy, ϕ_σ, as defined by eq. (2.77). Quantitative results cannot be easily obtained by this method because of the complex form of the stress tensor of the dislocations. A more direct method to obtain explicit results starts from the so-called Peach–Köhler relation [5.6, 5.7] where the stress tensor, σ^m, of the dw as defined by eqs. (2.60) and (4.27) is used, and the properties of the dislocations are exclusively described by Burgers vector b and dislocation line element dl [2.59, 5.1–5.4]. The force acting on the dw then may be written as

$$P = \int (\sigma^m \cdot b) \times dl, \tag{5.14}$$

where the integral extends over the dislocation line. In the following, all tensors and vectors refer to the coordinate system of the dw. The tensor of elastic stresses of the dw is then written as

$$\sigma^m = \begin{pmatrix} \sigma_{11}^m & \sigma_{12}^m & 0 \\ \sigma_{12}^m & \sigma_{22}^m & 0 \\ 0 & 0 & 0 \end{pmatrix}. \tag{5.15}$$

In the case of a 180°-wall the $(+)x$-coordinate (1) corresponds to the orientation of M_s within the domain (easy direction) and the z-coordinate (3) to the dw normal.

For the force, P_3, acting parallel or antiparallel to the dw normal we now obtain from eq. (5.14)

$$P_3 = \int \sum_{k-1,2} (\sigma_{1k}^m b_k \, dl_2 - \sigma_{2k}^m b_k \, dl_1) \qquad k = 1, 2$$

$$= \int \{(b_1 \sigma_{11}^m + b_2 \sigma_{12}^m) \, dl_2 - (b_1 \sigma_{12}^m + b_2 \sigma_{22}^m) \, dl_1\}. \tag{5.16}$$

5.6.1 Straight dislocation lines

The vectors b and dl in the following are described by their polar angles $\theta_{b,l}$ with respect to the z-axis and the azimuthal angles $\varphi_{b,l}$ with respect to the x-axis (Fig. 5.4). Then we may write

$$b = \begin{pmatrix} b_1 \\ b_2 \\ b_3 \end{pmatrix} = b \begin{pmatrix} \cos \varphi_b \sin \theta_b \\ \sin \varphi_b \sin \theta_b \\ \cos \theta_b \end{pmatrix},$$

$$dl = \begin{pmatrix} dl_1 \\ dl_2 \\ dl_3 \end{pmatrix} = dl \begin{pmatrix} \cos \varphi_l \sin \theta_l \\ \sin \varphi_l \sin \theta_l \\ \cos \theta_l \end{pmatrix}. \tag{5.17}$$

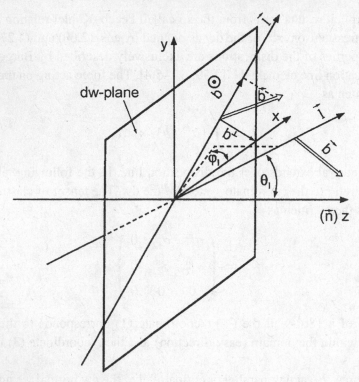

Fig. 5.4. Geometry of dw plane and dislocation lines. Screw dislocations: $b^{\odot} \parallel l$. Edge dislocations: $b^{\perp} \perp l$.

5.6.1.1 Dislocations of length l parallel to the domain wall plane (x, y)

In this case, in eq. (5.16) the integrand is a constant and the total force is proportional to the length of the dislocation:

$$P_3 = -\big(b_1\sigma_{12}^{m}(\varphi) + b_2\sigma_{22}^{m}(\varphi)\big)l \cos\varphi_1$$
$$+\big(b_2\sigma_{12}^{m}(\varphi) + b_1\sigma_{11}^{m}(\varphi)\big)l \sin\varphi_1. \qquad (5.18)$$

In the case of a screw dislocation, i.e., $b \parallel l$, this gives $\big(\theta_b = \frac{\pi}{2},\ \varphi_b = \varphi_1\big)$

$$P_3^{\odot} = -bl \cos^2\varphi_1\big(\sigma_{12}^{m}(\varphi) + \mathrm{tg}\,\varphi_1\,\sigma_{22}^{m}(\varphi)\big)$$
$$+bl \sin^2\varphi_1\big(\sigma_{12}^{m}(\varphi) + \mathrm{ctg}\,\varphi_1\,\sigma_{11}^{m}(\varphi)\big)$$
$$= -bl \cos 2\varphi_1 \cdot \sigma_{12}^{m}(\varphi) + \frac{1}{2}bl \sin 2\varphi_1\big(\sigma_{11}^{m}(\varphi) - \sigma_{22}^{m}(\varphi)\big). \qquad (5.19)$$

In the case of edge dislocations it is of interest to consider the two limiting cases with $\theta_b = 0$, i.e., $b \perp (x, y)$-plane and $\theta_b = \frac{\pi}{2}$, i.e., $\varphi_b = \varphi_1 + \frac{\pi}{2}$. For $\theta_b = 0$ we

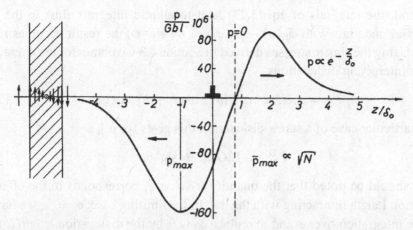

Fig. 5.5. Interaction force between a $(\bar{1}\bar{1}2)$ 180°-dw and a 60°-dislocation parallel to the $[\bar{1}10]$ direction. Depending on the sign of the Burgers vector the dislocation lines attract or repel the dw.

find a vanishing interaction force

$$P_3^\perp = 0, \tag{5.20}$$

and for $\theta_b = \varphi_1 + \frac{\pi}{2}$ the result is

$$P_3^\perp = bl\sigma_{12}^m(\varphi)\sin^2\varphi_1 - bl\sigma_{11}^m(\varphi)\sin^2\varphi_1 + \sigma_{22}^m(\varphi)\cos^2\varphi_1. \tag{5.21}$$

As an example, Fig. 5.5 shows the force acting on a $(\bar{1}\bar{1}2) -180°$ Bloch wall by a 60°-dislocation oriented parallel to the dw plane along the $[\bar{1}10]$-direction.

5.6.1.2 Dislocations intersecting the domain walls

In the case of intersecting dislocations we have to perform the integrations along the dislocation line as indicated by eq. (5.16). Taking care of eq. (5.17), the line element components dl_1 and dl_2 are given by $dl_1 = dz \, \text{tg} \, \theta_1 \cos\varphi_1$ and $dl_2 = dz \, \text{tg} \, \theta_1 \sin\varphi_1$ giving

$$P_3 = \int_{-\infty}^{\infty} \left\{ \left(b_1\sigma_{11}^m(\varphi) + b_2\sigma_{12}^m(\varphi)\right)\sin\varphi_1 - \left(b_1\sigma_{12}^m(\varphi) + b_2\sigma_{22}^m(\varphi)\right)\cos\varphi_1 \right\} \text{tg} \, \theta_1 \, dz.$$

$$\tag{5.22}$$

It is useful to substitute dz by $d\varphi$ using eq. (4.4), which gives us for a 180°-wall

$$P_3 = \sqrt{A} \, \text{tg} \, \theta_1 \int_0^{\pi} \left\{ \left(b_1\sigma_{11}^m(\varphi) + b_2\sigma_{12}^m(\varphi)\right)\sin\varphi_1 - \left(b_1\sigma_{12}^m(\varphi) + b_2\sigma_{22}^m(\varphi)\right)\cos\varphi_1 \right\}$$

$$\times \frac{d\varphi}{\left(\phi_B(\varphi) - \phi_B(\varphi_1)\right)^{1/2}}. \tag{5.23}$$

In general, the integrals of eq. (5.23) lead to elliptic integrals. Just in the case of uniaxial materials with $\phi_B(\varphi) - \phi_B(\varphi_1) = K_1 \sin^2 \varphi$, the result becomes rather simple. Using the elastic stresses derived in Section 4.3 we obtain for the interaction force of interacting dislocations

$$P_3 = 6G\lambda_s\delta_0 \, \mathrm{tg}\,\theta_1\{b_1 \sin\varphi_1 + b_2 \cos\varphi_1\}. \qquad (5.24)$$

In the particular case of a screw dislocation this gives for $b \parallel l$

$$P_3^{\ominus} = 12G\lambda_s\delta_0 b \sin^3\theta_1. \qquad (5.25)$$

Here it should be noted that the quantity $\pi\delta_0/\cos\theta_1$ corresponds to the effective dislocation length interacting with the dw. In the limiting case, $\theta_1 = \frac{\pi}{2}$, we have to omit the integration over φ and to replace $\delta_0 \, \mathrm{tg}\,\theta_1$ by the dislocation length l.

5.6.2 Straight dislocation dipoles

In many cases the dislocation structure of plastically deformed materials is composed of dislocation dipoles, which may be formed by cyclic deformation [5.8] or agglomeration of vacancies or interstitial atoms. In order to treat these dislocation loops it is useful to transform the line integral of eq. (5.14) into a surface integral according to

$$P = \int_l (\sigma^m \cdot b) \times \mathrm{d}l = -\int\int_F [(\mathrm{d}f \times \nabla) \times (\sigma^m \cdot b)], \qquad (5.26)$$

where $\mathrm{d}f$ denotes the vector surface element of an area F spreading over the dislocation loop. Since σ^m depends only on the z-coordinate we may put in eq. (5.26) $\nabla_1 = \nabla_2 \equiv 0$, thus leading to

$$P_3 = \int \sum_{k=1,2} (\nabla_3\sigma_{1k}^m b_k \, \mathrm{d}f_1 + \nabla_3\sigma_{2k}^m b_k \, \mathrm{d}f_2). \qquad (5.27)$$

According to Dehlinger and Kröner [5.9] and Eshelby [5.10] the products $b_k \, \mathrm{d}f_i$ may be interpreted as differential displacement dipole moments,

$$\mathrm{d}Q_{ik} = \mathrm{d}f_i \, b_k, \qquad (5.28)$$

(compare also the definition of the displacement dipole moment of point defects in eq. (5.4)). Equation (5.27) now may be written as

$$P_3 = -\int \sum_{k=1,2} (\nabla_3\sigma_{1k}^m \, \mathrm{d}Q_{1k} + \nabla_3\sigma_{2k}^m \, \mathrm{d}Q_{2k}). \qquad (5.29)$$

Applying eq. (5.29) to the case of straight dislocation dipoles intersecting the domain walls leads to vanishing interaction forces for edge and screw dislocations

$P_3^\perp = P_3^\odot = 0$. For dislocation dipoles lying parallel to the dw with the dislocations lying at positions z_1, φ_1 and z_2, φ_2 the interaction force for the screw and edge dislocations is given by

$$P_3^{\text{dip},\odot} = -bl \cos 2\varphi_1 \big(\sigma_{12}^{\text{m}}(\varphi_2) - \sigma_{12}^{\text{m}}(\varphi_1)\big)$$
$$+ \frac{1}{2} bl \sin 2\varphi_1 \big(\sigma_{11}^{\text{m}}(\varphi_2) - \sigma_{11}^{\text{m}}(\varphi_1) - \sigma_{22}^{\text{m}}(\varphi_2) + \sigma_{22}^{\text{m}}(\varphi_2)\big), \quad (5.30)$$

$$P_3^{\text{dip},\perp} = -bl \cos 2\varphi_1 \big(\sigma_{12}^{\text{m}}(\varphi_2) - \sigma_{12}^{\text{m}}(\varphi_1)\big)$$
$$- bl \big\{ \big(\sigma_{11}^{\text{m}}(\varphi_2) - \sigma_{11}^{\text{m}}(\varphi_1)\big) \sin^2 \varphi_1 + \big(\sigma_{22}^{\text{m}}(\varphi_2) - \sigma_{22}^{\text{m}}(\varphi_1)\big) \cos^2 \varphi_1 \big\}. \tag{5.31}$$

In the special case where the distance, D, between the two dipole dislocations is small in comparison to δ_B, the stress components, $\sigma_{ij}(\varphi_2)$, may be developed into a Taylor series giving

$$P_3^{\text{dip},\odot} = -blD \, \cos 2\varphi_1 \, \cos \theta_D \, \nabla_3 \sigma_{12}^{\text{m}}$$
$$+ \frac{1}{2} blD \, \sin 2\varphi_1 \, \cos \theta_D \, \nabla_3 (\sigma_{11}^{\text{m}} - \sigma_{22}^{\text{m}}), \tag{5.32}$$

$$P_3^{\text{dip},\perp} = -blD \, \sin 2\varphi_1 \, \cos \theta_D \, \nabla_3 \sigma_{12}^{\text{m}}$$
$$- blD \, \cos \theta_D \, \big\{ \sin^2 \varphi_1 \, \nabla_3 \sigma_{11}^{\text{m}} + \cos^2 \varphi_1 \, \nabla_3 \sigma_{22}^{\text{m}} \big\}. \tag{5.33}$$

Here θ_D denotes the polar angle between the dislocation dipole distance vector D and the z-axis.

5.6.3 Dislocation loops

For a calculation of the interaction force of small dislocation loops of area $F_{\text{dip}} = lD$ with lateral extensions $< \delta_B$ we obtain from eq. (5.29)

$$P_3 = -F_{\text{dip}} \sin \theta_f \sum_{k=1,2} \big\{ b_k \cos \varphi_f \, \nabla_3 \sigma_{1k}^{\text{m}} + b_k \sin \varphi_f \, \nabla_3 \sigma_{2k}^{\text{m}} \big\}, \tag{5.34}$$

where φ_f and θ_f denote the azimuthal and polar angle of the normal vector of the dislocation area F_B. Equation (5.34) leads to vanishing forces for dislocation loops lying parallel to the dw ($\theta_f = 0$) or for loops with a Burgers vector perpendicular to the dw ($b_1, b_2 = 0$).

5.7 Interaction of domain walls with planar defects

5.7.1 Pinning by thin planar defects

The interaction of dws with planar defects originates from the local perturbations of exchange interactions and of the magnetocrystalline anisotropy. Planar defects of

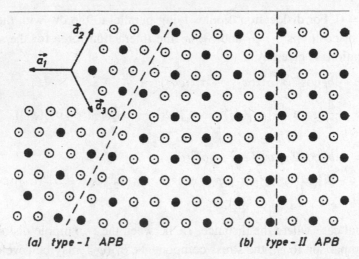

(a) type - I APB   (b) type - II APB

Fig. 5.6. Atomic configurations of antiphase boundaries in SmCo$_5$ showing the changes of nearest neighbour interactions within the basal plane. (a) Type-I APB, $(a/3)$ $\{1\bar{0}10\}\langle 1\bar{2}10\rangle$; (b) Type-II APB, $(a/3)$ $\{\bar{2}110\}\langle 2\bar{1}\bar{1}0\rangle$.

atomic width are stacking faults, antiphase boundaries in intermetallics and grain boundaries between grains of different orientations and different phases. As an example, Fig. 5.6 shows the atomic structure of two types of antiphase boundaries in SmCo$_5$ [5.15]. At the phase boundaries the exchange interactions and the magnetocrystalline energy are modified. These interactions are of importance for polycrystalline soft magnetic materials (nanocrystalline alloys) as well as hard magnetic materials, e.g., Sm$_2$Co$_{17}$-based permanent magnets. The pinning of dws by these planar defects may be treated by the micromagnetic continuum theory [5.10–5.13] or by the discrete Heisenberg model [5.14–5.16].

Consider a planar defect consisting of N atomic layers of interatomic distance d. The width of the planar defect is given by $D = (N - 1)d$. A dw lying parallel to the planar defect is divided into three regions as shown in Fig. 5.7. The external regions, $0 < \varphi < \varphi_1$ and $\varphi_N < \varphi < \pi$, are treated by the continuum approach. The internal region II includes the planar defect and will be treated by the discrete Heisenberg model. Within the framework of this model the wall energy per unit area of a uniaxial material is given by

$$\gamma = \frac{1}{F_P} \sum_{\alpha \neq \beta} J^{\alpha\beta} S^\alpha S^\beta \{1 - \cos(\varphi^\alpha - \varphi^\beta)\}$$

$$+ \sum_\alpha d\left(K_1^\alpha \sin^2 \varphi^\alpha + \mu_0 H_{ext} M_s \cos \varphi^\alpha\right), \qquad (5.35)$$

where φ^α and φ^β denote the angles of rotation of the spins S^α and S^β and $J^{\alpha\beta}$ corresponds to the exchange integral between these neighbouring spins. The summation

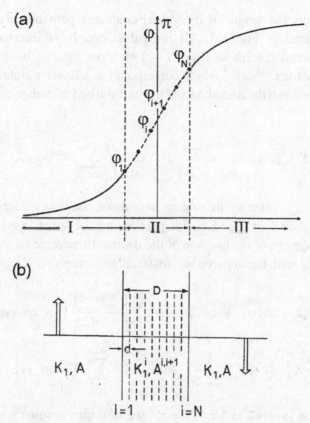

Fig. 5.7. Model of a planar defect with varying local anisotropy and exchange constants. (a) Definition of regions I, II, III. (b) Discretization of the planar defect.

extends over neighbouring spins of unit cells arranged parallel to the dw normal. F_P is the area of the unit cell lying parallel to the dw. For a more suitable representation we replace the exchange coupling between individual spins by the integral interaction between the adjacent atomic layers i and $i+1$ thus giving

$$\gamma = \sum_i \left\{ 2(A^{i,i+1}/d)\,[1 - \cos(\varphi_{i+1} - \varphi_i)] + dK_1^i \sin^2 \varphi_i + d\mu_0 H_{\text{ext}} M_s \cos \varphi_i \right\}.$$

$$(5.36)$$

The exchange constants $A^{i,i+1}$ contain all individual exchange couplings between adjacent layers i and $i+1$. In the case of an ideal crystal the macroscopic exchange constant A is defined as (see eq. (2.12))

$$A = \frac{d}{2F_P} \sum_{\alpha \neq \beta} J^{\alpha\beta} S^\alpha S^\beta (z^{\alpha\beta}/d)^2,$$

$$(5.37)$$

where $z^{\alpha\beta}$ denotes the length of individual couplings perpendicular to the wall plane as illustrated in Fig. 5.6. All individual couplings intersecting a fictitious plane between the ith and the $(i+1)$th layer have to be connected and weighted by a factor $(z^{\alpha\beta}/d)^{-1}$ which corresponds to a linear weighting. The ratio $A^{i,i+1}/A_{\text{undist.}}$ between the disturbed and the undisturbed exchange constant then is given by

$$A^{i,i+1}/A_{\text{undist.}} = \sum_n J^{i,n} \cdot z_{i,n} \bigg/ \sum_{\text{undist.}} J^{i,n} \cdot z_{i,n}, \qquad (5.38)$$

where $J^{i,n}$ and $z_{i,n}$ refer to all nearest neighbour couplings intersected by the fictitious plane shown in Fig. 5.6. For performing the sums in eq. (5.36) more explicitly we make use of the division of the dw into three regions. In regions I and III we are dealing with the unperturbed material parameters A, K_1^1, K_n^1 thus giving

$$\gamma = \frac{2A}{d}\frac{d}{\delta_0}(2 - \cos\varphi_1 + \cos\varphi_N) + \sum_{i=1}^{N-1} \frac{2A^{i,i+1}}{d}[1 - \cos(\varphi_{i+1} - \varphi_i)]$$

$$+ d\left\{\frac{1}{2}K_1^1 \sin^2\varphi_1 + \frac{1}{2}K_1^N \sin^2\varphi_N + \sum_{i=2}^{N-1} K_1^i \sin^2\varphi_i\right\}. \qquad (5.39)$$

For the transition layers $i = 1$ and $i = N$ we take into account only half of the magnetocrystalline energy. Furthermore, the effect of the applied field on the spin structure of the dw has been neglected. Minimizing the wall energy of eq. (5.39) with respect to the discrete angles φ_i leads to the following system of nonlinear equations, which for $\varphi_N - \varphi_1 \ll 1$ is written as

$$\varphi_{i+1} - \varphi_i = \frac{A}{A^{i,i+1}}\frac{d}{\delta_0}\sin\varphi_i. \qquad (5.40)$$

Making use of the micromagnetic dw equation, $\sin^2\varphi = \text{ch}^{-2}(z/\delta_0)$, which is valid for $D < \delta_B$, we obtain from eq. (5.39) the change of the dw energy

$$\Delta\gamma = \gamma_{\text{dist.}} - \gamma_{\text{undist.}} = K_1 d \sum_i^{N-1} \left(\frac{A}{A^{i,i+1}} - \frac{K_1^i}{K_1}\right)\text{ch}^{-2}(z/\delta_0). \qquad (5.41)$$

The force P per unit area acting on the dw is now given by

$$P_3'(z) = -\frac{d(\Delta\gamma)}{dz} = K_1\frac{d}{\delta_0}\sum_i^{N-1}\left(\frac{A}{A^{i,i+1}} - \frac{K_1^i}{K_1}\right)\frac{2\,\text{sh}(z/\delta_0)}{\text{ch}^3(z/\delta_0)}, \qquad (5.42)$$

and the maximum force acting on the dw is written as

$$P'_{3,\text{max}} = -\left.\frac{d(\Delta\gamma)}{dz}\right|_{\text{max}} = \frac{1}{3\cdot\sqrt{3}}\, 4K_1 \frac{d}{\delta_0} \sum_{i}^{N-1}\left(\frac{A}{A^{i,i+1}} - \frac{K_1^i}{K_1}\right). \tag{5.43}$$

If we assume that the perturbed layers all have the same material parameters $A^{i,i+1} = A'$ and $K_1^i = K_1'$, eq. (5.41) may be written as

$$\Delta\gamma = K_1 D\left[\frac{A}{A'} - \frac{K_1'}{K_1}\right]\text{ch}^{-2}(z/\delta_0), \tag{5.44}$$

with

$$\frac{1}{A'} = \frac{1}{N-1}\sum_{i=1}^{N-1}\frac{1}{A^{i,i+1}},$$

$$K_1' = \frac{1}{N-1}\left[\sum_{i=1}^{N-1} K_1^i + \frac{1}{2}\left(K_1^1 + K_1^N\right)\right], \tag{5.45}$$

and

$$P_{3,\text{max}} = \frac{1}{3\cdot\sqrt{3}}\, 4K_1 \frac{D}{\delta_0}\left|\frac{A}{A'} - \frac{K_1'}{K_1}\right|. \tag{5.46}$$

5.7.2 Pinning by extended planar defects

If the planar defect is wider than the dw width, the force per unit area acting on the dw derives from the space dependent wall energy $\gamma(z)$,

$$P'(z) = -\frac{d\gamma(z)}{dz}, \tag{5.47}$$

where the specific wall energy of a uniaxial crystal according to eq. (4.13) is given by $\gamma(z) = 4\sqrt{A(z)\,K_1(z)}$.

Let us assume that only $K_1(z)$ shows an appreciable inhomogeneity which may be described by

$$K_1(z) = K_1(\infty) - \frac{\Delta K}{\text{ch}^2(z/D)}. \tag{5.48}$$

$K_1(\infty)$ denotes K_1 in the perfect material, ΔK the maximum change of K_1 within the planar defect of halfwidth D. Inserting eq. (5.48) into eq. (5.47) and determining the maximum interaction force we obtain

$$P'_{\text{max}}(z) = 2K_1(\infty)\cdot\frac{4\delta_B}{3\pi D}. \tag{5.49}$$

It is of interest to note that in the case of wide defects, $D > \delta_B$, the pinning force decreases according to a $1/D$-law whereas in the case of thin defects, $D < \delta_B$, the pinning force increases linearly with D.

5.7.3 Pinning by phase boundaries

In contrast to the planar microstructures treated in Section 5.7.1, phase boundaries correspond to abrupt changes of the material constants, J_s, A and K_1. The micromagnetic model of a phase boundary is shown in Fig. 5.8. The material constants in phase I and II in the following are denoted as J_I, A_I, K_I and J_{II}, A_{II}, K_{II}. A 180°-dw is pressed by an applied magnetic field against the phase boundary. At position z_0 of the phase boundary, the azimuthal angle of J_s is denoted by φ_0. At a certain critical angle φ_0^{crit} the position of the dw becomes unstable and moves spontaneously from phase I into phase II. The micromagnetic calculation starts from the magnetic

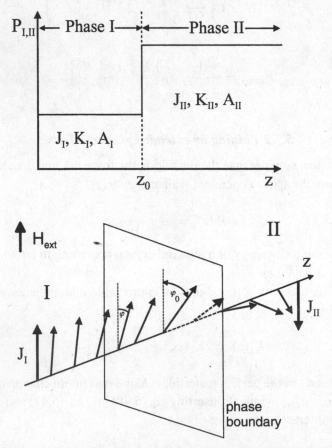

Fig. 5.8. Model of a phase boundary with abrupt changes of the material constants.

Gibbs free energy in phase I and II [5.17]:

$$\phi = \int_0^{\varphi_0} \left\{ A_{\mathrm{I}} \left(\frac{\mathrm{d}\varphi}{\mathrm{d}z} \right)^2 + K_{\mathrm{I}} \sin^2 \varphi \ - \ H_{\mathrm{ext}} J_{\mathrm{I}} \cos \varphi \right\} \mathrm{d}\varphi$$

$$+ \int_{\varphi_0}^{\pi} \left\{ A_{\mathrm{II}} \left(\frac{\mathrm{d}\varphi}{\mathrm{d}z} \right)^2 + K_{\mathrm{II}} \sin^2 \varphi - H_{\mathrm{ext}} J_{\mathrm{II}} \cos \varphi \right\} \mathrm{d}\varphi. \qquad (5.50)$$

Minimization of ϕ with respect to φ leads to the following micromagnetic equations in phases I and II:

$$A_{\mathrm{I}} \left(\frac{\mathrm{d}\varphi}{\mathrm{d}z} \right)^2 = K_{\mathrm{I}} \sin^2 \varphi - H_{\mathrm{ext}} J_{\mathrm{I}} \cos \varphi + H_{\mathrm{ext}} J_{\mathrm{I}}, \qquad \varphi < \varphi_0, \qquad (5.51)$$

$$A_{\mathrm{II}} \left(\frac{\mathrm{d}\varphi}{\mathrm{d}z} \right)^2 = K_{\mathrm{II}} \sin^2 \varphi - H_{\mathrm{ext}} J_{\mathrm{II}} \cos \varphi - H_{\mathrm{ext}} \cdot J_{\mathrm{II}}, \qquad \varphi > \varphi_0. \qquad (5.52)$$

At $\varphi = \varphi_0$ the following boundary condition holds:

$$A_{\mathrm{I}} \frac{\mathrm{d}\varphi}{\mathrm{d}z} \bigg|_{\varphi=\varphi_0} = A_{\mathrm{II}} \frac{\mathrm{d}\varphi}{\mathrm{d}z} \bigg|_{\varphi=\varphi_0}. \qquad (5.53)$$

Subtracting eq. (5.52) from eq. (5.51) and taking care of the boundary condition in eq. (5.53) we obtain for $\varphi = \varphi_0$

$$\left(\frac{A_{\mathrm{I}}}{A_{\mathrm{II}}} K_{\mathrm{I}} - K_{\mathrm{II}} \right) \sin^2 \varphi_0 - H_{\mathrm{ext}} \left(\frac{A_{\mathrm{I}}}{A_{\mathrm{II}}} J_{\mathrm{I}} - J_{\mathrm{II}} \right) \cos \varphi_0 + \left(\frac{A_{\mathrm{I}}}{A_{\mathrm{II}}} J_{\mathrm{I}} + J_{\mathrm{II}} \right) H_{\mathrm{ext}} = 0. \qquad (5.54)$$

Equation (5.54) gives φ_0 as a function of H_{ext}. The critical value $\varphi_0^{\mathrm{crit}}$, where the dw moves spontaneously into phase II is determined by the condition $\mathrm{d}H_{\mathrm{ext}}/\mathrm{d}\varphi_0 = 0$, giving

$$\cos \varphi_0^{\mathrm{crit}} = - \frac{H_{\mathrm{ext}} \left(\frac{A_{\mathrm{I}}}{A_{\mathrm{II}}} J_{\mathrm{I}} - J_{\mathrm{II}} \right)}{2 \left(\frac{A_{\mathrm{I}}}{A_{\mathrm{II}}} K_{\mathrm{I}} - K_{\mathrm{II}} \right)}. \qquad (5.55)$$

Inserting eq. (5.55) into eq. (5.54) gives a quadratic equation for $H_{\mathrm{ext}}^{\mathrm{crit}} = H_{\mathrm{c}}$:

$$H_{\mathrm{c}}^2 \frac{\left(\frac{A_{\mathrm{I}}}{A_{\mathrm{II}}} J_{\mathrm{I}} - J_{\mathrm{II}} \right)^2}{4 \left(\frac{A_{\mathrm{I}}}{A_{\mathrm{II}}} K_{\mathrm{I}} - K_{\mathrm{II}} \right)} + \left(\frac{A_{\mathrm{I}}}{A_{\mathrm{II}}} J_{\mathrm{I}} + J_{\mathrm{II}} \right) H_{\mathrm{c}} + \left(K_{\mathrm{I}} \frac{A_{\mathrm{I}}}{A_{\mathrm{II}}} - K_{\mathrm{II}} \right) = 0. \qquad (5.56)$$

With the ratios

$$\varepsilon_{\mathrm{A}} = A_{\mathrm{I}}/A_{\mathrm{II}}; \qquad \varepsilon_{\mathrm{J}} = J_{\mathrm{I}}/J_{\mathrm{II}}; \qquad \varepsilon_{\mathrm{K}} = K_{\mathrm{I}}/K_{\mathrm{II}},$$

the physical solution of eq. (5.56) is given by

$$H_c = \frac{2K_{II}}{J_{II}} \frac{1 - \varepsilon_K \varepsilon_A}{\left(1 + \sqrt{\varepsilon_A \varepsilon_J}\right)^2},$$

$$\cos \varphi_0^{crit} = \frac{\sqrt{\varepsilon_A \varepsilon_J} - 1}{\sqrt{\varepsilon_A \varepsilon_J} + 1}. \tag{5.57}$$

As to be expected, H_c vanishes for $\varepsilon_K = \varepsilon_A = 1$, and has an optimum value for $\varepsilon_K = 0$ ($K_I = 0$) with

$$H_c = \frac{2K_{II}}{J_{II}} \frac{1}{(1 + \sqrt{\varepsilon_A \varepsilon_J})^2}. \tag{5.58}$$

In hard magnetic materials large spontaneous magnetizations are desirable in order to realize large stored energies, and also large Curie temperatures are favoured in order to guarantee excellent temperature stability. Therefore $\varepsilon_J = \varepsilon_A \simeq 1$ are suitable conditions leading to a coercive field of

$$H_c = \frac{K_{II}}{2J_{II}}(1 - \varepsilon_K). \tag{5.59}$$

The optimum coercive field expected under these conditions for a soft phase I and a hard phase II is just $H_c = K_{II}/2J_{II}$ – one quarter of the ideal nucleation field.

As a practical case we determine the coercive field of phase boundaries between Sm_2Co_{17} and $SmCo_5$ using the material parameters given in Tables 2.1 and 2.2. For room temperature the calculation gives $\mu_0 H_c = 13.2$ T. This value is around a factor of three smaller than the ideal nucleation field of $SmCo_5$.

References

[5.1] Seeger, A., 1955, 'Kristallplastizität'. In *Handbuch der Physik*, Vol. 7/1, (Springer-Verlag, Berlin–Göttingen–Heidelberg) p. 383.

[5.2] Amelinckx, S., 1979. In *Dislocations in Solids* (Ed. F.R.N. Nabarro, North-Holland, Amsterdam, Vol. 2) p. 67.

[5.3] Haasen, P., 1974, *Physikalische Metallkunde* (Springer-Verlag, Berlin–Heidelberg–New York).

[5.4] Hirth, J.P., 1983, 'Dislocations'. In *Physical Metallurgy* (Eds. R.W. Cahn and P. Haasen, Elsevier Science Publ. BV) p. 1223.

[5.5] Vicena, F., 1954, *Czech. J. Phys.* **4**, 419; *ibid.* **5**, 11, 480.

[5.6] Peach, M., and Köhler, J.S., 1950, *Phys. Rev.* **80**, 436.

[5.7] Seeger, A., Kronmüller, H., Rieger, H., and Träuble, H., 1964, *J. Appl. Phys.* **35**, 740.

[5.8] Mughrabi, H., 1993, 'Microstructure and Mechanical Properties'. In *Plastic Deformation and Fracture of Materials*, Vol. 6, (Ed. H. Mughrabi, VCH Publishing, Weinheim–New York–Basel–Cambridge) p. 1.

[5.9] Dehlinger, U., and Kröner, E., 1960, *Z. Metallkd.* **51**, 457.

[5.10] Eshelby, J.D., 1956, 'The Continuum Theory of Lattice Defects'. In *Solid State Physics, Advances in Research and Applications*, Vol. 3 (Eds. F. Seitz and D. Turnbull) p. 79.

[5.11] Kronmüller, H., 1973, *AIP Conf. Proc.* **10**, 1006.

[5.12] Mitsek, A.I., and Semyannikov, S.S., 1969, *Sov. Phys. Solid State* **11**, 899.

[5.13] Friedberg, R., and Paul, D., 1975, *Phys. Rev. Lett.* **34**, 1234.

[5.14] Hilzinger, H.-R., and Kronmüller, H., 1975, *Phys. Lett.* **51A**, 59.

[5.15] Hilzinger, H.-R., 1977, *Appl. Phys.* **12**, 253.

[5.16] Hilzinger, H.-R., and Kronmüller, H., 1975, *IEEE Trans. Magn.* **11**, 81.

[5.17] Kronmüller, H., and Goll, D., 2002. *Physica B* **319**, 122.

6

Coercivity of modern magnetic materials

6.1 Introduction

Since the early times of the commercial use of permanent magnets (pms) the coercive field has been continuously increased from several 10 kA/m up to 10 MA/m. As a consequence of this development the volume of pms required for certain applications has been decreased dramatically. The characteristic properties of the hysteresis loop which are relevant for applications are shown in Figs. 6.1 and 6.2. These properties are the coercive field, H_c, the remanent magnetization, M_r, the saturation field, H_s, for optimum coercivity and the maximum energy product, $(BH)_{max}$. This latter quantity is given by

$$(BH)_{max} = \frac{1}{4\mu_0} J_r^2, \tag{6.1}$$

and presents the largest rectangular area inscribed in the second quadrant of the $B–H$-diagram. The maximum energy product, however, is only achievable if the hysteresis loop is rectangular and the coercive field obeys the condition

$$\mu_0 H_c > \frac{1}{2} J_r. \tag{6.2}$$

Equations (6.1) and (6.2) are the basis for the development of high-quality pms. Both quantities, J_r as well as H_c, should be large in order to produce outstanding pms.

During the first half of the last century, magnetic hardening was achieved by the pinning of domain walls; later on, with the development of Alnico magnets [6.1–6.4] and hard ferrites [6.5–6.10] in the 1940s, the principle of single domain particles became dominant. Presently, there exist pms based on both types of hardening mechanisms. For example, in Sm_2Co_{17}-based magnets the dw pinning mechanism determines H_c, whereas the (Ba,Sr)-ferrites, $SmCo_5$ and $Nd_2Fe_{14}B$ are governed by nucleation mechanisms. Both mechanisms are intimately connected with the

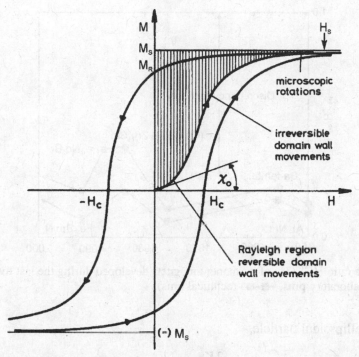

Fig. 6.1. Hysteresis loop $M(H)$ with characteristic properties, remanence M_R, coercive field H_s, saturation field H_s, and initial susceptibility χ_0.

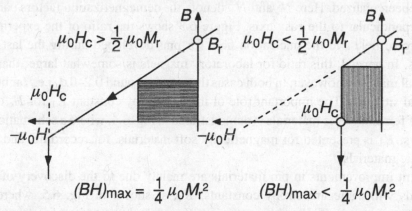

Fig. 6.2. Hysteresis loop $B(H)$ with maximum energy product $(BH)_{\max}$ as the shaded region. For $\mu_0 H_c < B_r = 0.5\mu_0 M_r$ the maximum energy product cannot be realized (right hand side).

microstructures of the hard magnetic materials. The understanding of pms therefore requires the analysis of the interaction between the microstructure and the magnetic properties. This is clearly demonstrated by the fact that in technical pms so far the ideal coercive field corresponding to the nucleation

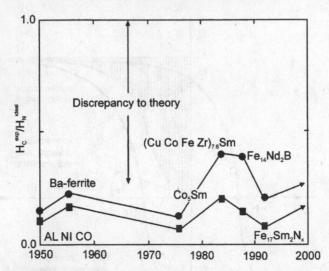

Fig. 6.3. The ratio H_c^{exp}/H_N of permanent magnets developed during the last five decades. ($-\bullet - \bullet-$ laboratory pms, $-\blacksquare-\blacksquare-$ technical pms).

field of an ellipsoidal particle,

$$\mu_0 H_N > \frac{2K_1}{M_s} - \left(N_\| - N_\perp\right)J_s, \tag{6.3}$$

has not been realized. Here $N_\|$ and $N_\perp$ denote the demagnetization factors parallel and perpendicular to the easy axis. Figure 6.3 shows the ratio of the experimental coercive field and the nucleation field for pms developed during the last four decades. In general, this ratio for laboratory magnets is somewhat larger than for technical magnets; however, in both cases the ratio is around 0.2–0.3, i.e., far below the ideal situation. The important role of the anisotropy constant K_1 for H_c of all types of ferromagnetic materials is demonstrated in Fig. 6.4, where a schematic plot of H_c vs. K_1 is presented for magnetically soft materials, for recording, and hard magnetic materials.

Recent improvements in pm materials are mainly due to the discovery of new materials with large anisotropy constants, K_1, as shown in Fig. 6.5, where the continuous increase of H_c with increasing K_1 values is presented for some hard magnetic intermetallic compounds. Improvements of the existing pms are possible by the development of suitable microstructures. Depending whether high-coercivity materials or high-remanence pms are required, the microstructure has to fulfil special properties. The basis for such a tailoring of the microstructure is a detailed knowledge of the interaction between magnetism and the microstructure. A rather successful tool for the treatment of these microstructure–property problems is the theory of micromagnetism. This latter theory was originally developed by Landau

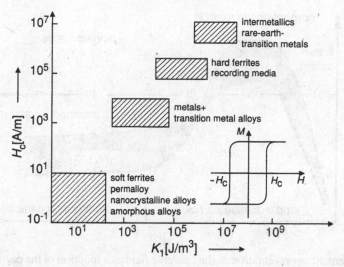

Fig. 6.4. Schematic plot of H_c vs. K_1 for prominent, soft, hard and extremely hard magnetic materials.

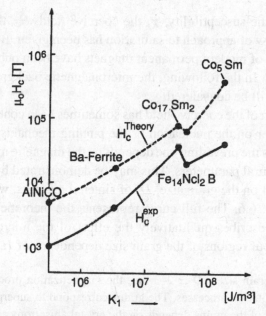

Fig. 6.5. Coercive field of prominent supermagnets as a function of anistropy constant K_1 and comparison with the ideal nucleation field $\mu_0 H_N = 2K_1/M_S$.

and Lifshitz [2.3], Brown [2.1-2.2], Néel [2.50-2.51] and Kittel [2.5]. An extension of this theory to the interaction of magnetization with lattice imperfections has been given by Kronmüller [2.7], [6.11–6.14] and Kronmüller *et al.* [6.15–6.17]. In these papers the basic micromagnetic concepts for an interpretation of structure sensitive

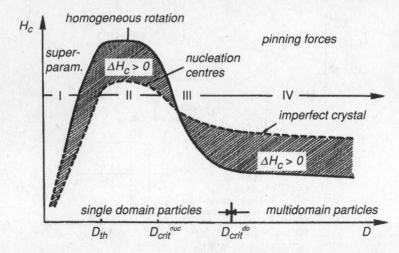

Fig. 6.6. Schematic representation of the coercive field as a function of the particle diameter *D* (————) perfect particle; (- - - - - -) imperfect particle.

properties such as the susceptibility, χ, the coercive field, H_c, the Rayleigh constant, α_R, and the law of approach to saturation has been given. Books and reviews on different aspects of modern permanent magnets have been published by several authors [6.18–6.42]. In the following, the micromagnetic background and the role of microstructure will be considered.

The interpretation of the coercive field has sometimes been controversial because pms are based either on the nucleation or the pinning mechanism. Which one of these mechanisms is the prevailing one depends on the magnetic material properties and the microstructural parameters. This may be demonstrated by the dependence of the coercive field on the grain size, *D*, of sintered magnets, which is presented qualitatively in Fig. 6.6. The full curve represents the theoretical prediction and the broken curve describes qualitatively the effect of the microstructure on H_c. Figure 6.6 shows four regions of the grain size dependence of H_c:

Region I: At small grain sizes, $D \simeq 2$–3 nm, the magnetization process takes place by thermally activated rotation processes. The grains correspond to superparamagnetic particles. The critical size of the grains depends on the crystal anisotropy and the spontaneous magnetization (see Section 6.2.4).

Region II: Reversion of magnetization by a homogeneous rotation process in grains up to diameters of 10 nm (see Section 6.2.1).

Region III: Reversion of magnetization by inhomogeneous rotation processes (curling process) with a decreasing coercive field with increasing grain sizes (see Section 6.2.2).

Region IV: Reversion of magnetization by domain wall displacements in multidomain grains.

According to Fig. 6.6 the role of the microstructure may be quite different. In regions I, II and III, i.e., in the single domain range, crystal imperfections reduce the coercive field, whereas at larger grain sizes, i.e., in the multidomain range, imperfections increase H_c due to the pinning of dws. The following sections deal with the interpretation of the hardening mechanisms in ideal and real hard magnetic materials.

6.2 Micromagnetism of hard magnetic materials

For a description of magnetization processes we consider the micromagnetic equations for uniaxial crystals. Starting from an ellipsoidal particle with uniaxial anisotropy, the total magnetic Gibbs free energy density according to

$$\phi'_t = \phi'_{ex} + \phi'_K + \phi'_s + \phi'_H, \qquad (6.4)$$

is composed of exchange, anisotropy, stray field and magnetostatic energies. In the case of one-dimensional problems, ϕ'_t depends only on the angle φ between M_s and the c-axis of the uniaxial crystals and the equilibrium condition for φ follows from a minimization of ϕ'_t with respect to φ as generally shown in Chapter 3.

The corresponding differential equations mainly depend on the type of magnetization mode considered, e.g., homogeneous rotation, curling or buckling, as shown in Fig. 6.7. Here the stray field plays a special role depending on whether magnetic volume or surface charges become important (see Fig. 6.7). In the following we give a review of the so-called nucleation fields as derived from the micromagnetic equations.

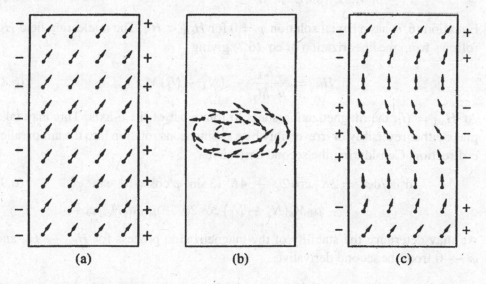

Fig. 6.7. Review of three different nucleation modes. (a) homogeneous rotations, (b) curling, (c) buckling.

6.2.1 Homogeneous rotation

We consider an ellipsoidal particle with its easy axis parallel to the rotational axis which is taken as the z-axis of a cartesian coordinate system. The rotation of the spontaneous magnetization is assumed to take place in the $(x–z)$-plane and is described by the rotation angle φ. In the case of a homogeneous rotation process the stray field components are

$$H_{s,x} = -N_\perp M_s \sin\varphi, \qquad H_{s,z} = -N_\parallel M_s \cos\varphi, \tag{6.5}$$

with the demagnetization factors $N_\parallel$ and $N_\perp$ for magnetization parallel and perpendicular to the c-axis. The total Gibbs free energy is now given by

$$\phi_t = \int \left\{ A(\nabla\varphi)^2 + K_1 \sin^2\varphi + K_2 \sin^4\varphi + \frac{1}{2}\mu_0 N_\perp M_s^2 \sin^2\varphi \right. \tag{6.6}$$
$$\left. + \frac{1}{2}\mu_0 N_\parallel M_s^2 \cos^2\varphi - \mu_0 H_{ext} M_s \cos(\psi_0 - \varphi) \right\} d^3r,$$

where ψ_0 denotes the angle between the positive c-axis and the applied magnetic field.

The homogeneous magnetization mode is characterized by a vanishing exchange energy. If the field is applied antiparallel to the c-axis, i.e., $\psi_0 = \pi$ variation of ϕ_t leads to the following equilibrium condition:

$$K_1 \sin 2\varphi + 2K_2 \sin^2\varphi \sin 2\varphi + \frac{1}{2}\mu_0 M_s^2 (N_\perp - N_\parallel) \sin 2\varphi$$
$$- \mu_0 M_s H_{ext} \sin\varphi = 0, \tag{6.7}$$

Equation (6.7) has a trivial solution $\varphi = 0$ for $H_{ext} < H_N$. The nucleation field H_N follows from the linearization of eq. (6.7) giving

$$H_N = \frac{2K_1}{\mu_0 M_s} - (N_\perp - N_\parallel) M_s. \tag{6.8}$$

At $H_{ext} = H_N$ the magnetization starts to rotate out of the c-axis. This may take place either reversibly or irreversibly by a spontaneous rotation into the antiparallel c-direction. Considering the second derivative

$$d^2\phi_t'/d\varphi^2 = 2K_1 \cos 2\varphi + 4K_2 (3\sin^2\varphi \cos^2\varphi - \sin^4\varphi) \tag{6.9}$$
$$- \mu_0 M_s (N_\perp - N_\parallel) \cos 2\varphi - \mu_0 M_s H_{ext} \cos\varphi,$$

we may determine the stability of the magnetization process for $H_{ext} \to H_N$ and $\varphi \to 0$ from the second derivative

$$d^2\phi_t'/d\phi^2|_{\varphi\to 0} = 3\varphi^2 \left(-K_1 - \frac{1}{2}\mu_0 M_s^2 (N_\perp - N_\parallel) + 4K_2 \right). \tag{6.10}$$

For $K_1 + \frac{1}{2}\mu_0 M_s^2 (N_\perp - N_\parallel) > 4K_2$ the condition $d^2\phi_t' / d\phi^2 < 0$ holds and consequently at $H_{\text{ext}} = H_N$ M_s rotates spontaneously into the opposite direction. For $K_1 + \frac{1}{2}\mu_0 M_s^2 (N_\perp - N_\parallel) < 4K_2$ the condition $d^2\phi_t' / d\phi^2 > 0$ holds, and therefore at $H_{\text{ext}} = H_N$ the magnetization rotates reversibly out of the c-axis. In this case the spontaneous reversion of M_s occurs at the instability field [6.14, 6.43, 6.44]

$$H_N' = \frac{4}{3 \cdot \sqrt{6}} \frac{K_2}{\mu_0 M_s} \left[2 + \frac{K_1 + \frac{1}{2}\mu_0 M_s^2 (N_\perp - N_\parallel)}{K_2} \right]^{3/2}, \qquad (6.11)$$

where the angle of irreversible rotation of $\underline{M}_s$ is given by

$$\sin^2 \varphi_N = -\frac{1}{6K_2} \left[K_1 - 4K_2 + \frac{1}{2}\mu_0 M_s^2 (N_\perp - N_\parallel) \right]. \qquad (6.12)$$

These relations follow from the condition $d^2\phi_t' / d\varphi^2 = 0$.

Due to the existence of two nucleation fields H_N and H_N', depending whether $K_1 \gtrless 4K_2$ holds, we observe different types of hysteresis loops as illustrated in Fig. 6.8. It is of interest to note that for vanishing K_1 and spherical particles,

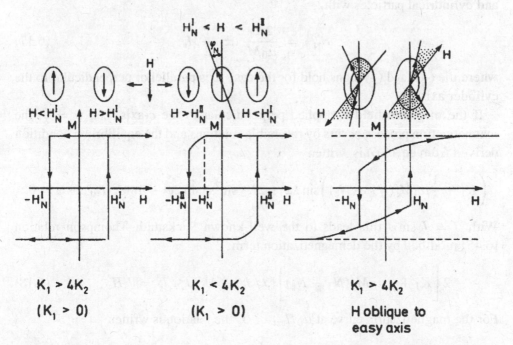

Fig. 6.8. Different types of hysteresis loops for uniaxial crystals for magnetic fields applied antiparallel and oblique with respect to the c-axis, for crystals with $K_1 \gtrless 4K_2$ [6.65].

$N_{\parallel} = N_{\perp}$, the nucleation field, H'_N, remains finite and is given by

$$H'_N = \frac{8}{9} \cdot \sqrt{3} \, \frac{K_2}{\mu_0 M_s}. \tag{6.13}$$

A further remarkable result is the independence of H_N on the particle size, because $N_{\perp}$ and $N_{\parallel}$ depend only on the ratio of the ellipsoid axes. The lower and upper bounds of the first nucleation field, H_N, are determined by the orientation of the easy axis with respect to the shape. For a plate with a perpendicular easy axis we obtain the lower bound,

$$H_N^{min} = \frac{2K_1}{\mu_0 M_s} - M_s, \tag{6.14}$$

and for an in-plane easy axis the upper bound,

$$H_N^{max} = \frac{2K_1}{\mu_0 M_s} + M_s. \tag{6.15}$$

Intermediate nucleation fields are obtained for the sphere with

$$H_N^{sph} = \frac{2K_1}{\mu_0 M_s}, \tag{6.16}$$

and cylindrical particles with

$$H_N^{cyl} = \frac{2K_1}{\mu_0 M_s} \pm \frac{1}{2} M_s, \tag{6.17}$$

where the (+)- and (−)-signs hold for the easy axis parallel or perpendicular to the cylinder axis.

If the magnetic field is applied perpendicular to the c-axis, i.e., $\psi = \frac{\pi}{2}$, the magnetization process occurs by reversible rotations and the equilibrium condition derived from eq. (6.6) is written

$$\left(K_1 + \frac{1}{2}\mu_0 M_s^2(N_{\perp} - N_{\parallel}) \right) \sin 2\varphi + 2K_2 \sin^2 \varphi \sin 2\varphi - \mu_0 M_s H_{ext} \cos \varphi = 0.$$

With $J = J_s \sin \varphi$ this leads to the well-known Sucksmith–Thompson relation [6.45] modified by the demagnetization term:

$$2\left[K_1 + \frac{1}{2}\mu_0 M_s^2(N_{\perp} - N_{\parallel}) \right](J/J_s) + 4K_2 J^3/J_s^3 = J_s H_{ext}. \tag{6.18}$$

For the magnetization curve at $\mu_0 H_{ext} \ll J_s$ the relation is written

$$J = \frac{J_s^2}{2K_1 + \mu_0 M_s^2(N_{\perp} - N_{II})} \cdot H_{ext}, \tag{6.19}$$

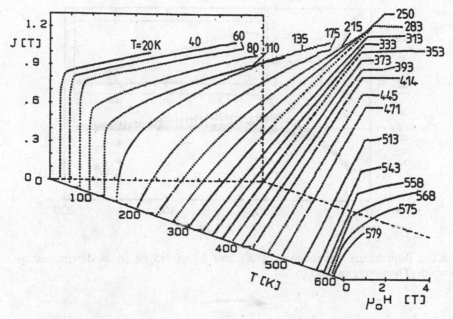

Fig. 6.9. Magnetization curves of a $Nd_2Fe_{14}B$ single crystals with H_{ext} applied perpendicular to the c-axis and a wide temperature range from 20 K to 579 K [6.47].

and the saturation field, H_{sat}, is given by

$$H_{sat} = (1/J_s)\big[2K_1 + \mu_0 M_s^2(N_\perp - N_\parallel) + 4K_2\big]. \tag{6.20}$$

It is important to note that the saturation field H_{sat} contains K_1 and K_2, whereas the nucleation field H_N and the initial susceptibility for $\psi = \pi/2$ contain only K_1. For an analysis of the coercive field therefore H_{sat} is not a suitable property.

As an application of eq. (6.18) to eq. (6.20), Fig. 6.9 presents magnetization curves of $Nd_2Fe_{14}B$ single crystals for $\psi = \frac{\pi}{2}$ and a wide temperature range [6.15]. This intermetallic tetragonal compound $Nd_2Fe_{14}B$ is of special interest because of its large anisotropy constants and large spontaneous magnetization at room temperature. The temperature dependence of the anisotropy constants K_1 and K_2 in Fig. 6.10 [6.46, 6.47] shows the well-known change in sign of K_1 at ≈ 130 K, which also coincides with a change of the shape of the magnetization curves. According to the results of Chapter 2, below $T = 130$ K the easy directions lie on a cone around the c-axis.

6.2.2 Inhomogeneous rotation by the curling mode

The ideal curling mode is characterized by a vanishing stray field, i.e., div $M_s = 0$ and $n \cdot M_s = 0$ [2.9, 2.10]. Under these conditions only exchange, crystal and

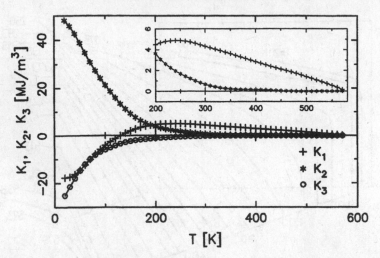

Fig. 6.10. Temperature dependence of K_1 and K_2 of $Nd_2Fe_{14}B$ as determined by the Sucksmith–Thompson method [6.46].

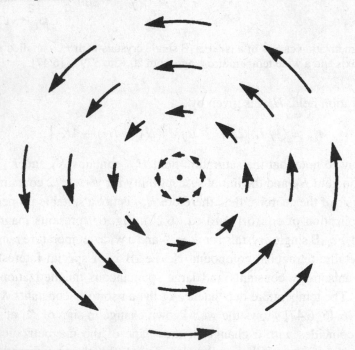

Fig. 6.11. Projection of magnetization of the curling mode on the cross-section of a cylindrical specimen.

magnetostatic energies have to be considered. In the case of a cylindrical particle of finite length, at the ends surface charges exist that produce a demagnetizing field $H_s = -N_\parallel M_s$, which has to be added to the inversely applied external field. Figure 6.11 shows the projection of M_s on the cross-section of a cylindrical specimen. Over the whole cross-section of radius R the radial component of M_s vanishes. The direction of M_s therefore is fully described by the angle φ between M_s and the c-axis. In cylindrical polar coordinates the linearized micromagnetic equations with the magnetic field applied antiparallel to the magnetization are now written as

$$
2A\left[\frac{d^2\varphi(r)}{dr^2} + \frac{1}{r}\frac{\partial\varphi(r)}{\partial r} - \frac{1}{r^2}\varphi(r)\right]
$$
$$
- \left(2K_1 - \mu_0 H_{ext}M_s - N_\parallel\mu_0 M_s^2\right)\varphi(r) = 0. \tag{6.21}
$$

The solution of eq. (6.21) is given by the Bessel function of first order

$$
\varphi(r) = \varphi_0 J_1\left[r\cdot\left(\frac{\mu_0 H_{ext}M_s + N_\parallel\mu_0 M_s^2 - 2K_1}{2A}\right)^{1/2}\right], \tag{6.22}
$$

where φ_0 denotes an undetermined amplitude. The nucleation field follows from the micromagnetic boundary condition

$$
\left.\frac{d\varphi}{dr}\right|_{r=R} = 0, \tag{6.23}
$$

which leads to the condition $J_1'(R) = 0$. The largest nucleation field is obtained for the first zero of $J_1'(R)$ leading to [2.10]

$$
H_N = \frac{2K_1}{\mu_0 M_s} - N_\parallel M_s + \frac{2A}{\mu_0 M_s}\left[\frac{1.84}{R}\right]^2. \tag{6.24}
$$

Equation (6.24) contains a term resulting from the exchange energy of the inhomogeneous magnetization which decreases according to a $1/R^2$-law. At small radii therefore, the nucleation field of the curling mode exceeds that of the homogeneous rotation mode. Since the smallest nucleation field always governs the demagnetization process at small radii, the rotation in unison always governs the reversal of magnetization. At large diameters of the particles, domain formation takes place and the coercive field is then determined by the pinning of dws at lattice imperfections. The critical diameter of single domain particles will be discussed in Section 6.2.4.

6.2.3 Inhomogeneous rotation by the buckling mode

The two nucleation modes discussed so far correspond to extreme cases either with vanishing exchange energy or vanishing stray field energy. Frei *et al.* [6.48] and Aharoni and Shtrikman [2.10][6.49] have shown that there may exist a buckling

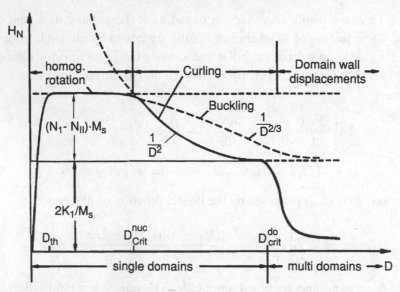

Fig. 6.12. Review of the nucleation fields of different hardening mechanisms. Critical diameters are discussed in Section 6.2.4.

mode where both energies play some role. This mode may be derived from the homogeneous mode by assuming a sinusoidal variation of the angular deviation φ from the easy axis. The increase of exchange energy is compensated by the reduced stray field energy due to the alternating surface charges (see Fig. 6.7). At larger radii the buckling mode leads to larger nucleation fields than the curling mode and therefore plays no important role for the practical cases. Figure 6.12 reviews H_N for the different nucleation modes.

In our considerations presented so far we have not correctly taken care of the surface charges at the ends, because these produce diverging stray fields at the circular edge. Holz [6.50, 6.51] has shown that the nucleation process at the ends is a combination of all three modes: homogeneous rotation, curling and buckling, corresponding to a butterfly-type ground state as shown by Fig. 6.13. The nucleation field therefore contains terms due to the crystal anisotropy, the exchange energy and the stray field:

$$ H_N = \frac{2K_1}{\mu_0 M_s} + \frac{2A}{\mu_0 M_s}\left[\frac{1.84}{R}\right]^2 - \frac{1}{2}M_s(1 - \varepsilon), \qquad (6.25) $$

where the parameter ε describes the effect of the alternating and reduced surface charges at the cylinder surface and at the ends. For diameters $R > 10$ nm ε decreases according to a $R^{-2/3}$-law. It is of interest to note that Holz's solution introduces a strong stray field term much larger than the term $N_\parallel M_s$ occuring in eq. (6.24), where it has been assumed that the magnetization remains rigid at the ends.

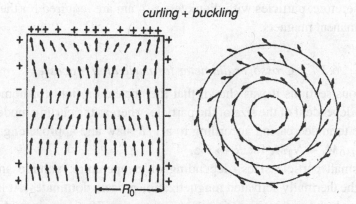

Fig. 6.13. Butterfly-type distribution of magnetization near the end of a cylindrical specimen.

6.2.4 Critical diameters of single domain particles

Small magnetic particles are characterized by at least three critical diameters where the type of magnetization process changes as indicated in Fig. 6.12. These diameters govern the thermal stability of magnetization, the transition from homogeneous to inhomogeneous nucleation and the formation of multidomain particles.

6.2.4.1 Thermal stability limit

For the application of small particles for the production of pms the stability of the magnetic state for time intervals of decades of years is an absolute prerequisite. At diameters $D < D_{crit}^{th}$, the nucleation field breaks down because reversion of M_s is induced by the thermally excited spin system. In very small particles the thermal fluctuation energy, kT, is sufficiently large to rotate M_s into the opposite direction. According to Néel [6.52–6.54] the lifetime of a magnetic particle of volume V is given by an Arrhenius law,

$$\tau = \tau_0 \exp[-K_{eff}V/2kT], \qquad (6.26)$$

where K_{eff} denotes an effective anisotropy energy, which for an ellipsoidal particle is given by $K_{eff} = K_1 + \frac{1}{2}(N_\perp - N_\parallel)\mu_0 M_s^2$. The pre-exponential factor, τ_0, corresponds to the resonance relaxation time of the spin system, which in a first approximation is given by $\tau_0^{-1} = (2\pi)^{-1}\gamma H_{eff}$ (γ = gyromagnetic ratio, $H_{eff} = 2K_{eff}/\mu_0 M_s$). If we consider a particle with material parameters $K_{eff} = 10^5$ J/m^3, $\tau_0 = 10^{-8}$ s, $T = 300$ K, we obtain $\tau = 0.1$ s for $D = 3.4$ nm and $\tau = 10^8$ s for $D = 4.4$ nm.

Accordingly, there exists a very narrow range of diameters where the transition from a thermally stable to a thermally unstable state takes place. In magnetically hard materials with $K_{eff} > 10^6$ J/m^3 the magnetic lifetime becomes longer than

10 years. Therefore, particles with diameters > 4 nm are required for the development of permanent magnets.

6.2.4.2 Crossover diameter for nucleation processes

In the previous sections it was shown that the nucleation field for homogeneous rotation is independent of the size of the particle, whereas the curling mode contains an exchange term decreasing according to a $1/R^2$-law and approaching the value $H_N = 2K_1/\mu_0 M_s - N_\parallel M_s$ for $R \to \infty$.

At very small particle sizes, depending on temperature, volume and crystal anisotropy, the thermally activated magnetization process dominates. At larger particle sizes the homogeneous rotation process governs H_N. At a critical diameter, D_{crit}^{nuc}, the homogeneous rotation is replaced by the curling mode. This crossover takes place at a diameter, D_{crit}^{nuc}, where both nucleation fields become equal. Comparison of eq. (6.8) and eq. (6.24) then gives

$$D_{crit}^{nuc} = 3.68 \sqrt{\frac{2A}{N_\perp \mu_0 M_s^2}}. \tag{6.27}$$

For appropriate material parameters, $A = 8 \cdot 10^{-12}$ J/m, $\mu_0 M_s = 1$ T, $N_\perp = \frac{1}{2}$, we find $D_{crit}^{nuc} \simeq 20$ nm. So the curling mode becomes dominant at rather small particle diameters if we deal with cylindrical particles. It should be noted, however, that D_{crit}^{nuc} increases to infinity for a platelet, where $N_\perp \to 0$ holds. Therefore, in real particles with a deteriorated surface of reduced anisotropy constant, where the nucleation process takes place in an extended planar area, the critical diameter may become rather larger.

6.2.4.3 Critical diameter for domain formation

There exists a critical diameter, D_{crit}^{do}, below which a single domain state is the energetically most favourable one [6.55]. With increasing size of the particle, however, the magnetostatic stray field energy ϕ_s increases. A lowering of ϕ_s takes place by the formation of a two-domain particle as shown in Fig. 6.14. The critical diameter, D_{crit}^{do}, for this transition follows from a comparison of the magnetic energies of these two states. State I is characterized by a homogeneously magnetized rotational ellipsoid with axes a and b. The stray field energy of state I is given by

$$\phi^I = \frac{1}{2}\mu_0 N_\parallel M_s^2 \cdot \frac{4\pi}{3} a^2 b. \tag{6.28}$$

State II corresponds to a two-domain state with a domain wall of energy $\pi ab\gamma_B$. Due to the two-domain state the stray field energy is reduced by a factor α. The

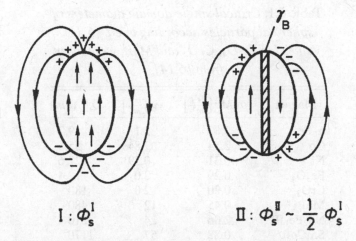

$$\text{I}: \Phi_s^{\text{I}} \qquad\qquad \text{II}: \Phi_s^{\text{II}} \sim \frac{1}{2}\,\Phi_s^{\text{I}}$$

Fig. 6.14. Magnetic stray fields of single and two-domain particles.

transition from a one- to a two-domain state is determined by the equation

$$\frac{1}{2}\mu_0 N_\parallel M_s^2 \cdot \frac{4\pi}{3}a^2 b = \pi a b \gamma_B + \alpha \frac{1}{2}\mu_0 N_\parallel M_s^2 \cdot \frac{4\pi}{3}a^2 b, \qquad (6.29)$$

which gives for the critical diameter

$$D_{\text{crit}}^{\text{do}} = 2a = \frac{3\gamma_B}{N_\parallel(1-\alpha)\mu_0 M_s^2}. \qquad (6.30)$$

For a sphere with $N_\parallel = 1/3$ and $\alpha = 1/2$ this gives

$$D_{\text{crit}}^{\text{do}} = \frac{18\gamma_B}{\mu_0 M_s^2} = \frac{72}{\mu_0 M_s^2}\sqrt{AK_1}. \qquad (6.31)$$

For a prolate ellipsoid with $N_\parallel \sim 1$ we find

$$D_{\text{crit}}^{\text{do}} = \frac{6\gamma_B}{\mu_0 M_s^2}. \qquad (6.32)$$

According to eqs. (6.31) and (6.32) spherical particles have a three times larger critical parameter than platelets with perpendicular anisotropy. Equation (6.30) furthermore shows that long needles with $N_\parallel \ll 1$ are characterized by large critical diameters. Needle-type particles therefore play an important role for magnetic recording systems. In Table 6.1 numerical results for D_{crit} are presented for spherical or polyhedral particles. For γ_B the expression for uniaxial crystals $\gamma_B = 4\sqrt{AK_1}$ has been inserted. According to Table 6.1 the largest critical diameter is found for Co_5Sm. The transition metals have rather small critical diameters and the important hard magnetic material $Nd_2Fe_{14}B$ also has a rather small critical diameter of only 0.2 μm.

Table 6.1. *Critical single domain diameters of*
spherical particles according to eq. (6.31).
Values for Fe₃O₄, CrO₂ and MnBi were taken
from [6.14].

Magnet	$\mu_0 M_s^2 \left[\frac{MJ}{m^3}\right]$	$\gamma_B \left[\frac{mJ}{m^2}\right]$	$D_{\text{crit}}^{\text{do}} [nm]$
α-Fe	3.82	2.1	9.7
Co	2.54	7.84	55.5
Ni	0.31	0.39	22.6
Fe_3O_4	0.29	2.0	12.4
CrO_2	0.20	2.0	180
MnBi	0.45	12	480
$Nd_2Fe_{14}B$	2.06	24	210
$SmCo_5$	0.88	57	1170
Sm_2O_{17}	1.33	31	420
$BaFe_{12}O_{19}$	0.183	6.3	62

Equations (6.31) and (6.32) represent excellent approximations for the single domain diameters. Recent numerical calculations of the critical diameter of a cube of edge length L by means of finite element techniques (see Chapter 13) led to the result that the transition from a so-called twisted flower state, with reduced stray field energy, for a particle with $Q = K_1/(J_s^2/2\mu_0) = 0.1$ takes place at $L = 8.6 l_s$ [6.56, 6.57]. Equation (6.31), which also may be written as $D_{\text{crit}}^{\text{do}} = 36\sqrt{Q} \cdot l_s$, gives $d_{\text{crit}} = 11.4 l_s$. These results are compatible with calculations of Aharoni [2.9], who determined upper and lower bounds for the single domain diameter.

Now the question arises why in sintered magnets the particles with diameters far above $D_{\text{crit}}^{\text{do}}$ act like single domain particles. Here it has to be taken into account that a transition from the single domain state to the multidomain state requires a nucleation process. This is only possible under the action of the nucleation field H_N as derived in the preceding sections. Therefore, the particles under zero field only transform into a multidomain state if $H_N < 0$ holds.

In hard magnetic materials the term $2K_1/\mu_0 M_s$ always exceeds the negative demagnetization field $-N_{\text{eff}} M_s$. Once magnetized at large fields, the particles remain in this single domain state until an inversely applied field approaches the negative nucleation field $(-)H_N$. Since the experimental H_N^{exp} in general is a factor of 3–4 smaller than the ideal values, multidomain particles appear at much lower fields. This discrepancy with respect to the theoretical predictions, which has to be attributed to the role of the microstructure, is known as Brown's paradox. Brown [2.6][6.58] was the first to realize this contradiction between theory and experiment. According to eq. (6.8) even for macroscopic ellipsoidal samples a transition into a multidomain state cannot take place without application of an inverse magnetic field

$H_{ext} > H_N$. Today it is generally accepted that micromagnetic theory developed for ideal homogeneous materials has to be expanded to the case of real materials in which, due to lattice imperfections, there are regions of reduced anisotropy where the nucleation fields are reduced. Besides soft magnetic regions, magnetostatic demagnetization fields also reduce the ideal nucleation fields because these may become rather large near the edges and corners of polyhedral grains. In the following sections the role of microstructures in nucleation fields and magnetization processes will be discussed in more detail.

6.2.5 Comparison with experiment

An experimental test of the dependence of H_N on the diameter of cylindrical particles requires thin, almost perfect wires with large aspect ratio. The first historical attempt of such a test was due to Luborsky and Morelock [6.59] who performed measurements on long iron whiskers with varying diameter. Their results shown in Fig. 6.15 clearly demonstrate that the predictions for the curling mode are approached for whiskers of diameters ≈ 50 nm. For diameters > 100 nm the coercive field deviates strongly from the theoretical limit $H_c = 2K_1/J_s = 540$ Oe. This discrepancy has to be attributed to the transition to multidomain configurations. More recently, Seberino and Bertram [6.60] and Suhl and Bertram [6.61] have analysed the nucleation field of Ni wires [6.62] with aspect ratios 5:1 and 20:1 where the

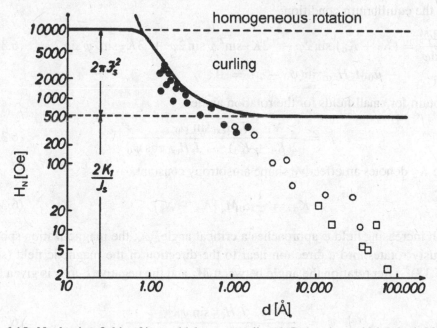

Fig. 6.15. Nucleation fields of iron whiskers according to Luborsky and Morelock [6.59].

finite length plays a decisive role. According to Fig. 6.16, for diameters of 20 nm H_N deviates by a factor of 3 from the theoretical predictions of the Stoner–Wohlfarth model [6.63]. On the basis of a micromagnetic model of Schabes and Bertram [6.64], the demagnetization process of the thin wires has been simulated leading to the full curves shown in Fig. 6.16. Here it becomes obvious that the demagnetization process starts from both ends of the wires and moves to the centre of the particles as demonstrated by Fig. 6.17.

6.3 Nucleation under oblique magnetic fields

6.3.1 Homogeneous rotation

The magnetization process in small particles under oblique applied fields has been treated by a number of authors [6.63, 6.65–6.67]. The original paper of Stoner and Wohlfarth [6.63] gives the basis for the following investigations. If the external field is applied at an angle ψ_0 with respect to the negative c-axis, the spontaneous magnetization rotates out of the easy axis, and for small fields the rotation angle φ follows from the equilibrium condition $d\phi_t'/d\varphi = 0$, where ϕ_t' is given by

$$\phi_t' = K_1 \sin^2 \varphi + K_2 \sin^4 \varphi + K_3 \sin^6 \varphi + \frac{1}{2}\mu_0 M_s N_\parallel \cos^2 \varphi \quad (6.33)$$

$$+ \frac{1}{2}\mu_0 M_s N_\perp \sin^2 \varphi - \mu_0 M_s H_{ext} \cos(\psi_0 - \varphi).$$

From the equilibrium condition

$$\frac{d\phi_t'}{d\varphi} = (K_1 + K_d) \sin 2\varphi + 2K_2 \sin^2 \varphi \sin 2\varphi + 3K_3 \sin^4 \varphi \sin 2\varphi \quad (6.34)$$

$$- \mu_0 M_s H_{ext} \sin(\psi_0 - \varphi) = 0,$$

we obtain for small fields for the rotation angle

$$\varphi = \frac{J_s H_{ext} \sin \psi_0}{2(K_1 + K_d) - J_s H_{ext} \cos \psi_0}, \quad (6.35)$$

where K_d denotes an effective shape anisotropy constant,

$$K_d = \frac{1}{2}\mu_0 M_s^2 (N_\perp - N_\parallel). \quad (6.36)$$

If with increasing field φ approaches a critical angle, φ_N, the magnetization spontaneously rotates into a direction near to the direction of the magnetic field (see Fig. 6.18). After rotation the angle between M_s and the negative c-axis is given by

$$\varphi = \frac{J_s H_{ext} \sin \psi_0}{2(K_1 + K_d) + J_s H_{ext} \cos \psi_0}. \quad (6.37)$$

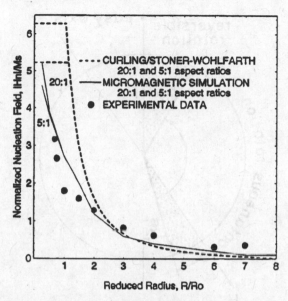

Fig. 6.16. Nucleation field H_N/M_s of Ni-wires with aspect ratios 5:1 and 20:1 versus reduced particle radius R for uniform nucleation, curling and micromagnetic simulations according to [6.60]. R_0 is defined as $R_0 = \sqrt{2\pi}\, l_s$.

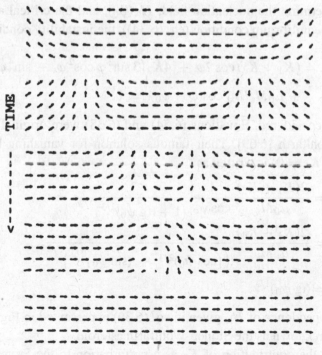

Fig. 6.17. Reversal mode simulation of magnetization of a wire with aspect ratio 5:1 according to [6.60].

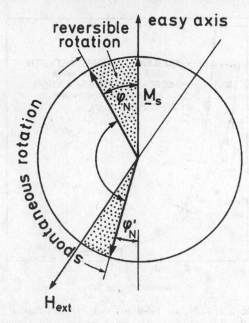

Fig. 6.18. Orientation of magnetization before and after spontaneous rotation in the case of an obliquely applied magnetic field [6.65].

In order to determine the nucleation field, $H_N(\psi_0)$, and the critical angle, φ_N, in addition to the equilibrium condition (eq. (6.34)), the instability condition,

$$\frac{\mathrm{d}^2\phi'_t}{\mathrm{d}\varphi^2} = (K_1 + K_\mathrm{d})\cos 2\varphi + 4K_2\big(3\sin^2\varphi\cos^2\varphi - \sin^4\varphi\big) \tag{6.38}$$
$$+ \mu_0 M_\mathrm{s} H_\mathrm{ext}\cos(\varphi - \psi_0) = 0,$$

has also to be considered. Equations (6.34) and (6.38) were originally solved by Stoner and Wohlfarth [6.63]. Their famous solution for vanishing higher order constants ($K_2, K_3 = 0$) is written as

$$H_N^{(0)}(\psi_0) = \frac{2(K_1 + K_\mathrm{d})}{\mu_0 M_\mathrm{s}} \frac{1}{\cos\psi_0} \frac{1}{\big(1 + (\mathrm{tg}\,\psi_0)^{2/3}\big)^{3/2}} \tag{6.39}$$
$$= \frac{2(K_1 + K_\mathrm{d})}{\mu_0 M_\mathrm{s}} \frac{1}{\big((\cos\psi_0)^{2/3} + (\sin\psi_0)^{2/3}\big)^{3/2}},$$

$$\mathrm{tg}\,\varphi_N = (\mathrm{tg}\,\psi_0)^{1/3}.$$

The angular dependence of H_N given in eq. (6.39) is presented in Fig. 6.19 and in a polar diagram known as the Stoner–Wohlfarth asteroid.

Considering the contribution of K_2 as a perturbation to the Stoner–Wohlfarth solution, the following extended expression for H_N is derived from eqs. (6.34)

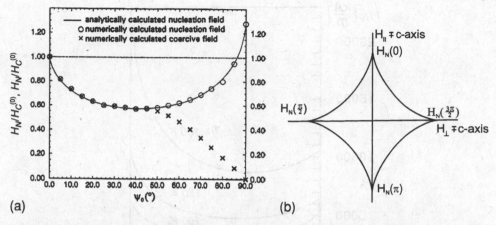

Fig. 6.19. Angular dependence of the nucleation and of the coercive field of a $Nd_2Fe_{14}B$ sphere in reduced units $\alpha_\psi = H_N(\psi)/H_N^{(0)}$. (a) Linear plot. (b) Stoner–Wohlfarth asteroid. $H_\parallel$ and $H_\perp$ denote the components of the applied field parallel or antiparallel to the negative c-axis, and perpendicular to the c-axis respectivly.

and (6.38):

$$H_N(\psi_0) = H_N^{(0)}(\psi_0)\left\{1 + \frac{2K_2}{K_1 + K_d}\frac{(\text{tg } \psi_0)^{2/3}}{1 + (\text{tg } \psi_0)^{2/3}}\right\}. \tag{6.40a}$$

The angle φ_N by which J_s is rotated before spontaneous nucleation takes place is given by

$$\varphi_N = \text{arctg } \sqrt[3]{\text{tg } \psi_0} + \frac{2}{3}\frac{K_2}{K_1 + K_d}. \tag{6.40b}$$

From eq. (6.40a) we obtain a minimum nucleation field for $\psi_0 = \pi/4$, which is given by

$$H_N^{\min}\left(\frac{\pi}{4}\right) = \frac{K_1 + K_d + K_2}{\mu_0 M_s}. \tag{6.41}$$

Equations (6.40a) and (6.41) describe the angular dependence of $H_N(\psi_0)$ quite nicely for the case $K_1, K_2 > 0$. The expression of eq. (6.40a), however, is less precise for $K_1 < 0$. Figure 6.20 shows precise numerical results which were obtained from a solution of the equilibrium and the stability conditions, eq. (6.34) and eq. (6.38) for $K_d, K_3 = 0$. For $M_s(T), K_1(T), K_2(T)$ the values determined by Hock and Kronmüller [6.47] have been used. It is noteworthy that for $H_N^{\min}(\psi_0)$ an exact expression may be derived in contrast to $H_N(\psi_0)$ which cannot be determined explicitly. From the equilibrium and the stability conditions the following

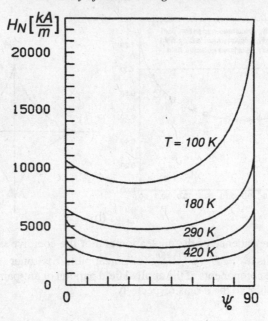

Fig. 6.20. Angular dependence of H_N of $Nd_2Fe_{14}B$ from eq. (6.40a) for various temperatures taking account of K_1, K_2 and K_3 [6.68].

expression has been derived by Martinek and Kronmüller [6.68]:

$$H_N^{min} = \frac{1}{4\mu_0 M_s} \sqrt{2}\left[K_1 + \frac{K_2}{4}\left(W - \frac{K_1}{K_2} + 3\right)\right]$$

$$\times \sqrt{\left\{W\left(\frac{K_1}{K_2} + 1\right) - \left(\frac{K_1}{K_2}\right)^2 + \frac{2K_1}{K_2} + 3\right\}}, \qquad (6.42)$$

with

$$W = (\pm)\sqrt{\left(\frac{K_1}{K_2} + 1\right)^2 + 8}, \qquad (6.43)$$

where the $(+)-$ sign holds for $K_2 > 0$, $K_1 > -2K_2$ and the $(-)-$ sign for $K_2 < 0$, $K_1 > 0$. The actual nucleation field for obliquely applied magnetic fields may now be written as

$$H_N(\psi_0) = \alpha_\psi \cdot H_N(0), \qquad (6.44)$$

with

$$\alpha_\psi = \frac{1}{\left((\cos \psi_0)^{2/3} + (\sin \psi_0)^{2/3}\right)^{3/2}}\left[1 + \frac{2K_2}{K_1 + K_2}\frac{(tg\,\psi_0)^{2/3}}{1 + (tg\,\psi_0)^{2/3}}\right]. \qquad (6.45)$$

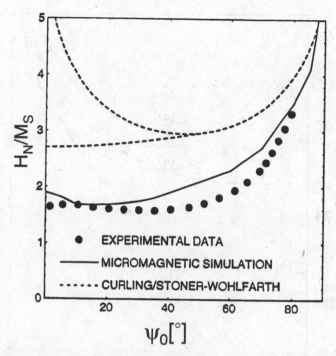

Fig. 6.21. Angular dependence of the nucleation field of a finite Ni-wire according to Seberino and Bertram [6.60]. The pointed curves present the theoretical angular dependence for homogeneous rotation and curling and a wire radius $R = 3.9l_s$.

It is of interest to note that for $K_2 = 0$ the angular dependence of H'_N and φ_N depend only on ψ_0 and not on material constants. The angular dependence of $H_N(\psi_0)$ is characterized by a symmetric function for $K_1 > 4K_2$ and becomes more and more asymmetric approaching a $1/\sin \psi_0$-law for $\psi_0 \rightarrow \pi/2$ and $K_1 < 4K_2$.

6.3.2 Curling mode

Experiments on the angular dependence of wires of finite length have been performed by a number of authors [6.62, 6.69, 6.70]. Considering the results of Seberino and Bertram [6.60] shown in Fig. 6.21, it is obvious that the conventional theory cannot explain the measured angular dependence of H_N. The angular dependence of H_N of the curling mode has been investigated extensively by Aharoni [6.66, 6.71] in a series of papers. His main numerical results for infinite cylinders and a prolate spheroid are shown in Fig. 6.22 and Fig. 6.23. In both cases, for small anisotropy constants the angular dependence of H_N is similar to that of the coercivity due to dw pinning ($1/\cos \psi_0$). For larger values of $Q = 2K_1/M_sJ_s$, the angular dependence of H_N approaches that of the Stoner–Wohlfarth model.

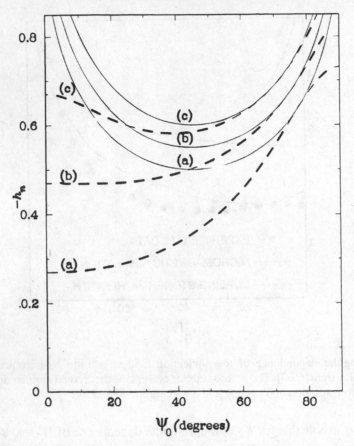

Fig. 6.22. Angular dependence of the reduced nucleation field, $h_n = 2H_N/M_s$, for the curling mode in an infinite cylinder according to Aharoni [6.66]. Calculations were performed for a reduced radius of $R/l_s = \sqrt{2}$ and parameters Q = 0 (a); Q = 0.1 (b); Q = 0.2 (c).

6.4 Nucleation in magnetically soft regions

In ideal homogeneous, ellipsoidal particles the nucleation fields are derived from linear homogeneous differential equations of second-order and constant coefficients (see eq. (6.7)). In the case of inhomogeneous materials we still have to consider second-order differential equations, however, with nonconstant coefficients. In principle all phenomenological material parameters, $J_s = \mu_0 M_s$, K_1, A, etc., may vary spatially. For an analytical treatment of the nucleation problem we consider a planar region of thickness $2r_0$, where the crystal anisotropy K_1 is reduced spatially thus forming a soft magnetic strip within a hard magnetic matrix. Figure 6.24 presents the geometry of this nucleation region, where under the action of a magnetic field a rotation process takes place leading to an inversely magnetized region. For the

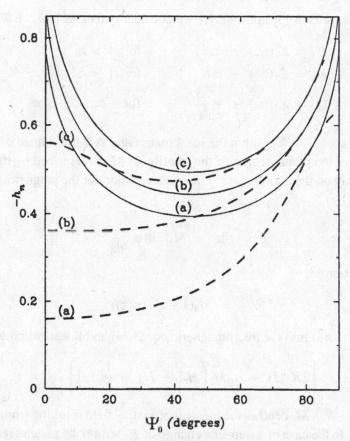

Fig. 6.23. Angular dependence of the reduced nucleation field $h_n = 2H_N/M_s$ for the curling mode in a prolate spheroid according to Aharoni [6.66]. Calculations were performed for a reduced radius of $R/l_s = \sqrt{2}$ and parameters $Q = 0$ (a); $Q = 0.1$ (b); $Q = 0.2$ (c).

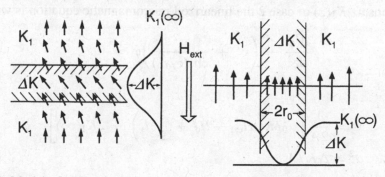

Fig. 6.24. Model of soft magnetic planar nucleus of width $2r_0$. Left side: $H_{ext} \perp$ stripe. Right side $H_{ext} \parallel$ stripe.

variation of the anisotropy constant K_1 we consider two cases [6.72, 6.73]:

$$K_1 = K_1(\infty) \qquad\qquad \text{for}|z| > r_0,$$
$$K_p = K_1(\infty) - \Delta K \qquad\qquad \text{for}|z| < r_0,$$
$$K_1(z) = K_1(\infty) - \frac{\Delta K}{\text{ch}^2(z/r_0)} \qquad \text{for} -\infty < z < \infty. \qquad (6.46)$$

$K_1(\infty)$ corresponds to K_1 within the ideal matrix and ΔK the change of K_1 at the centre $z = 0$ of the planar region. If the rotation of M_s is described by the angle φ depending only on the coordinate z, Poisson's equation for the magnetostatic stray field is given by

$$dH_{s,z}/dz = M_s \sin\varphi \, \frac{d\varphi}{dz}, \qquad (6.47)$$

with the solution

$$H_{s,z} = M_s(1 - \cos\varphi(z)). \qquad (6.48)$$

Inserting eq. (6.48) into the micromagnetic eq. (3.24) and linearization gives

$$2A \, \frac{d^2\varphi}{dz^2} - \left\{ 2K_1(z) - \mu_0 M_s \left(H_{\text{ext}} - H_d + \frac{1}{2} M_s \right) \right\} \cdot \varphi = 0, \qquad (6.49)$$

where $H_d = -N_d \cdot M_s$ denotes the macroscopic stray field resulting from fixed surface charges. In the case of a step-like change of K_1 within the platelet (eq. (6.46)), the nucleation field derived from eq. (6.49) is given by

$$H_N = \frac{2K_p}{\mu_0 M_s} - \frac{1}{2} M_s + \frac{A\pi^2}{2\mu_0 M_s r_0^2} + H_d. \qquad (6.50)$$

With the ansatz $K_1(z)$ of case 2 the linearized micromagnetic equation is written

$$\frac{d^2\varphi}{dz^2} + \left(\kappa^2 + \frac{\alpha^2}{\text{ch}^2(z/r_0)} \right)\varphi = 0, \qquad (6.51)$$

with

$$\kappa^2 = \frac{1}{2A} \left[\mu_0 M_s \left(H_{\text{ext}} - H_d + \frac{1}{2} M_s \right) - 2K_1(\infty) \right], \qquad (6.52)$$
$$\alpha^2 = \Delta K / A.$$

Equation (6.52) may be transformed into the hypergeometric differential equation with the solution in hypergeometric functions, F_n, where the lowest nucleation field follows from the eigenvalue equation $r_0^2 \kappa^2 = s^2$ with $s(s + 1) = r_0^2 \alpha^2$. From these relations we obtain the nucleation field

$$H_N = H_c = \frac{2K_1(\infty)}{\mu_0 M_s} \alpha_K - N_{eff} M_s,$$ (6.53)

with

$$N_{eff} = \frac{1}{2} + N_d,$$ (6.54)

$$\alpha_K = 1 - \frac{1}{4\pi^2} \frac{\delta_B^2}{r_0^2} \left[1 - \sqrt{1 + \frac{4\Delta K r_0^2}{A}} \right]^2.$$

Equation (6.54) allows the derivation of three limiting cases for H_N, where the fictitious wall width $\delta_B' = \pi \sqrt{A/\Delta K}$ corresponds to the characteristic length scale.

1. Narrow inhomogeneities $r_0 < \delta_B'$ lead to the ideal nucleation field modified by stray field terms:

$$H_N = \frac{2K_1(\infty)}{\mu_0 M_s} + H_d - \frac{1}{2} M_s.$$ (6.55)

2. Inhomogneities of average thickness $2\pi r_0 > \delta_B'$:

$$H_N = \frac{2K_1(\infty)}{\mu_0 M_s} \left(\frac{\delta_B^2}{\pi \delta_B' r_0} \right) + \frac{2(K_1(\infty) - \Delta K)}{\mu_0 M_s} + H_d - \frac{1}{2} M_s.$$ (6.56)

3. Extended inhomogeneities $2r_0 \gg \delta_B'$:

$$H_N = \frac{2(K_1(\infty) - \Delta K)}{\mu_0 M_s} + H_d - \frac{1}{2} M_s.$$ (6.57)

For $\Delta K = K_1$ the α_K in case 2 reduces to $(\delta_B/\pi r_0)$.

In case 1 obviously the region of reduced anisotropy has no influence on H_N because the exchange energy averages over the narrow inhomogeneity. In case 3 the nucleation field is determined by the minimum anisotropy $K_1(\infty) - \Delta K$. Case 2 describes the intermediate situation where H_N decreases according to a $1/r_0$-law and approaches the limit of case 3. Figure 6.25 presents the full dependence of α_K on r_0 as a function of the parameter ΔK.

6.5 Nucleation in inhomogeneous misaligned grains

In the preceding sections we have determined the nucleation fields for misaligned grains and inhomogeneous regions separately. The reduction of H_N due to these properties has been described by the parameters α_K and α_ψ. Now the question arises what is the role of α_K and α_ψ if both effects are superimposed.

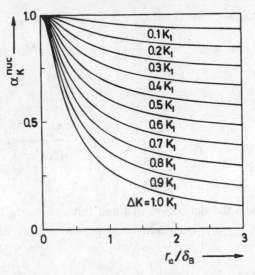

Fig. 6.25. Variation of the microstructural parameter α_K as a function of r_0/δ_B for various parameters ΔK [6.102].

Under an oblique magnetic field M_s rotates in the inhomogeneous region as well as in the bulk hard magnetic material. As in Section 6.3 we have to determine the nucleation field after a rotation of M_s up to a critical angle φ_c. For small deviations of M_s from this angle, φ_c, the linearized micromagnetic equation is given by [6.14]

$$2A\frac{d^2\varphi}{dz^2} - \left\{2K_1(z)\cos 2\varphi_c - \mu_0 M_s\left(H_{ext} - H_d + \frac{1}{2}M_s\right)\right\}$$
$$\times \cos(\varphi_c + \psi_0)\varphi = 0. \tag{6.58}$$

Equation (6.58) is similar to eq. (6.49) derived for the ideally oriented particle. We therefore can use the same solution derived in the preceding section by substituting M_s by $M_s\cos(\varphi_c + \psi_0)$ and $K_1(z)$ by $K_1(z)\cos 2\varphi_c$. With the same spatial dependence for $K_1(z)$ used in section 5.4 (eq. (6.46)) we obtain

$$H_N = \frac{2K_1(\infty)}{\mu_0 M_s}\frac{\cos 2\varphi_c}{\cos(\varphi_c + \psi_0)} + H_d - \frac{1}{2}M_s \tag{6.59}$$

$$-\frac{A}{2\mu_0 M_s r_0^2}\frac{1}{\cos(\varphi_c + \psi_0)}\cdot\left[-1 + \sqrt{1 + \frac{4\Delta K r_0^2\cos 2\varphi_c}{A}}\right]^2.$$

In Section 5.3 it has been shown that for vanishing anisotropy constant K_2 the critical angle, where the spontaneous reversion of M_s takes place, is given by $\operatorname{tg}\varphi_c = (\operatorname{tg}\psi_0)^{1/3}$, i.e., φ_c is independent of $K_1(z)$. Accordingly, it is a straightforward approximation to replace the first term in eq. (6.59) by the nucleation field $H_N(\psi_0)$. Similarly, we may replace the term $1/\cos(\varphi_c + \psi_0)$ in eq. (6.59) by $H_N(\psi_0)\, M_s/(2K_1(\infty)\cos 2\varphi_c)$. Thus we finally obtain

$$
H_N(\psi_0, r_0) = H_N(\psi_0) \left\{ 1 - \frac{1}{4\pi^2 \cos 2\varphi_c} \frac{\delta_B^2}{r_0^2} \right.
$$

$$
\left. \times \left[-1 + \sqrt{1 + \frac{4\Delta K r_0^2 \cos 2\varphi_c}{A}} \right]^2 \right\} + H_d - \frac{1}{2} M_s.
$$

$$(6.60)$$

Equation (6.60) can be rearranged according to

$$
H_N(\psi_0, r_0) = \frac{2K_1(\infty)}{\mu_0 M_s} \alpha_K(\psi_0, r_0)\alpha_\psi - N_{\text{eff}} M_s, \qquad (6.61)
$$

where α_ψ is given by eq. (6.39) and $\bar{\alpha}_K(\psi_0, r_0)$ follows from eq. (6.60),

$$
\alpha_K(\psi_0, r_0) = 1 - \frac{1}{4\pi^2 \cos 2\varphi_c} \frac{\delta_B^2}{r_0^2} \cdot \left[-1 + \sqrt{1 + \frac{4\Delta K r_0^2 \cos 2\varphi_c}{A}} \right]^2 ,
$$

$$(6.62)$$

with $\operatorname{tg}\varphi_c = (\operatorname{tg}\psi_0)^{1/3}$. Equation (6.62) now contains the main microstructural effects: misalignment (φ_0, φ_c), inhomogeneous magnetic regions (ΔK, r_0), and demagnetizing stray fields (N_{eff}).

6.6 Micromagnetic analysis of the coercive field of modern permanent magnets

6.6.1 Nucleation versus pinning

Large coercive fields are obtained either by dw pinning or by nucleation hardening. The first of these mechanisms has been realized for the classical iron–steel based pms and in the modern Sm_2Co_{17} based pms. The nucleation mechanism has been found to be realized in Alnico, in the hard ferrites and in the rare earth metal–transition metal intermetallic compounds. In a number of papers the role of thermally activated nucleation and the expansion of the nuclei have been considered to be the leading coercivity mechanisms [6.74–6.78]. Here it should be noted, however, that the effective field due to thermal fluctuations is only of the order of 10^{-1} T and therefore irrelevant for the interpretation of coercive fields of several teslas. This also cannot be removed by assuming that the nucleus expansion determines the coercive

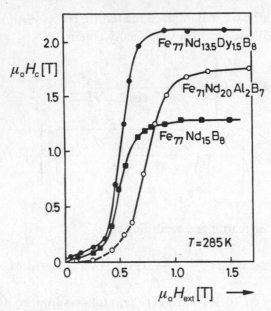

Fig. 6.26. H_c as a function of the applied maximum field for nucleation hardened, sintered magnets.

field. This has been discussed in detail previously. The following two experimental results clearly demonstrate the relevance of the nucleation mechanism:

1. The coercive field of minor hysteresis loops shows characteristic differences between pinning and nucleation hardened pms. In the case of a pinning hardened pm the coercive field of a minor hysteresis loop is not larger than the maximum applied external field, whereas in the case of the nucleation hardened pm the coercive field in general is larger than the applied external field if this is larger than the pinning coercive field which naturally also exists in the nucleation hardened pms. On approaching this critical field in nucleation hardened pms an abrupt increase of H_c is observed which is considerably larger than the applied field. As shown in Fig. 6.26 for three sintered pms the H_c-values induced by an applied field may be larger by a factor of 4 as compared with the applied field. Such behaviour cannot be understood by a pinning mechanism because in this case H_c can be only as large as the applied field.

2. By means of the magneto-optical Kerr effect the nucleation of reversed domains has been observed in an aligned sintered NdFeB pm. Since the grains of a sintered pm are of the order 5–20 μm their size is considerably larger than the critical single domain diameter of 0.2 μm. Figure 6.27 shows the domain pattern of a thermally demagnetized, aligned sintered pm where the characteristic deficiencies of all types of sintered magnets are observable: misaligned grains, coupled grains, and non-ferromagnetic grains. The existence of coupled grains due to the lack of an intergranular paramagnetic phase is

Fig. 6.27. Domain pattern of a thermally demagnetized, sintered aligned NdFeB magnet obtained by the magneto-optical Kerr effect.

responsible for cascades of magnetic reversal processes if these grains are also misoriented. Accordingly magnetically coupled misoriented grains are one of the main sources for a reduction of the coercive field. In eq. (6.61) we therefore have to introduce $\alpha_\psi(\psi) = \alpha_\psi(\pi/4)$. Figure 6.28 shows the domain pattern of an ensemble of aligned grains with one of the grains being misaligned, contacting a paramagnetic grain of $NdFe_4B_4$ in its neighbourhood. After applying a magnetic field of 1.4 T all domains have vanished. Reducing the magnetic field to 0.4 T we observe in the misaligned grain at the right hand surface the formation of some small spike-like reversed domains. This clearly demonstrates the damaging effects of misaligned grains contacting paramagnetic grains.

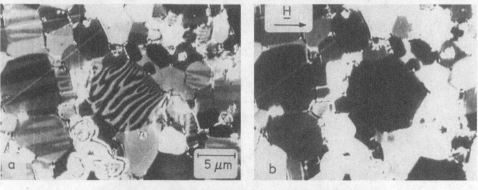

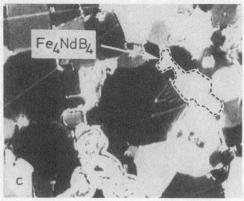

Fig. 6.28. Domain pattern within a misaligned grain of a $Nd_{16}Fe_{77}B_6$ pm under different applied magnetic fields [6.14]. (a) $\mu_0 H_{ext} = 0$ T (thermally demagnetized), (b) $\mu_0 H_{ext} = 1.4$ T (saturated state), (c) $\mu_0 H_{ext} = 0.4$ T (just nucleated reversed domains).

6.6.2 Analysis of the temperature dependence of the coercive field

A sensitive test of the nucleation model is the temperature dependence of H_c. A series of measured temperature dependences are shown in Fig. 6.29. All pms show a monotonous decrease of H_c, however, with characteristic differences in the low and high temperature range which depend on the special composition of the pms:

1. Magnets containing Dy in general have larger H_c-values because the antiferromagnetic coupling of the Dy moment reduces M_s, i.e., increases H_c.
2. Nd-rich pms being composed mainly of two phases – the hard magnetic phase $Nd_2Fe_{14}B$, and a nonferromagnetic intergranular Nd-rich phase.
3. B- and Fe-rich pms being composed mainly of the phases $Nd_2Fe_{14}B$, $NdFe_4B_4$, Fe_2B, Fe_3B.

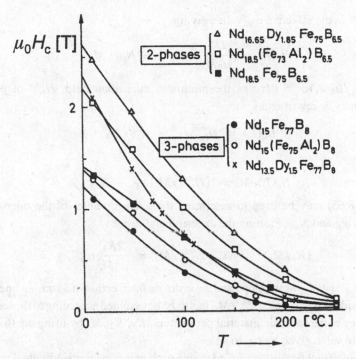

Fig. 6.29. Temperature dependence of H_c of two-phase and three-phase NdFeB magnets.

The better temperature stability of H_c in the case of the two-phase pms is due to the smaller demagnetizing local stray fields if nonferromagnetic grains are absent. Here it must be noted that the stray field term, $(-)N_{\text{eff}} \cdot M_s$, is less temperature dependent than $2K_1/\mu_0 M_s$ and consequently at temperatures near T_c the term $(-)N_{\text{eff}} \cdot M_s$ may compensate $2K_1/\mu_0 M_s$. To avoid large local stray fields therefore the two-phase compositions containing a narrow intergranular phase are a prerequisite.

For a quantitative interpretation of $H_c(T)$ we have to take into account the damaging effects of the microstructure: misaligned grains, magnetically coupled grains (exchange or dipolar coupling), magnetically perturbed grain surfaces, large local demagnetizing stray fields at sharp corners and edges of polyhedral grains. Equation (6.61) takes care of these damaging effects by the parameters α_K, α_ψ and N_{eff}. Equation (6.61) has been derived for an individual particle. In order to apply eq. (6.61) to an ensemble of grains, effective values for α_ψ have to be considered:

1. In the case of magnetically coupled grains it is obvious that the grains with the smallest nucleation field determines H_c. Once such a grain reverses its magnetization, reversal processes also take place in neighbouring grains producing a cascade of demagnetization processes. Since grains misoriented by 45° have the lowest nucleation field we may

choose $\alpha_\psi^{\min}$ as the effective α_ψ-value, giving

$$H_c(T) = \frac{2K_1}{\mu_0 M_s}\alpha_K\alpha_\psi^{\min} - N_{\text{eff}}\cdot M_s. \tag{6.63}$$

Since $(2K_1/\mu_0 M_s)\alpha_\psi^{\min}$ denotes the minimum nucleation field, $H_N^{\min}$ of eq. (6.39), eq. (6.63) may be rewritten as

$$H_c(T) = H_N^{\min}\alpha_K - N_{\text{eff}}\cdot M_s, \tag{6.64}$$

$$H_c(T)/M_s = (H_N^{\min}/M_s)\alpha_K - N_{\text{eff}}. \tag{6.65}$$

Equation (6.65) may be used to determine the average values of the microstructural parameters α_K and N_{eff}. Plotting the experimental result

$$(H_c/M_s)^{\exp} \quad \text{vs.} \quad H_N^{\min}/M_s = \frac{2K_1}{\mu_0 M_s^2}\alpha_\psi^{\min}, \tag{6.66}$$

we obtain α_K and N_{eff} as the slope and the ordinate intersection of a straight line. Here the right hand side of eq. (6.65), $H_N^{\min}/M_s$, has to be determined according to the temperature dependences of the intrinsic material parameters, K_1, K_2, M_s by using eq. (6.42) or the approximate value given by eq. (6.41).

2. In the case of magnetically decoupled grains each grain acts with individual microstructural parameters. Therefore if no correlation exists between α_ψ and α_K, the averages $\langle\alpha_\psi\rangle$, $\langle\alpha_K\rangle$ and $\langle N_{\text{eff}}\rangle$ should be considered in eq. (6.62) giving

$$H_c(T) = \langle H_N\rangle\langle\alpha_K\rangle - \langle N_{\text{eff}}\rangle\cdot M_s, \tag{6.67}$$

where

$$\langle H_N\rangle = (2K_1/\mu_0 M_s)\langle\alpha_\psi\rangle. \tag{6.68}$$

$\langle H_N\rangle$ has to be determined according to the actual distribution function of the misaligned grains, which in general may be described by Gaussian functions [6.79]. $\langle\alpha_K\rangle$ and $\langle N_{\text{eff}}\rangle$ again are obtained from a plot of

$$(H_c(T)/M_s)^{\exp} \quad \text{vs.} \quad \langle H_N\rangle/M_s. \tag{6.69}$$

Since coupled grains are always observed in sintered magnets, the analysis of $H_c(T)$ has been performed using eq. (6.66). This analysis is shown in Fig. 6.30 for three types of sintered pms produced by Vacuumschmelze GmbH ($Nd_{15}Fe_{77}B_8$), Sumitomo ($Nd_{15}Fe_{77}B_6$), and the Max-Planck-Institut für Metallforschung ($Nd_{20}Fe_{71}B_7Al_2$). Within a large temperature range between 200 K and 500 K a linear relation is observed. The deviations at low temperatures are due to the spin reorientation in $Nd_2Fe_{14}B$ and at high temperature to a transition from the nucleation to the pinning mechanism. From the slope of the linear plots of Fig. 6.30 we obtain average α_K-values of 0.93 for the VAC-magnet, 0.90 for the Sumitomo magnet and 0.89 for the MPI-MF magnet. These results were obtained despite the

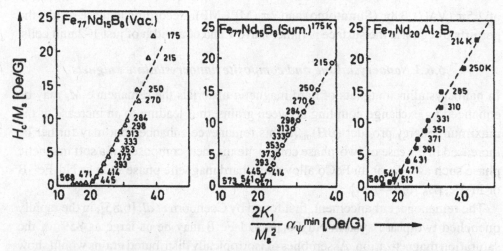

Fig. 6.30. H_c/M_s vs. $(2K_1/M_s^2)\alpha_\psi^{min}$ for three sintered magnets [6.14] produced by Vakuumschmelze, Sumitomo and Max-Planck-Institut für Metallforschung. From the slopes of these plots, lower bounds for the width r_0 of the magnetic inhomogeneity may be derived from eq. (6.54) ($r_0 = 1.2$ and 1.0 nm for the $Nd_{15}Fe_{77}B_6$ pms, and $r_0 = 1.4$ nm for the $Nd_{20}Fe_{71}Al_2B_7$ pms) [6.102].

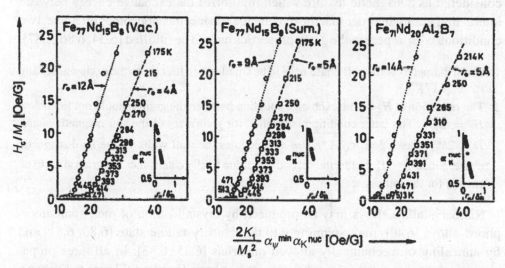

Fig. 6.31. H_c/M_s vs. $(2K_1/M_s^2)\alpha_\psi^{min}\alpha_K^{nuc}$ for the three sintered pms of Fig. 6.30, where upper and lower bounds for the α_n^{nuc} values have been used [6.14]. These insets show the temperature dependence of the parameter α_K^{nuc}.

fact that α_K should be temperature dependent. Figure 6.31 shows a more refined plot of $(H_c(T)/M_s)^{exp}$ vs. $H_N^{min}(T) \cdot \alpha_K(T)$. Here it has been assumed that $\Delta K = K_1$ and for each magnet two values, r_0, of the unperturbed surface region have been considered. In this more refined plot a spectrum of r_0-values varying between 4Å and 14 Å has been found. The effective average demagnetization factors are of the order of 2.5π (VAC), 3.8π (Sumitomo) and 2π (MPI-MF). According to these results the perturbed region at the surface is rather narrow and of a width of just 1–2 unit cells.

6.6.3 Nanocrystalline and composite nanocrystalline magnets

In nanocrystalline magnets of hard magnetic materials the remanence M_r may be enhanced by exchange coupling between grains, thus leading to an increase of the maximum energy product $(BH)_{max}$. This remanence enhancement may further be increased in the case of two-phase composite magnets composed of a soft magnetic phase such as α-Fe or an FeCo alloy and a hard magnetic phase such as $Nd_2Fe_{14}B$ or $Pr_2Fe_{14}B_6$ [6.80–6.84].

The remanence enhancement, first found by Coehoorn *et al.* [6.85], in the rapidly quenched two-phase system $Nd_2Fe_{14}B$ and Fe_3B may be as large as 85% of the saturation magnetization. Assemblies of isotropically distributed grains would show a remanence of $0.5M_s$. Due to the exchange interaction between coupled grains the spontaneous magnetization in unfavourably oriented grains is rotated into the alignment direction of the saturation field.

This tendency of neighbouring grains to align J_s parallel to each other may be considered as a magnetic texture which minimizes the exchange energy between interacting grains. In order that the remanence enhancement becomes effective, two conditions with respect to the grain properties have to be fulfilled [6.34, 6.86, 6.87]:

1. Grain diameters of the soft phase have to be smaller than four times the exchange length, $l_K^s = \sqrt{A/K_1}$,
2. The reduction of H_c by the exchange coupling between the grains should not fall below $H_c = M_r/2$. This latter condition is fulfilled for grain sizes of the soft magnetic phase, $D < 2\delta_B^h$, where $\delta_B^h = \pi\sqrt{A/K_1^h} = \pi l_K^h$ denotes the wall width of the hard magnetic phase. Since $\delta_B^h < l_K^s$ holds, the second condition finally determines the grain sizes to be chosen for the soft phase.

Nanocrystalline alloys may be produced by crystallization of melt-spun amorphous alloys [6.40], melt-spinning into the nanocrystalline state [6.80, 6.81] and by annealing of mechanically alloyed materials [6.25, 6.33]. In all three preparation techniques remanence enhancement has been observed. Figure 6.32 shows hysteresis loops of melt-spun $Pr_2Fe_{14}B$ and α-Fe showing the change of J_r and of H_c with increasing α-Fe content. Three types of hysteresis loops may be detected from Fig. 6.32 the microstructures of which are visualized in Fig. 6.33:

1. High-coercivity pms for grains decoupled by a paramagnetic Pr-rich intergranular phase.
2. High-remanence–high-coercivity pms in the case of strongly exchange coupled grains in highly stoichiometric alloys.
3. Composite high-remanence pms in alloys with overstoichiometric α-Fe.

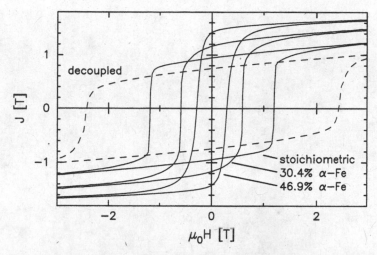

Fig. 6.32. Room temperature hysteresis loops of PrFeB-based pms; $Pr_{15}Fe_{78}B_7$ (decoupled grains), $Pr_{12}Fe_{82}B_6$ (stoichiometric, exchange coupled grains), $Pr_8Fe_{87}B_5$ (30.4% α-Fe), $Pr_6Fe_{90}B_4$ (46.9% α-Fe) (composite exchange coupled grains) [6.81].

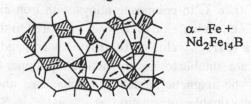

Fig. 6.33. Schematic microstructures of nanocrystalline melt-spun pms. (1) Exchange decoupled grains of a high-coercivity two-phase pm. (2) Exchange coupled single-phase stoichiometric, high-remanence pm. (3) Composite exchange coupled high-remanence two-phase pm.

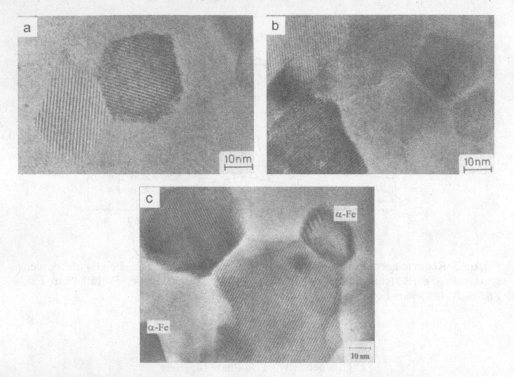

Fig. 6.34. High-resolution transmission electron microscopy micrographs of NdFeB pms.
(a) Single-phase $Nd_2Fe_{14}B$ pm. (b) Two-phase NdFeBGaNb pm with Ga segregated at the
grain boundaries. (c) Composite pm of $Nd_2Fe_{14}B$ and 15% α-Fe (dark grains).

In the latter case remanence enhancement is due to the exchange coupling as well
as the large magnetic moment of α-Fe. In Fig. 6.34a–c micrographs of transmission
electron microscopy of these three types of magnets are presented for NdFeB-based
nanocrystalline pms [6.15, 6.17]. Similar results have been obtained for PrFeB-
based pms [6.24].

Figure 6.35 summarizes the dependence of J_r, H_c and $(BH)_{max}$ on iron content.
Stoichiometric $Pr_2Fe_{14}B$ shows a remanence $J_r = 0.95$ T which is larger than
the isotropic value of 0.78 T. In composite alloys with iron excess of 30%, the
observed J_r is 1.18 T, i.e., an increase of 52% as compared to the isotropic value.
With this remanence a $(BH)_{max}$ value of 181 kJ/m^3 is realized. In particular the
nanocrystalline alloys are suitable for the production of polymer-bounded magnets.

For the analysis of the magnetic hardening mechanisms, the temperature dependence of H_c gives valuable information as discussed in Section 6.6.2. Due
to the exchange coupling between the grains in nanocrystalline magnets, the pa-
rameter $\alpha = \alpha_K \cdot \alpha_\psi^{min}$ is expanded by an additional parameter α_{ex}, which takes
care of the cooperative demagnetization process where several grains are involved.

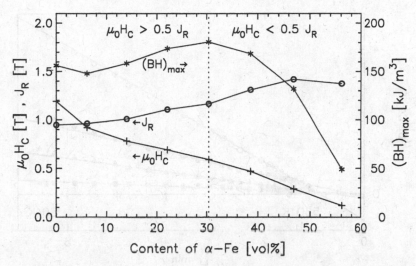

Fig. 6.35. Magnetic properties of composite PrFeB-based pms measured at RT for different amounts of α-Fe (according to [6.81]).

Equation (6.63) is now rewritten as

$$H_c(T) = \frac{2K_1}{\mu_0 M_s} \cdot \alpha_K \cdot \alpha_\psi^{min} \cdot \alpha_{ex} - N_{eff} M_s$$
$$= H_N^{min} \cdot \alpha_K \alpha_{ex} - N_{eff} M_s. \tag{6.70}$$

Figure 6.36 compares plots of $\mu_0 H_c(T)/J_s$ for decoupled, stoichiometric and composite PrFeB magnets, and Fig. 6.37 gives a review over different types of nanocrystalline pms. In Fig. 6.36 the parameters $\alpha_K \alpha_{ex}$ vary from 0.8 for the decoupled material to 0.18 and 0.06 for the composite materials, with percentages of 30.4 and 46.9% excess α-Fe, respectively. Since the average grain sizes of the α-Fe grains are of the order of 15 nm [6.81] (see Fig. 6.38) these are perfectly exchange coupled to the PrFeB grains. Nevertheless, rotations of J_s within the α-Fe grains are easier than in the hard magnetic grains and therefore induce, via the exchange coupling, an enhanced rotation within the hard grain, thus approaching their nucleation field at lower fields, which leads to a decrease of H_c as also seen in the hysteresis loop of Fig. 6.32 and which is also demonstrated by the drastic decrease of $\alpha_K \cdot \alpha_{ex}$. It is of interest that N_{eff} for the exchange coupled nanocrystalline composite magnets with $N_{eff} \simeq 0.1$ is much smaller than $N_{eff} \simeq 0.74$ for the decoupled grains. This has to be attributed to the more polyhedral shapes for the decoupled grains, whereas the exchange coupled grains are submitted to a cooperative demagnetization process of clusters of grains taking place at a minimum stray field. The parameters α_{ex} and α_K cannot be measured separately for a given magnet. A separation is possible by using results of the computational micromagnetism

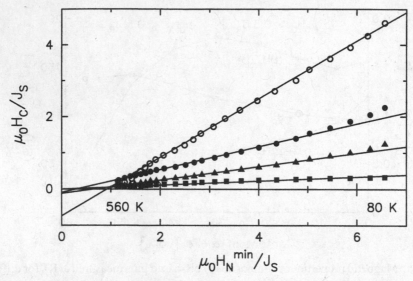

Fig. 6.36. Analysis of the temperature dependence of H_c by plots of $\mu_0 H_c / J_s$ vs. $\mu_0 H_N^{min} / J_s$ for composite pms with different amounts of α-Fe. ○, decoupled pm: $\alpha_K \alpha_{ex} = 0.80$, $N_{eff} = 0.74$; ●, stoichiometric pm: $\alpha_K \alpha_{ex} = 0.32$, $N_{eff} = 0.09$; ▲, 30.4% α-Fe: $\alpha_K \alpha_{ex} = 0.18$, $N_{eff} = 0.12$; ■, 46.9% α-Fe: $\alpha_K \alpha_{ex} = 0.06$, $N_{eff} = 0.00$ [6.81].

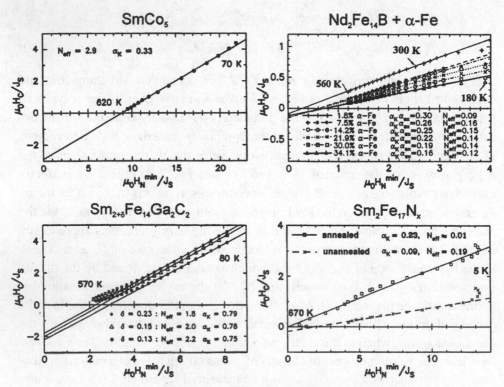

Fig. 6.37. $\mu_0 H_c / J_s$ vs. $\mu_0 H_N^{min} / J_s$ for different types of pms.

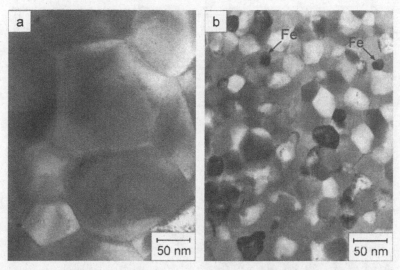

Fig. 6.38. Transmission electron microscopy micrographs of a melt-spun $Pr_8Fe_{87}B_8$ ribbon flake with (a) decoupled grains in a Pr-rich pm ($Pr_{15}Fe_{78}B_7$) and (b) exchange coupled spherical grains of a composite pm ($Pr_8Fe_{87}B_5$).

as performed by Fischer and Kronmüller [6.88] for stoichiometric nanocrystalline pms with ideal grain boundaries, i.e, $\alpha_K = 1$. In this case α_{ex} has been found to be 0.65 and $N_{eff} = 0.018$ for 20 nm grains. Using these results and $\alpha_K\alpha_{ex} = 0.32$ as obtained for the stoichiometric magnet [6.81], we find $\alpha_K = 0.49$. Consequently in nanocrystalline materials ideal grain boundaries do not exist. This is also demonstrated by the TEM micrograph shown in Fig. 6.39 where the grain boundary shows a destruction of the net planes approaching a more disordered structure. Obviously grain boundaries act as nuclei for the reversal of magnetization. The role of deteriorated grain boundaries has been treated by micromagnetic model calculations [6.73, 6.89, 6.90] and numerical simulations by the finite element method (see Chapter 13). These investigations, to be discussed in Chapter 13, have shown that $\alpha_K \simeq 0.5$ requires a reduction of K_1 and A to one tenth of their ideal values within grain boundaries of width 3 nm.

6.6.4 Nanostructured, nanocrystalline Sm_2Co_{17}-based permanent magnets

The largest coercive fields so far obtained have been for SmCo-based pms: 8 T for $SmCo_5$ [6.91] and 4 T for Sm_2Co_{17} [6.92, 6.93]. Because of their excellent temperature stability up to 500 °C [6.94, 6.95] the 2:17-based pms are of high technical interest. The coercivity of Sm_2Co_{17}-based pms with additives of Cu, Fe and Zr is usually attributed to the pinning of dws at the cell walls of a 1:5 structure separating pyramical cells of a 2:17 structure as shown by the TEM micrograph in Fig. 6.40.

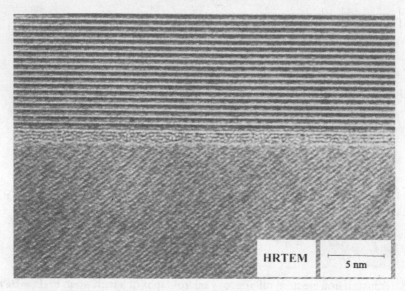

Fig. 6.39. High resolution TEM of a Pr-rich nanocrystalline PrFeB pm showing a strongly disordered intergranular phase.

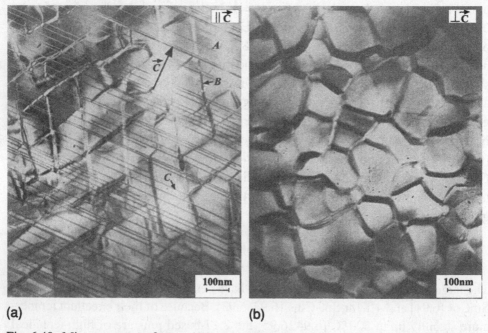

(a)   (b)

Fig. 6.40. Microstructure of an optimized $Sm(Co_{bal}Cu_{0.05}Fe_{0.10}Zr_{0.03})_{8.5}$ pm. TEM micrograph observed (a) parallel and (b) perpendicular to the c-axis [6.96].

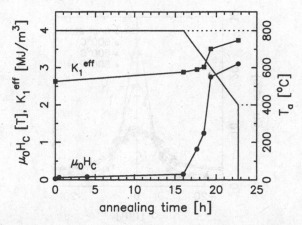

Fig. 6.41. Evaluation of the RT coercivity $\mu_0 H_c$ and the effective magnetocrystalline constant K_1 as a function of the ageing programme for a sintered $Sm(Co_{bal}Cu_{0.07}Fe_{0.22}Zr_{0.04})_{7.4}$ pm [6.31]. K_1^{eff} denotes the effective anisotropy constant.

According to Fig. 6.40 the morphology of the $Sm(Co_{bal}Cm_{0.05}Fe_{0.10}Zr_{0.03})_{8.5}$ magnet is composed of three phases:

1. The rhombohedral Fe-rich 2:17 structure of the cells.
2. The hexagonal Cu-rich 1:5 structure of the cell walls.
3. The hexagonal 2:17 structure of the so-called lamellar Zr-rich Z-phase.

One of the striking phenomena of the 2:17 pms is the fact that after an annealing treatment at 800 °C for 16 h the morphology of the three phases is fully developed; however, the coercive field is only of the order of 0.1 T. This result refutes the suggestion [6.97] that the lamellar Z-phase is responsible for the high coercivity of 2:17 pms. According to Fig. 6.41, the coercivity and the effective anisotropy constant develop only during a slow cooling process from 800 °C to 400 °C with a cooling rate of 0.5–1.0 °C/min, followed by quenching to RT. Therefore, the development of H_c has to be attributed to the chemical and structural modifications of the cell walls during the cooling process. Further information is obtained by applying high-resolution TEM-EDX [6.96]. Figure 6.42 presents the development of the Cu-profile within the cell and the cell wall during the cooling process and Fig. 6.43 shows the distribution of all components in cell and cell wall. The development of the Cu-profile is related to a sharpening of the transition region between the 2:17 and the 1:5 phase. The sharp transition region between the cell and cell wall is also demonstrated by the high-resolution transmission electron microscopy micrograph of Fig. 6.44 showing clearly the large coherence between the two phases. Obviously the existing internal stresses are relaxed by the existence of the lamellar Z-phase. Actually in Zr-free samples the lamellar phase does not exist and only Cu-rich ellipsoidal precipitates are formed resulting in rather low coercivities.

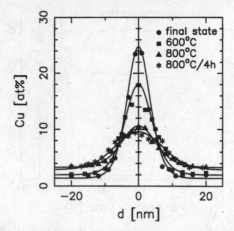

Fig. 6.42. EDX profiles of copper within the 1:5 cell walls and the 2:17 cells as a function of the annealing programme [6.31].

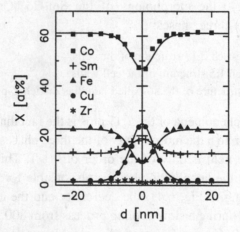

Fig. 6.43. EDX profiles of the elements Sm, Co, Cu, Fe, Zr within the 1:5 cell wall and the 2:17 cell in the optimized state after cooling down to 400 °C [6.31].

From measurements of magnetocrystalline constant of the Cu-doped $SmCo_5$ phase [6.98] the measured Cu-profile in Fig. 6.42 can be transformed into a K_1-profile as shown in Fig. 6.45 [6.99]. The anisotropy constant varies within a narrow width of 2 nm from a value $K_1^{2:17} = 2.9\,MJ/m^3$ to $K_1^{1:5} = 8.1\,MJ/m^3$. The increase of H_c during the cooling treatment has two main sources. After the 800 °C isothermal treatment the cell walls are not in the 1:5 structure. Due to the increasing Cu content in the cell wall during cooling, a transformation from the rhombohedral Sm_2Co_7 and the hexagonal Sm_5Co_{19} into the hexagonal $SmCo_5$ phase takes place favoured by the increased Cu content.

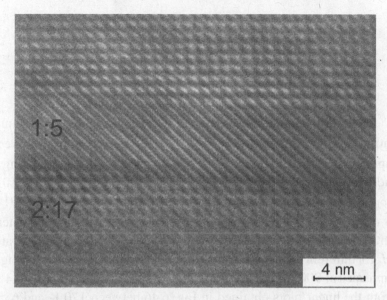

Fig. 6.44. High-resolution TEM micrograph of the plane boundaries between the cell and the cell walls [6.31].

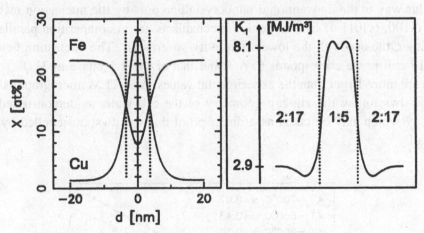

Fig. 6.45. Calculated profile of the magnetocrystalline anisotropy constant K_1 using the results of EDX-profiles and K_1 measurements of Lectard *et al.* [6.107].

With the knowledge of the K_1 profile the interaction between a dw and the cell wall can be determined by the results derived in Section 5.7.1 for planar defects. The relation between the coercive field and the repulsive force $P_{3,\mathrm{max}}$ given by eq. (5-43), is written

$$H_c = \frac{1}{2 J_s \cos \psi_0} P_{3,\mathrm{max}}$$

$$= \frac{\pi}{3 \cdot \sqrt{3}} \frac{2K_1}{J_s} \frac{1}{\cos \psi_0} \frac{d}{\delta_B} \left| \sum_{1=1}^{N-1} \frac{A}{A^{i,i+1}} - \frac{K_1^i}{K_1} \right|$$

$$= \frac{2K_1}{J_s} \alpha_K^{pin}, \tag{6.71}$$

where α_K^{pin} determines the pinning strength of the phase boundary. For a numerical determination of α_K^{pin} we may assume $A^{i,i+1} \approx A$ because both phases have similar Curie temperatures. An approximate value of α_K^{pin} can be estimated assuming further a linear increase of K_1 from $K_1^{2:17}$ to $3K_1^{2:17} = K^{1:5}$ extending over $N = 10$ layers. With the δ_B of 5.74 nm from Table 4.1, we then obtain with $K_1^{2:17} = 2.91$ MJ/m^3 a value of $\alpha_K^{pin} = 0.35$, where we have used $d = 0.2$ nm. α_K^{pin} may be determined experimentally by plotting $\mu_0 H_c / J_s|^{exp}$ vs. $2\mu_0 K_1 / J_s^2|^{theor.}$ for the temperature de-pendent H_c for different isothermal annealing and cooling treatments (see Fig. 6.46). For pms annealed at 800 °C and 900 °C the largest α_K^{pin} parameters of ~ 0.4 are ob-tained. From the linear plots obtained in Fig. 6.46 between 170 K and 635 K we may conclude that α_K^{pin} is more or less temperature independent. Sometimes it has been proposed that the coercivity of Sm_2Co_{17}-based pms should be interpreted in a similar way to the conventional nanocrystalline pms by the nucleation mecha-nism [6.100, 6.101]. If two hardening mechanisms are in competition parallel to each other, the one with the lower coercivity governs H_c. The nucleation field of the 2:17 cell phase corresponds to 6 T and that of the 1:5 phase to 31 T. These values are much larger than the experimental values. Also TEM micrographs show that the dws follow the zig-zag geometry of the cell walls as demonstrated by Fig. 6.47 which may be taken as a direct proof that the dws are repelled by the cell walls.

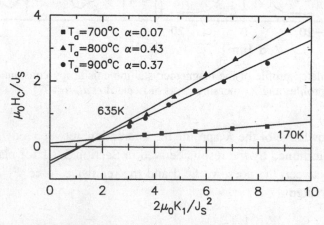

Fig. 6.46. Analysis of the temperature dependence of H_c by plotting $\mu_0 H_c / J_s$ vs. $2\mu_0 K_1 / J_s^2$ for different annealing treatments of $Sm(Co_{bal}Cu_{0.07}Fe_{0.22}Zr_{0.04})_{7.4}$ pms [6.107].

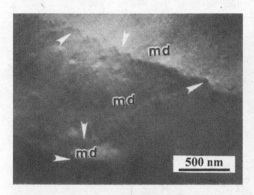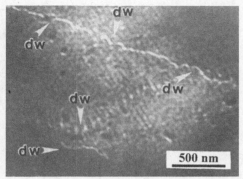

Fig. 6.47. Lorentz microscopy TEM images of an optimized $Sm(Co_{bal}Cu_{0.07}Fe_{0.22}Zr_{0.09})_{7.4}$ sintered pms showing the zig-zag pinning of dws along the 1:5 cell walls. Foucoult mode (left), Fresnel mode (right) [6.31].

The situation may be different for compositions with large Cu-content at temperatures above 550 K. According to measurements of Tellez-Blanco *et al.* [6.100] $\mu_0 H_c / J_s$ plots show a cross-over behaviour with large α-values above 550 K, indicating possibly a transition from the pinning to the nucleation mechanism. For a large Cu content a transition to a pinning behaviour may also take place due to the lower Curie temperature of the 1:5 cell wall phase.

The question whether domain wall pinning or nucleation determines the coercivity can be classfied by comparing the parameters α_K^{nuc} and α_K^{pin} as represented in Fig. 6.48. Here α_K^{nuc} decreases according to a $1/r_0$-law for $r_0 > 0.25\delta_B$ whereas α_K^{pin} increases linearly with r_0. If domain wall bowing is possible the pinning coercivity also decreases according to a $1/r_0$-law, but lies below α_K^{nuc}. If no macroscopic domain walls are available the nucleation mechanisms governs H_c throughout. If macroscopic domain walls are present, a crossover occurs from the pinning to the nucleation mechanism at around $r_0 = 0.7\delta_B$. The maximum coercivity achievable then is $H_c \sim 0.28 H_N$ in good agreement with eq. (5.60) which describes the pinning at phase boundaries.

A very important feature of the 2:17 based magnets is their high Curie temperatures up to 850 °C. Therefore those pms are suitable for high temperature magnets [6.103–6.105]. Due to the strong temperature dependence of J_s and K_1, however, in conventional 2:17 pms the coercivity decreases remarkably above 300°C, thus frustrating the condition $H_c \geq 0.5 M_r$ which guarantees the achievement of the maximum energy product. By the development of special compositions with reduced Fe-content and modified Cu-content, e.g., $Sm(Co_{bal}Cu_{008}Fe_{0.10}Zr_{0.03})_{85}$ [6.96], the high-temperature qualities of 2:17 pms have been improved considerably leading to coercivities of 0.5 Tesla at 500 °C, so far not achieved by other pms. Figure 6.49 shows temperature dependences of $H_c(T)$ for magnets of different compositions,

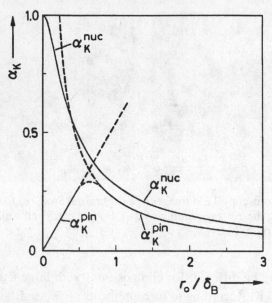

Fig. 6.48. The α_K parameters for pinning and nucleation in planar perturbations of width r_0 as a function of r_0/δ_B. α_K^{nuc} has been determined for $\Delta K = K_1$, and α_K^{pin} for $r_0/\delta_B > \delta_B$ corresponds to the bowing mechanism of eq. (7.47) [6.102].

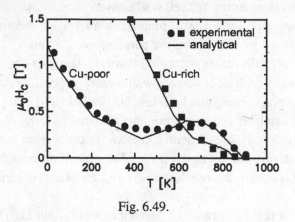

Fig. 6.49.

in particular for varying Cu-contents. Whereas the pm with the largest Cu-content shows a monotonous decrease of H_c with increasing temperature, the pms with lower Cu and Fe contents show a nonmonotonous temperature dependence with a minimum of $H_c(T)$ at 450 K and a maximum at 750 K. Whereas the monotonous decrease of H_c is nicely described by eq. (6.71) and supported by eq. (6.46), the nonmonotonous temperature behaviour requires a different treatment where a detailed consideration of the temperature dependence of the material parameters of the cells and the cell walls has to be taken into account. Here experimental results

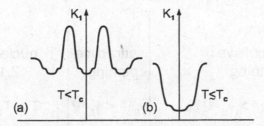

Fig. 6.50. Schematic K_1 profiles within the cell walls for different temperature ranges. (a) $RT < T < T_c^{1:5}$, (b) $T \lesssim T_c^{1:5}$.

obtained by Lectard *et al.* [6.98] and Goll [6.106] for the alloys $Sm(Co_{1-x}Cu_x)_5$ may be used. Using these results the K_1-profile shown in Fig. 6.45 has been determined [6.107]. Since the Curie temperatures are different for the cells ($T_c \approx 850$ K) and the cell walls ($T_c \approx 700$ K) a crossover of the absolute values of K_1 of these two phases takes place. At room temperature the K_1 value of the 1:5 cell walls is larger than that of 2:17 cells. With increasing temperature, the increase in K_1 of the 1:5 cell walls is stronger than that of the 2:17 cells due to their lower Curie temperature. At an intermediate state Fig. 6.50 shows a K_1 profile where in the cells and the centre of the cell walls equal K_1 values exist. Within the transition region between the two phases increased K_1 values still exist, leading to a two peak potential. At larger temperatures these two peaks vanish and the K_1 values of the 1:5 phase become smaller than that of the 2:17 phase. The cell walls now act as attractive pinning potentials. At the Curie temperatures of the 1:5 cell walls the cells behave as single domain particles separated from each other by the paramagnetic cell walls. The demagnetization process is now governed by the nucleation mechanism. According to these results three characteristic temperature ranges [6.107] govern $H_c(T)$. These are represented in Fig. 6.51:

- At low temperatures, $T \leq 200\,°C$, the cell walls act as repulsive barriers for domain walls.
- In the intermediate temperature range, $200\,°C < T < T_c$, the cell walls become attractive trapping centres for domain walls.
- Above $T_c^{1:5}$, the Curie temperature of the 1:5 phase, the nucleation mechanism governs $H_c(T)$.

An equation which describes this seemingly complex temperature dependence of $H_c(T)$ has been derived for the interaction of a dw with a phase boundary in Section 5.7.3. Equation (5.57) contains all six material parameters J_s, K_1 and A of the two phases and taking into account the temperature dependence of these material parameters in addition to their Curie temperatures $T_c^{1:5}$ and $T_c^{2:17}$, $H_c(T)$ may be described quantitatively.

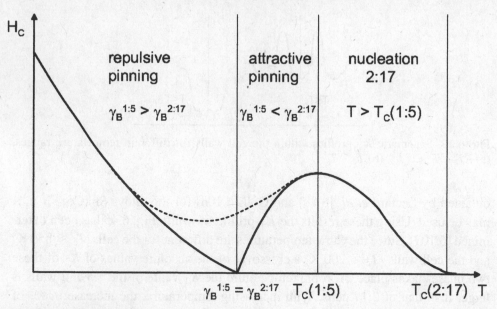

Fig. 6.51. Schematic temperature dependence of $H_c(T)$ according to eq. (5.75) showing the three ranges of repulsive and attractive pinning as well as the high-temperature nucleation range ($T_c^{1:5} < T_c^{2:17}$).

Lack of a detailed knowledge of the high-temperature dependence of the material parameters of the 2:17 and 1:5 phases, the theoretical predictions for the phase transition of ferromagnetic crystals are used. For simplicity, in the following upper indices I and II are introduced for the 2:17 and the 1:5 phase, respectively. With the critical exponent β, the materials law for uniaxial crystals near T_c is written (see eqs. 12.6, 2.13, 2.34):

$$J_S^{I,II}(T) = c_J^{I,II}(1 - T/T_c^{I,II})^\beta,$$
$$K_1^{I,II}(T) = c_K^{I,II}\left(1 - T/T_c^{I,II}\right)^{3\beta},$$
$$A^{I,II}(T) = c_A^{I,II}(1 - T/T_c^{I,II})^{2\beta}. \tag{6.72}$$

Here, $\beta = 0.5$ for the molecular field theory and $\beta = 0.365$ for the three dimensional Heisenberg model. The critical amplitudes $c_X^{I,II}$ are related to the material constants at $T = 0$. Inserting eq. (6.72) into eq. (5.57) gives

$$\mu_0 H_c(T) = \mu_0 \frac{2C_K^{II}}{C_J^{II}}\left(1 - T/T_c^{II}\right)^{2\beta} \frac{1 - \varepsilon_{KA}\left(\frac{1-T/T_c^{I}}{1-T/T_c^{II}}\right)^\beta}{\left[1 + \left\{\varepsilon_{AJ}\left(\frac{1-T/T_c^{I}}{1-T/T_c^{II}}\right)^{3\beta}\right\}^{1/2}\right]^2}, \tag{6.73}$$

with the parameters $\varepsilon_{KA} = c_K^I c_A^I / c_K^{II} c_A^{II}$ and $\varepsilon_{AJ} = c_A^I c_J^I / c_A^{II} c_J^{II}$. Equation (6.73) allows a quantitative discussion of the complex temperature dependence of $H_c(T)$. Since in general $\varepsilon_{KA} < 1$, $\varepsilon_{AJ} > 1$ and $T_c^I > T_c^{II}$ holds, two crossover temperatures exist. At low temperatures H_c is mainly determined by the first temperature coefficient describing the repulsive temperature range where H_c decreases with increasing temperature, and if the condition $\varepsilon_K \varepsilon_A = 1$ is fulfilled H_c vanishes because $T_c^I > T_c^{II}$ holds. Above this temperature attractive pinning occurs and the superscripts I and II have to be exchanged in eq. (6.72) and eq. (6.73). In this temperature range II H_c increases up to a maximum value at $T = T_c^I$, the Curie temperature of the 1:5 cell wall phase, and H_c becomes equal to the nucleation field $(2K_1^{2:17}/J_s^{2:17})\alpha$ of the 2:17 cell phase which also governs the decreasing branch of $H_c(T)$ in temperature range III of Fig. 6.51. In Fig. 6.49, by the superposition of three $H_c(T)$ curves for varying parameters ε_{KA} and ε_{AJ} and average Curie temperatures of $T_c^{II} = 700$ K for the Cu-poor and $T_c^{II} = 790$ K for the Cu-rich pm the experimental result of a positive and negative temperature gradient dH_c/dT could be nicely explained. In particular the positive temperature gradient is now explained by the transition from repulsive to attractive pinning and the magnetic phase transition in the 1:5 cell phase at $T = T_c^I$.

6.7 Alternative coercivity models – the nucleus expansion model

Besides the spontaneous nucleation model discussed in Sections 6.3 and 6.4, a so-called phenomenological 'global model' has been discussed by several authors [6.74–6.78]. In this model it is assumed that the expansion of a preformed nucleus takes place by means of a thermally excited process within the lifetime of the reversed nucleus which is given by an Arrhenius equation $\tau = \tau_0 \exp[-\Delta E/kT]$. The pre-exponential factor is of the order of magnitude $\tau_0 \cong 10^{-11}$ s and ΔE denotes the activation enthalpy required to expand the nucleus of volume v spontaneously. For the usual measuring times of $t = 1-10^3$ s the activation enthalpy $\Delta E(t) = kT \ln(t/\tau_0)$ is of the order of $\Delta E \cong 25kT$ [6.77].

The activation energy ΔE is required in order to overcome the magnetic enthalpy required for the expansion of the nucleus. As shown in Fig. 6.52, at the coercive field H_c the three terms of the magnetic enthalpy are balanced by ΔE which leads to

$$s\gamma_B' - \mu_0 M_s H_c v - \mu_0 N_{\text{eff}} M_s^2 v = \Delta E = 25kT, \tag{6.74}$$

with the domain wall energy, the magnetostatic energy of the external field at H_c and the demagnetization energy. Since it is suggested that the nucleus is formed in a perturbed region, the specific wall energy γ_B is reduced as compared to the value of the perfect crystal, giving $\gamma_B' = \alpha_B \gamma_B = \alpha_B \cdot 4\sqrt{AK_1}$, where α_B corresponds to a microstructural parameter < 1. The surface s, of the nucleus may be related to

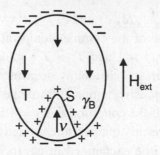

Fig. 6.52. Spike-type nucleus of volume v, surface area s and wall energy γ'_B.

the volume v by the relation $s = \alpha_s v^{2/3}$ where α_s corresponds to a geometrical parameter relating the nucleus surface to the nucleus volume. From Eq. (6.74) we now obtain

$$\mu_0 H_c = \frac{\alpha_s \alpha_B \gamma_B}{M_s v^{1/3}} - \mu_0 M_s \cdot N_{\text{eff}} - \frac{25kT}{v M_s}. \qquad (6.75)$$

Measurements of the activation volume v by means of relaxation curves have shown that v obeys a similar temperature dependence as δ_B^3. Therefore, with $v = \alpha_v^3 \delta_B^3$ eq. (6.75) can be rewritten as

$$\mu_0 H_c = \frac{\alpha_s \alpha_B}{\alpha_v} \frac{1}{M_s} \frac{\gamma_B}{\delta_B} - \mu_0 M_s \cdot N_{\text{eff}} - \frac{25kT}{v M_s}$$

$$= \frac{2K_1}{M_s} \frac{2\alpha_s \alpha_B}{\pi \alpha_v} - \mu_0 M_s \cdot N_{\text{eff}} - \frac{25kT}{v M_s}. \qquad (6.76)$$

Here it becomes obvious that the coercive field of the global model is described by the same eq. (6.53) or eq. (6.63) as in the case of the nucleation model. The only difference seems to be that the nucleation model is based on the micromagnetic equations whereas the global model starts from an energetic approach, i.e., the integrated micromagnetic equations. Since the thermal fluctuation field $\mu_0 H_f = 25kT/v M_s$ corresponds only to 5–10% of the coercive field, the micromagnetic energy terms are the dominant ones. Naturally the fluctuation field can also be introduced in the nucleation model as a term reducing the nucleation field. The microstructural parameters $\alpha_s \cdot \alpha_B$ can be derived from the plot

$$\left. \frac{H_c + H_f}{M_s} \right|_{\text{exp}} \quad \text{vs } \gamma_B / \mu_0 M_s^2 v^{1/3} \text{ or vs. } \mu_0 \gamma_B / J_s^2 v^{1/3}, \qquad (6.77)$$

where H_c, v and H_f are determined experimentally and M_s and γ_B are obtained from the intrinsic material parameters. Figure 6.53 presents the temperature dependence of the activation volume of three types of magnets (melt-spun, sintered and annealed and as-sintered). The corresponding plots according to eq. (6.77) are shown in Fig. 6.54 [6.108]. For all three magnets linear plots are obtained and the

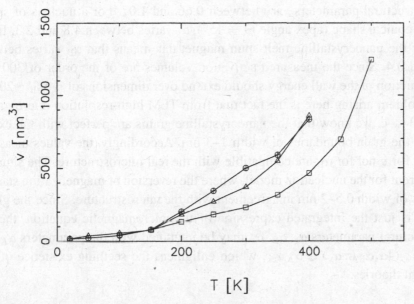

Fig. 6.53. Temperature dependence of the activation volume of three types of magnets, melt-spun □, sintered and annealed ⊕, and as-sintered △ [6.108].

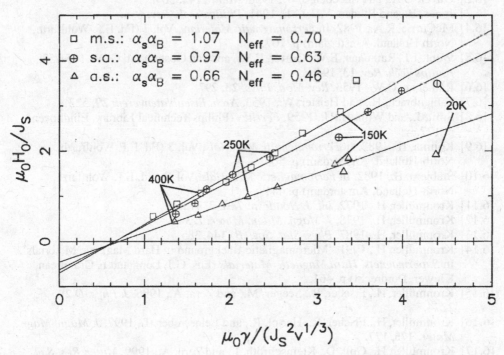

Fig. 6.54. Plots for the determination of $\alpha_s \cdot \alpha_B$, and N_{eff} for the global model of the pms of Fig. 4.53 [6.108].

microstructural parameters vary between 0.66 and 1.07. For a nucleus of spherical or conical shape (apex angle $\Theta = 15°$), α_s varies between 4.8 and 2.3. In the case of the nanocrystalline melt-spun magnet this means that α_B varies between 0.2 and 0.4. Since the measured activation volumes are of the order of 300 nm^3 the reduction of the wall energy should extend over dimensions of $\sim 5\delta_B \sim 20$ nm. The problem arising here is the fact that from TEM high-resolution micrographs, Fig. 6.34a–c, we know that the nanocrystalline grains are perfect with the exception of the grain boundaries of width 1–3 nm. Accordingly, the values measured neither for v nor for α_B are compatible with the real microstructure. The situation is different for the nucleation model, where the reversion of magnetization starts in regions of width 0.5–2 nm in agreement with the microstructure. Since the global theory is just the integrated expression of the micromagnetic equation, the microstructural parameters $\alpha_s, \alpha_B, \alpha_v$ may be identified with the parameters α_K and α_{ex}, i.e., $(4\alpha_s\alpha_B/\pi\alpha_v) \equiv \alpha_K\alpha_{ex}$, which enlightens the seeming existence of two different theories.

References

[6.1] Mishima, T., 1932, *Ohm* **19**, 353.

[6.2] Oliver, D.A., and Shedden, J.W., 1938, *Nature* **142**, 309.

[6.3] Jones, B., and Emden, V.H.J.M., 1941, *Philips Tech. Rev.* **6**, 8.

[6.4] McCurrie, R.A., 1982. In *Ferromagnetic Materials*, Vol. 3 (Ed. E.P. Wohlfarth, North Holland, Amsterdam) p. 107.

[6.5] Went, J.J., Rathenau, E.W., Gorter, E.W., and van Oosterhout, G.W., 1951/52, *Philips Tech. Rev.* **13**, 194.

[6.6] Rathenau, G.W., 1953, *Rev. Mod. Phys.* **25**, 297.

[6.7] Fahlenbrach, H., and Heister, W., 1953, *Arch. Eisenhüttenverein* **29**, 523.

[6.8] Smit, J., and Wijn, H.P.J., 1959, *Ferrites* (Philips Technical Library, Eindhoven) p. 177.

[6.9] Kojima, H., 1982. In *Ferromagnetic Materials*, Vol. 3 (Ed. E.P. Wohlfarth, North-Holland, Amsterdam) p. 305.

[6.10] Stäblein, H., 1982. In *Ferromagnetic Materials*, Vol. 3 (Ed. E.P. Wohlfarth, North-Holland, Amsterdam) p. 441.

[6.11] Kronmüller, H., 1972, *Int. J. Nondestr. Test.* **3**, 315.

[6.12] Kronmüller, H., 1978, *J. Magn. Magn. Mater.* **7**, 341.

[6.13] Kronmüller, H., 1987, *Phys. Stat. Sol. (B)* **144**, 385.

[6.14] Kronmüller, H., 1991, 'Micromagnetic Background in Hard Magnetic Materials'. In *Supermagnets, Hard Magnetic Materials* (Eds. G.J. Long and F. Grandjean, Kluwer, Dordrecht) p. 461.

[6.15] Kronmüller, H., Fischer, R., Seeger, M., and Zern, A., 1996, *J. Phys. D* **29**, 2274.

[6.16] Kronmüller, H., Fischer, R., Hertel, R., and Leineweber, T., 1997, *J. Magn. Magn. Mater.* **175**, 177.

[6.17] Kronmüller, H., Goll, D., Kleinschroth, I., and Zern, A., 1999, *Mater. Res. Soc. Symp. Proc.* 577 (Mat. Res. Soc.) p. 307.

[6.18] Coey, J.M.D., 1996, *Rare-Earth Iron Permanent Magnets* (Clarendon Press, Oxford).

[6.19] Coey, J.M.D., 1991, 'Intermetallic Compounds and Crystal Field Interactions'. In *Science and Technology of Nanostructured Magnetic Materials* (Eds. G.C. Hadjipanayis and G.A. Prinz, Plenum Press, New York) p. 439.

[6.20] Coey, J.M.D., and Smith, P.A.I., 1999, *J. Magn. Magn. Mater.* **200**, 405.

[6.21] Givord, D., Lu, Q., and Rossignol, M.F., 1991, 'Coercivity and Hard Magnetic Materials'. In *Science and Technology of Nanostructured Magnetic Materials* (Eds. G.C. Hadjipanayis and G.A. Prinz, Plenum Press, New York) p. 635.

[6.22] Buschow, K.H.J., 1997, 'Magnetism and Processing of Permanent Magnet Materials'. In *Handbook of Magnetic Materials*, Vol. 10 (Ed. K.H.J. Buschow, North-Holland, Amsterdam) p. 463.

[6.23] Buschow, K.H.J., 1991, 'Novel Permanent Magnet Materials'. In *Supermagnets, Hard Magnetic Materials* (Eds. G.J. Long and F. Grandjean, Kluwer Academic Publishers, Dordrecht) p. 49. *ibid.* p. 527, *ibid.* p. 553.

[6.24] Goll, D., and Kronmüller, H., 2000, *Naturwissenschaften* **87**, 423.

[6.25] Gutfleisch, O., 2000, *J. Phys. D: Appl. Phys.* **33**, R 157.

[6.26] Hadjipanayis, G.C., 1999, *J. Magn. Magn. Mater.* **200**, 373.

[6.27] Hadjipanayis, G.C., 1982, *Proc. 6th Int. Workshop on RE-Co Permanent Magnets* (Ed. J. Fidler, Techn. Univ. Vienna) p. 609.

[6.28] Hadjipanayis, G.C., Tang, W., Zhang, Y., Chui, S.T., Liu, J.F., Chen, C., and Kronmüller, H., 2000, *IEEE Trans. Magn.* **36**, 3382.

[6.29] Fidler, J., and Schrefl, T., 2000, *J. Phys. D: Appl. Phys.* **33**, R 135.

[6.30] Fidler, J., 1982, *J. Magn. Magn. Mater.* **30**, 58.

[6.31] Goll, D., Sigle, W., Hadjipanayis, G.C., and Kronmüller, H., 2001, *Mater. Res. Soc. Symp. Proc.*, Mat. Res. Soc. Vol. 674.

[6.32] Coehoorn, R., 1991, 'Electron Structure Calculations for Rare-Earth Transition Metal Compounds'. In *Supermagnets, Hard Magnetic Materials* (Eds. G.J. Long and F. Grandjean, Kluwer, Dordrecht) p. 133.

[6.33] Schultz, L., 1991, 'Preparation and Properties of Mechanically Alloyed Nd-Fe-B Magnets'. In *Supermagnets, Hard Magnetic Materials* (Eds. G.J. Long and F. Grandjean, Kluwer, Dordrecht) p. 573.

[6.34] Skomski, R., and Coey, J.M.D., 1993, *Phys. Rev. B* **48**, 15812.

[6.35] Skomski, R., and Coey, J.M.D., 1999, *Permanent Magnetism* (Institute of Physics, Bristol).

[6.36] Strnat, K.J., 1978, *J. Magn. Magn. Mater.* **7**, 351.

[6.37] Strnat, K.J., 1988. In *Ferromagnetic Materials*, Vol. 4, (Eds. E.P. Wohlfarth and K.H.J. Buschow, North-Holland) p. 131.

[6.38] Sagawa, M., and Hirosawa, S., 1988, *J. Physique* **49**, C8, 617.

[6.39] Sagawa, M., Fujimori, S., Togawa, M., and Matsuna, Y., 1984, *J. Appl. Phys.* **55**, 2083.

[6.40] Croat, J.J., Herbst, J.F., Lee, R.W., and Pinkerton, F.E., 1984, *J. Appl. Phys.* **55**, 2083.

[6.41] Feutrill, E.H., McCormick, P.G., and Street, R., 1996, *J. Phys. D: Appl. Phys.* **29**, 2320.

[6.42] Ding, J., McCormick, P.G., and Street, R., 1994, *Trans. Mater. Res. Soc., Japan* **B 14**, 1059.

[6.43] Kronmüller, H., 1985, *Phys. Stat. Sol. (B)* **130**, 197.

[6.44] Herzer, G., Fernengel, W., and Adler, E., 1986, *J. Magn. Magn. Mater.* **58**, 48.

[6.45] Sucksmith, W., and Thompson, F.E., 1954, *Proc. R. Soc. London,* **A 225**, 362.

[6.46] Hock, St., 1988, Dr.-Thesis Univ. Stuttgart.

[6.47] Hock, St., and Kronmüller, H., 1987, *Proc. 5th Int. Conf. on Magnetic Anisotropy and Coercivity in RE-Transition Metal Alloys* (Eds. C. Herget, H. Kronmüller, and P. Poerschke, DPG-GmbH, Bad Honnef) p. 275.

[6.48] Frei, E.H., Shtrikman, S., and Treves, D., 1957, *Phys. Rev.* **106**, 446.

[6.49] Aharoni, A., 1986, *J. Appl. Phys.* **60**, 1118.

[6.50] Holz, A., 1967, *Z. Angew. Physik* **23**, 170.

[6.51] Holz, A., 1968, *Phys. Stat. Sol.* **26**, 751.

[6.52] Néel, L., 1949, *Compt. Rend. Paris* **228**, 604.

[6.53] Néel, L., 1949, *Ann. Geophys.* **5**, 99.

[6.54] Néel, L., 1950, *J. Phys. Rad.* **11**, 49.

[6.55] Kittel, C., 1946, *Phys. Rev.* **70**, 965.

[6.56] Hertel, R., and Kronmüller, H., 2002, *J. Magn. Magn. Mater.* **238**, 185.

[6.57] Rave, W., Fabian, K., and Hubert, A., 1998, *J. Magn. Magn. Mater.* **190**, 332.

[6.58] Brown, W.F., Jr., 1945, *Rev. Mod. Phys.* **17**, 15.

[6.59] Luborsky, F.E., and Morelock, C.R., 1964, *J. Appl. Phys.* **35**, 2055.

[6.60] Seberino, C., and Bertram, H.N., 1997, *IEEE, Trans. Magn.* **33**, 3055.

[6.61] Suhl, H., and Bertram, H.N., 1997, *J. Appl. Phys.* **82**, 6128.

[6.62] O' Barr, R., Ledermann, M., and Schultz, S., 1995, *IEEE Trans. Magn.* **31**, 3793.

[6.63] Stoner, E.C., and Wohlfarth, E.P., 1948, *Philos. Trans. R. Soc. (London)* **240**, 599.

[6.64] Schabes, M.E., and Bertram, H.N., 1988, *J. Appl. Phys.* **64**, 5832.

[6.65] Kronmüller, H., Durst, K.-D., and Martinek, G., 1987, *J. Magn. Magn. Mater.* **69**, 149.

[6.66] Aharoni, A., 1997, *J. Appl. Phys.* **82**, 1281.

[6.67] Aharoni, A., 1969, *IEEE Trans. Magn.* **5**, 207.

[6.68] Martinek, G., and Kronmüller, H., 1990, *J. Magn. Magn. Mater.* **86**, 177.

[6.69] O'Barr, R., Ledermann, M. Schultz, S., Chu, W., Scherer, A., and Tonuncci, R.J., 1996, *J. Appl. Phys.* **79**, 5303.

[6.70] Wernsdorfer, W., Doudin, B., Mailly, D., Hasselbach, A., Benoit, J., Meier, J., Ansermet, Ph., and Barbara, B., 1996, *Phys. Rev. Lett.* **77**, 1873.

[6.71] Aharoni, A., 1966, *Phys. Stat. Sol.* **16**, 1.

[6.72] Kronmüller, H., Hilzinger, H.-R., 1976, *J. Magn. Magn. Mater.* **2**, 3.

[6.73] Kronmüller, H., 1987, *Phys. Stat. Sol. (B)* **144**, 385.

[6.74] Givord, D., Lienard, A., Tennaud, P., and Viadieu, T., 1987, *J. Magn. Magn. Mater.* **67**, L 281.

[6.75] Givord, D., Tenaud, P., and Viadieu, T., 1988, *IEEE Trans. Magn.* **24**, 1921.

[6.76] Givord, D., Lu, Q., Missell, E.P., Rossignol, M.F., Taylor, D.W., and Villa-Boas, V., 1992, *J. Magn. Magn. Mater.* **104–107**, 1129.

[6.77] Givord, D., Rossignol, M.F., Taylor, D.W., and Ray, A.E., 1992, *J. Magn. Magn. Mater.* **104–107**, 1126.

[6.78] Givord, D., Rossignol, M.F., and Taylor, D.W., 1992, *J. Phys. (Paris) IV,* **2**, 95.

[6.79] Rieger, G., Seeger, M., and Kronmüller, H., 1999, *Phys. Stat. Sol. (A)* **171**, 583.

[6.80] Bauer, J., Seeger, M., Zern, A., and Kronmüller, H., 1996, *J. Appl. Phys.* **80**, 1667.

[6.81] Goll, D., Seeger, M., and Kronmüller, H., 1998, *J. Magn. Magn. Mater.* **185**, 49.

[6.82] Clemente, G.B., Keem J.E., and Bradley, J.P., 1988, *J. Appl. Phys.* **64**, 5299.

[6.83] Matsumoto, F., Sakamoto, H., Komiya, M., and Fujikura, M., 1988, *J. Appl. Phys.* **63**, 3507.

[6.84] Manaf, A., Buckley, R.A., Davis, H.A., and Leonowicz, M., 1990, *J. Magn. Magn. Mater.* **101**, 360.

[6.85] Coehoorn, R., de Mooji, D.B., Duchateau, J.P., and Buschow, K.H.J., 1988, *J. Phys. (Paris)* **49**, C8, 669.

[6.86] Kneller, H.E., and Hawig, R., 1991, *IEEE Trans. Magn.* **27**, 3588.

[6.87] Schrefl, Th., Kronmüller, H., and Fidler, F., 1994, *Phys. Rev. B* **49**, 6100.

[6.88] Fischer, R., and Kronmüller, H., 1998, *Phys. Stat. Sol. (A)* **166**, 489.

[6.89] Fischer, R., and Kronmüller, H., 1996, *Phys. Rev. B.* **54**, 5469.

[6.90] Fischer, R., and Kronmüller, H., 1998, *J. Appl. Phys.* **83**, 3271.

[6.91] Ding, J., McCormick, P.G., and Street, R., 1994, *Trans. Mater. Res. Soc. Japan B* **14**, 1059.

[6.92] Livingston, J.D., and Martin, D.L., 1977, *J. Appl. Phys.* **48**, 1350.

[6.93] Ray, A.E., 1984, *J. Appl. Phys.* **55**, 2094.

[6.94] Hadjipanayis, G.C., 1982. In *Proc. 6th Int. Workshop Rare-Earth-Cobalt Permanent Magnets* (Ed. J. Fidler, Technical Univ. Vienna,) p. 609.

[6.95] Durst, K.D., Kronmüller, H., and Ervens, W., 1988, *Phys. Stat. Sol. (A)* **108**, 403; *ibid.*, 705.

[6.96] Goll, D., Kleinschroth, I., Sigle, W., and Kronmüller, H., 2000, *Appl. Phys. Lett.* **76**, 1054.

[6.97] Katter, M., Weber, J., Assmus, W., Schrey, P., and Rodewald, W., 1996, *IEEE Trans. Magn.* **32**, 4815.

[6.98] Lectard, E., Allibert, C.H., and Ballou, R., 1994, *J. Appl. Phys.* **75**, 6277.

[6.99] Kronmüller, H., Durst, K.P., Ervens, W., and Fernengel, W., 1984, *IEEE Trans. Magn.* **20**, 1569.

[6.100] Tellez-Blanco, J.C., Kou, X.C., Grössinger, R., Estevez-Rams, E., Fidler, F., and Ma, B.M., 1996. In *Proc. 14th Int. Workshop Rare Earth Magnets and Their Applications* (Eds. F.P. Missell, *et al.*, World Scientific, Singapore) p. 707.

[6.101] Tang, W., Gabay, M., Zhang, Y., Hadjipanayis, G.C., and Kronmüller, H., 2001, *IEEE Trans. Magn.* **37**, 2515.

[6.102] Kronmüller, H., Durst, K.-D., and Sagawa, M., 1988, *J. Magn. Magn. Mater.* **74**, 291.

[6.103] Chen, C.H., Walmer, M.S., Walmer, M.H., Liu, S., Kuhl, E., and Simon, G., 1998, *J. Appl. Phys.* **83**, 6706.

[6.104] Liu, J.F., Ding, Y., Zhang, Y., Dimitar, F., Zheng, F., and Hadjipanayis, G.C., 1994, *J. Appl. Phys.* **85**, 5660.

[6.105] Corte Real, M.M., De Cupos, M.F., Zhang, Y., Hadjipanayis, G.C. and Liu, J.F., 2002, *Phys. Stat. Sol. A* **193**, 302.

[6.106] Goll, D., 2001, Doctor Thesis, University Stuttgart.

[6.107] Kronmüller, H., and Goll, D., 2002, *Scripta Materialia* **47**, 545, *ibid.*, Kronmüller, H., and Goll, D., 2003, **48**, 833.

[6.108] Becher, M., Seeger, M., Zern, A., and Kronmüller, H., 1998, 'Magnetic viscosity measurements on nanocrystalline NdFeB and PrFeB magnets'. In *Proc. 10th Int. Symp. Magn. Anisotropy and Coercivity in Rare-Earth Transition Metal Alloys* (Eds. L. Schultz and K.-H. Müller, Werkstoff-InformationsgesellschaftmbH, Frankfurt, Germany) p. 307.

7

Statistical theory of domain wall pinning

7.1 Statistical pinning potential

In multidomain systems at small magnetic fields, the initial magnetization curve is determined by displacements of dws. The domains, within which the magnetization is aligned parallel to one of the easy directions, have dimensions of the order of 10^{-3} cm to 10^{-1} cm. The volume of the domains varies between 10^{-9} cm^3 and 10^{-3} cm^3. In real materials the number of defects in such volumes is rather large. Accordingly, the dws interact with large numbers of defects which means the whole interaction force has to be determined by statistical methods. The quantities to be determined by statistics are the parameters of the Rayleigh law [7.1],

$$J = \chi_0 \mu_0 H + \alpha_R (\mu_0 H)^2, \qquad (7.1)$$

and the coercive field, H_c. The susceptibility, χ_0, describes the reversible part of the magnetization and the Rayleigh constant, α_R, takes care of the irreversible Barkhausen jumps [7.2] of dws at fields smaller than the coercive field. The statistical problem becomes complicated if in the case of strong pinning centres the dws are bowing out. In a first approximation, we assume rigid dws[1] and one type of defects with positions z_j with respect to the dw, the position of which in the following is denoted by z. The force acting on the dw parallel to the dw normal (z-axis) is given by

$$p(z - z_j) = -\frac{\mathrm{d}}{\mathrm{d}z} \phi(z - z_j), \qquad (7.2)$$

where ϕ denotes the interaction energy between a defect at position z_j and the dw. In the case of a multitude of defects the total force acting on the dw is given by

$$P(z) = \sum_j p(z - z_j), \qquad (7.3)$$

---

[1] The case of flexible dws is treated in Appendix A.

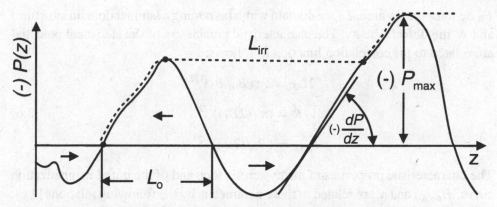

Fig. 7.1. Characteristic parameters of the statistical field of force of a domain wall. Reversible and irreversible displacements of the dw are indicated by the broken line.

where the sum extends over all defects in the neighbouring domains. The statistical field of force, $P(z)$, is characterized by three parameters as shown in Fig. 7.1:

1. The average wavelength, $2L_0$, defined as the average distance between next nearest intersections of $P(z)$ with $P(z) = 0$.
2. The average value $\overline{P_{max}}$ of the maxima of $P(z)$.
3. The average value $1/R = \overline{1/(dP/dz)}$ of the reciprocal slopes of $P(z)$ at $P(z) = 0$.

These parameters have been determined in previous papers [2.8, 5.5, 7.3–7.6] by means of correlation functions under the assumption that the total forces $P(z)$ obey a Gaussian distribution function. The probabilities, f, to find an interaction force between P and $P + dP$ or a slope $R = dP/dz$ between R and $R + dR$ are given by Gaussian distributions:

$$f(P) = \frac{1}{(2\pi B_0)^{1/2}} \exp\left[-\frac{P^2}{2B_0}\right], \tag{7.4}$$

$$f(R) = \frac{1}{(2\pi B_1)^{1/2}} \exp\left[-\frac{R^2}{2B_1}\right].$$

The correlation functions B_0 and B_1 are related to the individual interaction forces p by the following integrals:

$$B_0 = \frac{F_B N}{L_3} \int\limits_{-L_3/2}^{L_3/2} \left[p^2(z) - \langle p(z)\rangle^2\right] dz, \tag{7.5}$$

$$B_1 = \frac{F_B N}{L_3} \int\limits_{-L_3/2}^{-L_3/2} \left\{\left(\frac{dp(z)}{dz}\right)^2 - \left\langle\frac{dp(z)}{dz}\right\rangle^2\right\} dz.$$

F_B denotes the dw area, L_3 the domain width (assuming a laminar domain structure) and N the defect density. The characteristic parameters of the statistical potential are related to the correlation function as follows:

$$2L_0 = 2\pi(B_0/B_1)^{1/2},$$

$$\overline{1/R} = (\pi/(2B_1))^{1/2}, \tag{7.6}$$

$$\overline{P_{max}} = (B_0/(2\pi))^{1/2}.$$

The characteristic properties of the hysteresis loop and of the initial magnetization curve, H_c, χ_0 and α are related to these parameters by the following relations [7.6]:

$$\mu_0 H_c = \frac{\pi^{1/2}}{M_s F_B |\cos\varphi_0|} \overline{P_{max}} \left(\ln\frac{L_3}{2L_0}\right)^{1/2},$$

$$\chi_0 = \frac{\mu_0(2M_s\cos\varphi_0)^2 F_B}{L_3} \overline{1/R}, \tag{7.7}$$

$$\alpha_R = \frac{\mu_0|2M_s\cos\varphi_0|^3 F_B^2}{(2\pi)^2} \frac{L_0}{L_3} \frac{1}{\left(\overline{P_{max}}\right)^2},$$

where φ_0 denotes the angle between the applied field and M_s within the domains. Combining eqs. (7.6) and eqs. (7.7) the following relations between H_c, χ_0 and α_R may be derived:

$$\chi_0 H_c/M_s = \frac{1}{2}|\cos\varphi_0|\pi^{1/2}\frac{L_0}{L_3}\left(\ln\frac{L_3}{2L_0}\right)^{1/2},$$

$$\mu_0\frac{\alpha_R H_c}{\chi_0} = \frac{8}{3\pi^{1/2}}\left(\ln\frac{L_3}{2L_0}\right)^{1/2}, \tag{7.8}$$

$$\frac{\chi_0^2}{\mu_0 M_s \alpha_R} = \frac{3\pi}{16}\frac{L_0}{L_3}|\cos\varphi_0|,$$

where we may put $L_0 \sim \delta_B$. Equations (7.5)–(7.7) also predict the dependences on the defect density which are given by

$$H_c \propto N^{1/2}, \qquad \chi_0 \propto 1/N^{1/2}, \qquad \alpha_R \propto 1/N. \tag{7.9}$$

The above mentioned relations have been derived under the assumption of rigid dws and a statistical distribution of defects. It is of interest that the wavelength $2L_0$ is independent of the defect density.

7.2 Applications of the statistical pinning theory

7.2.1 Dislocations in crystalline metals

In crystalline soft magnetic materials, dislocations, impurity atoms and grain boundaries are the major sources for the pinning of domain walls. In the case of dislocations, the pinning effect is due to the magnetoelastic interaction between the dw and the elastic stresses of the dislocation. Using eq. (2.75) for the magnetoelastic coupling energy the calculation of the interaction energy between the dw and the dislocation becomes an elaborate task. An elegant method to determine this complex interaction starts from the so-called Peach–Köhler formula [5.6]. If we describe a dislocation by the line element, $\mathrm{d}\boldsymbol{l}^j$, and by the Burgers vector $\boldsymbol{b}^j$, the force exerted by the dislocation line element on the dw at position z is written

$$p(z - z_j) = \int \mathrm{d}\boldsymbol{l}^j \times \left(\boldsymbol{\sigma}(z_j) \cdot \boldsymbol{b}^j\right). \tag{7.10}$$

Here $\boldsymbol{\sigma}(z_j)$ denotes the elastic magnetostrictive stress tensor of the domain wall at position z_j of the dislocation. Choosing a coordinate system with the x_1-axis parallel to the easy direction, the x_3-axis parallel to the dw normal and the x_2-axis perpendicular to the x_1- and x_3-axes, the magnetostrictive stresses of a 180°-dw for a magnetostrictive isotropic crystal are given by

$$\sigma_{11} = -\sigma_{22} = 3G\lambda_s \sin^2 \varphi, \qquad \sigma_{12} = -\frac{3}{2}G\lambda_s \sin 2\varphi, \tag{7.11}$$

where G is the shear modulus and λ_s the isotropic magnetostriction constant. The interaction force exerted parallel to the dw normal by a dislocation lying parallel to the dw plane according to eq. (5.16) is obtained as

$$p_3(z) = l_1 b_2 \sigma_{22}(z) - l_2 b_1 \sigma_{11}(z) + (l_1 b_1 - l_2 b_2)\sigma_{12}(z), \tag{7.12}$$

where l_i and b_i correspond to the components of the dislocation line and of the Burgers vector, respectively. With $l_1, b_2 = 0$ this gives

$$p_3(z) = 3b_1 l_2 G\lambda_s \sin^2 \varphi(z). \tag{7.13}$$

Inserting this result into eq. (7.5) and using eqs. (7.6) and eqs. (7.7) with $b_1 = b$ and $l_2 = l$, gives

$$2L_0 = \sqrt{5}\delta_B, \tag{7.14}$$

$$\mu_0 H_c = \frac{Gbl\lambda_s}{\pi^{1/2} F_B M_s |\cos \varphi_0|} (6N F_B \delta_B)^{1/2} \left[\ln\left(\frac{L_3}{2L_0}\right)\right]^{1/2},$$

$$\chi_0 = \mu_0 \frac{\sqrt{5}\pi M_s^2 \cos^2 \varphi_0 F_B^{1/2} \delta_B^{1/2}}{2(6N)^{1/2} Gbl\lambda_s L_3},$$ (7.15)

$$\alpha_R = \mu_0 \frac{2\sqrt{5}\pi |M_s \cos \varphi_0|^3 F_B}{3N L_3 (Gbl\lambda_s)^2}.$$

Equations (7.15) contain all the important microstructural parameters of a plasti-cally deformed crystal. The dependence of H_c, χ_0 and α_R on the dislocation density may be used as a test of the statistical theory. According to previous investigations the dislocation density in Ni is related to the applied stress, τ, by [7.7]

$$\tau - \tau_0 = 0.36Gb(Nl)^{1/2},$$ (7.16)

with τ the initial flow stress. As a function of the applied flow stress according to eq. (7.9) and eq. (7.16) we expect a linear relation for H_c, $1/\chi_0$ and $1/\alpha_R^{1/2}$ as demonstrated by Fig. 7.2 for a plastically deformed Ni single crystal [7.8, 7.9].

Another interesting result is the dependence of H_c on the magnetic intrinsic material properties, being given by

$$\mu_0 H_c(T) \propto \frac{\lambda_s}{M_s} \delta_B^{1/2} \propto K_1^{-1/4},$$ (7.17)

$$\chi_0(T) \propto \frac{M_s^2}{\lambda_s} \delta_B^{1/2} \propto K_1^{-1/4}.$$

According to this result the dependence of χ_0 and H_c on δ_B, i.e., on K_1, is the same. However, with increasing magnetostriction and dislocation density, χ_0 decreases

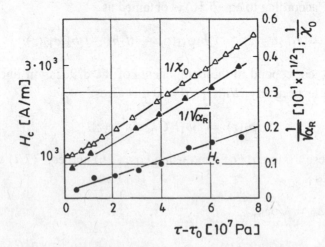

Fig. 7.2. Test of the dependences of H_c, χ_0 and α_R on the density of dislocations of a plastically deformed [100]-Ni single crystal [7.4, 7.8, 7.9].

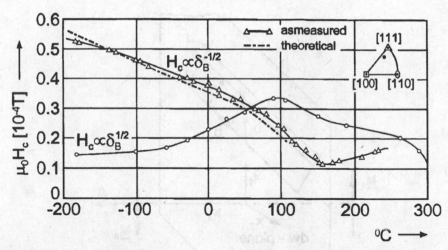

Fig. 7.3. Temperature dependence of the coercive field of a Ni single crystal due to dislocations ($\delta_B^{1/2}$) in the as-grown crystal (-o-o-o-) and due to dislocation dipoles ($\delta_B^{-1/2}$) after heavy neutron irradiation (-$\triangle$-$\triangle$-$\triangle$-) ($\phi = 3 \cdot 10^{18}$ n/cm^2) at 90 °C [2.7].The broken line corresponds to the theoretical prediction for $H_c(T)$ of dislocation dipoles for a (112) − 180°-Bloch wall.

whereas H_c increases. In general, K_1 increases with decreasing temperature and therefore H_c decreases in plastically deformed materials with decreasing temperature, as shown in Fig. 7.3 for Ni single crystals.

This behaviour may be modified if λ_s also shows a strong temperature dependence. It is of interest to note that for rigid dws the Rayleigh constant is independent of δ_B and its temperature dependence due to magnetic material constants is given by

$$\alpha_R = M_s^3/\lambda_s^2. \qquad (7.18)$$

The Rayleigh constant α_R is inversely proportional to L_3, i.e., with increasing dw density α_R increases.

7.2.2 Dislocation dipoles

In many cases the dislocation structure of plastically deformed materials or of rapidly quenched materials is composed of dislocation dipoles. In particular, in cyclic deformation or by the agglomeration of vacancies or interstitial atoms, dislocation dipoles may be formed. For a calculation of the interaction forces, $p(z)$, the Peach–Köhler formula given by eq. (7.10) can be rearranged as follows:

$$p_3(z) = -\int_f \mathrm{d}\boldsymbol{Q} \cdot \nabla_3 \sigma = -\nabla_3 \int_f \sigma_{ik} \, \mathrm{d}Q_{ik}, \qquad (7.19)$$

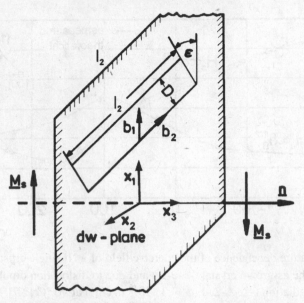

Fig. 7.4. Model of a dislocation dipole interacting with a dw.

where the dyadic product $dQ = df \cdot b$ corresponds to the differential displacement tensor of the dipole [5.9], with df corresponding to the vector surface element of the area spread out by the dislocation and b denoting the Burgers vector. We consider a dislocation loop of width D, extending parallel to the x_2-axis over a length l_2, and a loop area inclined by an angle ε with respect to the x_1-axis (Fig. 7.4) [7.10]. The interaction force for a material of isotropic magnetostriction is then written, with $dQ_{11} = Db_1 \sin \varepsilon \, dx_2$, $dQ_{12} = Db_2 \sin \varepsilon \, dx_2$,

$$p_3(z) = -3\pi D l_2 \sin \varepsilon \, G\lambda_s \, \delta_B^{-1} \sin \varphi (b_1 \sin 2\varphi + b_2 \cos 2\varphi). \qquad (7.20)$$

Inserting eq. (7.20) into eqs. (7.5)–(7.7) gives, for $\varepsilon = \pi/2$,

$$\mu_0 H_c = \frac{12\sqrt{\pi}}{|M_s \cos \varphi_0|} \frac{1}{(30F_B)^{1/2}} |G\lambda_s \, \Delta V| \left(\frac{N_{dip}}{\delta_B}\right)^{1/2} \left(\ln \frac{L_3}{2L_0}\right)^{1/2}, \qquad (7.21)$$

where $\Delta V = (Dl_3 b_2)$ corresponds to the volume expansion or contraction due to the dislocation dipole and N_{dip} denotes the volume density of dislocation dipoles. The leading terms which determine the temperature dependence of H_c give

$$\mu_0 H_c(T) \propto \frac{\lambda_s}{M_s} \delta_B^{-1/2} \propto K_1^{1/4}. \qquad (7.22)$$

According to this result, H_c of dislocation dipoles increases with increasing K_1 in contrast to the behaviour of dislocations. It is also of interest that χ_0 follows a $\delta_B^{3/2}$ law, i.e., with increasing K_1, χ_0 decreases.

As an example of the temperature dependence of dislocation dipoles, Fig. 7.3 shows $H_c(T)$ for a Ni single crystal irradiated by neutrons, at a flux of $\phi = 2 \cdot 10^{18} \, \text{cm}^{-2}$. It is well known that irradiation above RT produces dislocation dipoles of high density on {111}-planes by agglomeration of vacancies [2.7, 2.11].

7.2.3 Point defects

The results obtained for dislocation dipoles in Section 7.2.2 can be easily applied to point defects if we interpret the dipole tensor, Q, as that of a point defect. Instead of an integral like eq. (7.19) we may write

$$p_3(z) = -Q \cdot \nabla_3 \sigma. \tag{7.23}$$

In the case of an isotropic defect with $Q = IQ_0$ (I = unit tensor), the force acting on a dw vanishes. For an anisotropic defect of orthorhombic symmetry, $Q_{11} \neq Q_{22} \neq Q_{33}$, the interaction force $p_3(z) = 3\pi G \lambda_s \delta_B^{-1}(Q_{22} - Q_{11}) \sin \varphi \sin 2\varphi$ leads to a coercive field of

$$\mu_0 H_c = \frac{12\pi^{1/2}}{M_s \cos \varphi_0} \frac{G|\lambda_s||Q_{22} - Q_{11}|}{(30 F_B)^{1/2}} \left(\frac{N_{\text{dip}}}{\delta_B} \right)^{1/2} \left(\ln \frac{L_3}{2 L_0} \right)^{1/2}. \tag{7.24}$$

It should be noted that eq. (7.24) corresponds to eq. (7.21), the volume dilatation being replaced by the dipole tensor Q_0.

7.2.4 Amorphous alloys

In amorphous alloys neither individual dislocations nor grain boundaries are thought to be present. Therefore, the most important pinning centres governing H_c in crystalline materials are absent in amorphous alloys. The only 'defect structures' are fluctuations of the mass density, known as free or antifree volumes [7.12-, 7.14], short-range ordering or agglomerates of 'free' or 'antifree' volumes. Since, in addition to the lack of conventional defect structures, a long-range crystal anisotropy is absent in amorphous alloys, they are predestined to be applied as soft magnetic materials. Nevertheless, the initial susceptibilities of amorphous alloys are of the same order of magnitude as those of crystalline alloys, e.g., permalloy. Obviously, magnetic inhomogeneities exist in amorphous alloys leading to a moderate intrinsic pinning of dws. Five types of intrinsic magnetic inhomogeneities may be of importance [7.5, 7.15]:

1. Fluctuations of the exchange energy δA,
2. Fluctuations of the local magnetic anisotropy δk,

3. Elastic coupling energy of elastic dipoles,
4. Clusters of atomic short-range order,
5. Surface irregularities.

All these different types of pinning centres have to be treated by the statistical pinning theory. Since in general the distances between the local fluctuations are much smaller than the wavelength $2L_0$ of the statistical potential, eqs. (7.5), eqs. (7.6) and eqs. (7.7) are the basis for the calculation of H_c. We just have to consider the sum $B_0 = \sum_i B_0^{(i)}$ where $B_0^{(i)}$ corresponds to the self-correlation functions of the above mentioned pinning processes.

7.2.4.1 Intrinsic fluctuations of exchange and local anisotropy energy

The fluctuation of the exchange energy is related to fluctuations of the local exchange integrals at the position r_i and may be written as

$$\delta\phi_{ex} = \delta A \left(\frac{d\varphi(r_i)}{dz}\right)^2 \langle\Omega\rangle, \tag{7.25}$$

where $\langle\Omega\rangle$ denotes the average atomic volume and φ the rotation angle of M_s within the dw, given by eq. (4.7) or eq. (4.29). Similarly the fluctuation of the anisotropy energy is given by

$$\delta\phi_K = \delta k \cos^2\Theta_i \langle\Omega\rangle, \tag{7.26}$$

where Θ_i denotes the angle between M_s and the local anisotropy axis at position r_i. Characterizing the orientation of the easy axes with respect to the coordinate system of the dw by the polar angles ϑ_i and the azimuthal angles φ_i we have

$$\cos\Theta = \sin\vartheta_i \sin(\varphi - \varphi_i). \tag{7.27}$$

The fluctuations $\delta\phi_{ex}$ and $\delta\phi_K$ produce a small deviation of the dw angle φ from its ideal orientation and are the origin of forces acting on the dw

$$p_{ex}^{(i)} = -\frac{\partial\delta\phi_{ex}(z - z_i)}{\partial z}, \tag{7.28}$$

$$p_K^{(i)} = -\frac{\partial\delta\phi_K(z - z_i)}{\partial z}.$$

Inserting eqs. (7.28) into eqs. (7.5), eqs. (7.6) and eqs. (7.7) gives for the intrinsic coercive field (ρ_M = density of magnetic ions)

$$\mu_0 H_c = \frac{\pi^{1/2}}{M_s} \frac{\rho_M^{1/2}\langle\Omega\rangle}{\sqrt{2F_B\delta_B}} \left[\frac{8}{15}\langle\delta k^2\rangle + \frac{11}{15}\langle\delta A^2\rangle \frac{\pi^4}{\delta_B^4}\right]^{1/2} \cdot \left[\ln\frac{L_3}{2L_0}\right]^{1/2}. \tag{7.29}$$

If we assume that the material properties, P, obey Gaussian distribution functions

$$w(P) = \frac{1}{\sqrt{2\pi\sigma_P}} \exp\left[-\frac{(P - \langle P \rangle)^2}{2\sigma_P}\right], \tag{7.30}$$

the volume averages of the quadratic fluctuations of the anisotropy energy are given by

$$\langle \delta k^2 \rangle = \langle k \rangle^2 + \sigma_k, \tag{7.31}$$

where σ_k corresponds to the dispersion of the average anisotropy constant $\langle k \rangle$.

The fluctuations of the exchange energy may be calculated on the basis of a nearest neighbour exchange interaction. In the case of FeNi(PB) amorphous alloys the Ni-atoms are considered as nonmagnetic. The total exchange interaction of a certain Fe-atom is then proportional to the number of nearest neighbour Fe-atoms. In the case of a dense random packing model with 12 nearest neighbours (nn), the probability, w_n, of finding n nn atoms is given by $w_n = (12/n) c_{Fe}^n (1 - c_{Fe})^{12-n}$, where c_{Fe} corresponds to the atomic density of Fe-atoms. The exchange energy of n Fe–Fe pairs is given by $(n/r_0) J_{FeFe}/3$, where r_0 corresponds to the nn distance and J_{FeFe} denotes the exchange integral between Fe-atoms. A detailed calculation gives, for $c = 1/2$ [7.15],

$$\langle \delta A^2 \rangle = \frac{1}{9r_0^2} \sum_{n=0}^{12} w_n n^2 J_{FeFe}^2 = \frac{13}{3r_0^2} J_{FeFe}^2. \tag{7.32}$$

Inserting eq. (7.31) and eq. (7.32) into eq. (7.29) gives

$$\mu_0 H_c = \frac{\pi^{1/2}}{M_s} \frac{\rho_M^{1/2} \langle \Omega \rangle}{\sqrt{2 F_B \delta_B}} \left[\frac{8}{15}(\langle k \rangle^2 + \sigma_k) + \frac{143\pi^4}{45 r_0^2} \frac{J_{FeFe}^2}{\delta_B^4}\right] \left(\ln \frac{L_3}{2L_0}\right)^{1/2}. \tag{7.33}$$

Equation (7.33) may be used to evaluate intrinsic coercive fields of amorphous alloys. Some material parameters of amorphous alloys are summarized in Table 7.1. The long-range anisotropy $\langle k \rangle$ is due to long-range internal stresses and is in general much smaller than $\sigma_k^{1/2}$ which has values of the order of the anisotropy constant of the corresponding crystalline material.

Using the values given for $\langle k \rangle$, σ_k, $J_{FeFe} = kT_c/8$ ($T_c = 1000$ K), $r_0 = 0.25$ nm, $F_B = 4 \cdot 10^{-5}$ cm^2, $L_3 = 100$ μm and $\rho_M = 0.5/\Omega$ gives $\mu_0 H_c = 3 \cdot 10^{-8}$ T. If we assume that the long-range anisotropy as well as the dispersion σ_k vanishes, the coercive field is determined by the fluctuations of the exchange energy only, leading to a value of $\mu_0 H_c = 3 \cdot 10^{-9}$ T, i.e., the minimum intrinsic coercive field achievable in amorphous alloys is of the order of nanoteslas.

Table 7.1. *Material parameters of amorphous and nanocrystalline alloys [7.5, 7.6, 7.10, 7.27].*

Ferromagnet	J_s [T]	$A\left[\frac{pJ}{m}\right]$	$\langle k \rangle \left[\frac{kJ}{m^3}\right]$	$\sigma_k^{1/2}\left[\frac{kJ}{m^3}\right]$	δ_B [nm]	$\gamma_B\left[\frac{mJ}{m^2}\right]$	λ_s [10^{-6}]
$Fe_{80}B_{20}$	1.6	5	6	10^2	87	0.69	45
$Fe_{40}Ni_{40}P_{14}B_6$	0.85	3.1	0.6	10^3	215	0.17	15
$Fe_{40}Ni_{40}B_{20}$	1.05	8.1	1.2	$2 \cdot 10^2$	258	0.40	25
$Co_{58}Ni_{10}Fe_5Si_{11}B_{16}$	0.55	4.4	0.04	$2 \cdot 10^2$	1000	0.06	0.4
$Fe_{73.5}Si_{13.5}Nb_3Cu_1B_9$ (nanocryst.)	1.2	10	0.024	20	6600	0.67	6

So far these low values of intrinsic coercive fields have not been observed. The lowest values actually measured are of the order of several 10^{-7} T. Obviously other microstructures are governing the coercive fields in amorphous alloys as will be discussed in the following.

7.2.4.2 Internal stress sources

In crystalline materials the coercive fields are predominantly determined by dislocations, grain boundaries and impurity atoms. In amorphous alloys these types of defects are not effective in the conventional way. Nevertheless, the coercive fields in amorphous alloys are a factor of 10–100 larger than those determined for the intrinsic coercive fields in the preceding section. An indication that the coercive fields of amorphous alloys are determined by magnetostrictive interactions follows from a comparison of the characteristic parameters of the hysteresis loops of $Fe_{80-x}Ni_xB_{20}$ alloys, which are characterized by a variation of λ_s from $5 \cdot 10^{-6}$ for $x = 60$ up to $\lambda_s = 45 \cdot 10^{-6}$ for $x = 0$. According to the results obtained in Section 7.2.1, H_c, χ_0 and α_R depend in a characteristic way on λ_s and M_s. In Fig. 7.5 we show experimental results for these properties as a function of λ_s. The coercive field ($H_c M_s$) shows the expected monotonous increase with λ_s as expected from eqs. (7.17). As predicted by eqs. (7.17) and eqs. (7.18), χ_0/M_s^2 and α_R/M_s^3 decrease as $1/\lambda_s$, and as $1/\lambda_s^2$ respectively, with increasing λ_s. In particular, α_R/M_s^3 shows a stronger decrease as χ_0/M_s^2. Also the dependence of the effective anisotropy constant, K_{eff}, which follows that of H_c, can be understood assuming that it results from elastic long-range stresses, i.e., from the magnetoelastic coupling energy. The type of stress sources existing in amorphous alloys are long-range quenching stresses and short-range stresses due to quasidislocation dipoles. The long-range stresses are the source for a magnetic anisotropy energy whereas the elastic dipoles act as pinning centres for dws. These latter defect structures have been derived from measurements of the high-field susceptibility and an analysis of the law of approach to ferromagnetic

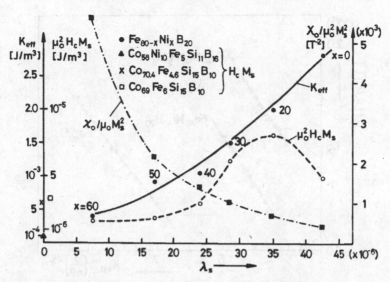

Fig. 7.5. Effective anisotropy constant, K_{eff}, coercive field $H_c M_s$ and $\chi_0/\mu_0^2 M_s^2$ for $Fe_{80-x}Ni_x B_{20}$ alloys as a function of λ_s. For nearly nonmagnetostrictive Co alloys the results for $\mu_0^2 H_c M_s$ are shown in the lower left side.

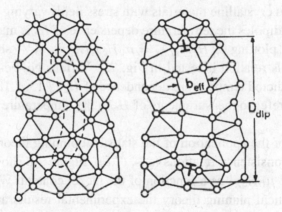

Fig. 7.6. Two-dimensional model of a quasidislocation dipole due to the agglomeration of free volumes in amorphous alloys.

saturation (see Chapter 8). In a number of papers it has been shown that elastic dipoles of extensions 1–10 nm are present in amorphous alloys at a high density of 10^{18} m^{-3} [7.16]. Quasi-dislocation dipoles in amorphous alloys may result from mass density fluctuation quenched-in during the rapid cooling process. A simple model of such a mass density defect is shown in Fig. 7.6. Here it is assumed that vacancy-type defects, so-called free volumes [7.13], have agglomerated during the quenching process, thus forming a diluted zone which leads to a local collapse of the amorphous structure. This local contraction induces elastic deformations similar

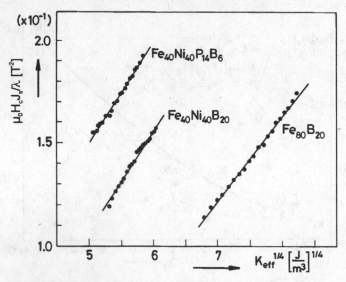

Fig. 7.7. Temperature dependence of the quantity $H_c J_s / \lambda$ of different amorphous alloys proving the validity of the $K^{1/4}$ law of eq. (7.22) [7.17].

to elastic dipoles in crystalline materials with stress fields varying as $1/r^2 - 1/r^3$. For circular quasi-dipoles the temperature dependence of H_c is mainly determined by eq. (7.22), i.e., plotting $\mu_0^2 H_c M_s / \lambda_s = \mu_0 H_c J_s / \lambda_s$ vs. $K^{1/4}$ should result in a linear relation. This relation is tested in Fig. 7.7. Various Ni–Fe-based alloys, in fact, show the predicted temperature dependence of H_c [7.15, 7.18]. The different points in Fig. 7.7 refer to measurements of H_c in the temperature range from 100 to 310 K.

A further test for the application of the statistical pinning theory to amorphous alloys is the self-consistency relations (eqs. (7.8)). Figure 7.8 shows $\mu_0 H_c \alpha_R$ as a function of χ_0 and $\mu_0 H_c \chi_0$ as a function of the nickel content. Within the framework of the statistical pinning theory the experimental results are in agreement with the predictions. The dashed line in Fig. 7.8 corresponds to the parameter $(8/(3 \cdot \sqrt{\pi}))(\ln(L_3/(2L_0)))^{1/2}$ of eqs. (7.8), where we have used a domain width of $L_3 = 5\,\mu m$ and $2L_0 = 0.4\,\mu m$. With the same parameters we obtain for the constant $|\cos \varphi_0| \, \pi^{1/2} (L_0/(2L_3)) \, (\ln(L_3/(2L_0)))^{1/2}$ of the $H_c \chi_0$-test a value of $1.2 \cdot 10^{-2}$ (putting $(1/2)|\cos \varphi_0| \pi^{1/2} = 0.5$) which is not far from the experimental result of $0.95 \cdot 10^{-2}$. According to these results it is quite obvious that the statistical pinning theory is a rather effective tool to describe the hysteresis loops of amorphous alloys. It should be noted, however, that the nonmagnetostrictive (Co, Ni, Fe)-based alloys do not fit the self-consistency relations.

Due to the self-consistency relations (7.8) the properties χ_0 and H_c depend on each other and may vary only between an upper and a lower ratio of L_3/δ_B. Since

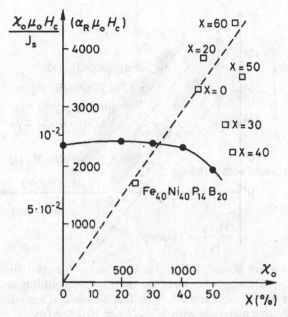

Fig. 7.8. Test of the self-consistency relations eqs. (7.8) for amorphous $Fe_{80-x}Ni_xB_{20}$ alloys ($\chi_0 H_c/M_s$ vs. $x\%$ Ni; $\alpha_R \mu_0 H_c$ vs. χ_0).

the square root of the logarithm $(\ln(L_3/(2L_0)))^{1/2}$ does not depend sensitively on $L_3/(2L_0)$ and is of the order of 2 ($L_3 \approx 10\delta_B - 1000\delta_B$), the self-consistency relation is written approximately as ($L_0 \approx \delta_B$)

$$\chi_0 \simeq (M_s/H_c) \cdot (\delta_B/L_3).$$

Ratios of δ_B/L_3 vary between 10^{-3} and 10^{-1} and give, with $M_s = 10^5$ A/m, values of 10 A/m $< M_sL_0/L_3 < 10^3$ A/m. The double logarithmic plot of Fig. 7.9 shows that the Co–Fe- and the Ni–Fe-based amorphous and crystalline alloys lie between these two limits, which means that L_3 should be of the order of $10-10^3 \delta_B$. Nanocrystalline materials produced by nanocrystallization of FeSiNbCuB amorphous alloys have a position with $M_sL_0/L_3 \simeq 10\,000$ A/m, which reflects the fact that in nanocrystalline materials, due to the random anisotropy effect, the dw widths are a factor 5–10 larger than in amorphous alloys. Figure 7.9 also includes the absolute limits for χ_0 and H_c which are expected in an ideally amorphous material with $K_1 \rightarrow 0$ and $\lambda_s \rightarrow 0$.

7.2.4.3 Coercive field due to surface irregularities

In low-coercivity materials the natural surface roughness may contribute significantly to the coercivity. Amorphous ribbons produced by the melt-spin technique

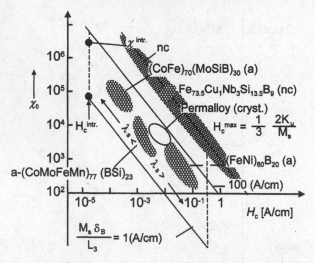

Fig. 7.9. Upper and lower bounds for χ_0 vs. H_c relations for crystalline, nanocrystalline and amorphous alloys. The upper bound of the intrinsic susceptibility and the lower bound for H_c are obtained for $K_1 \to 0$ and $\lambda_s \to 0$. The fine-shaded region indicates an upper bound for nanocrystalline materials with $M_s \delta_B / L_3 = 10\,000$ A/m.

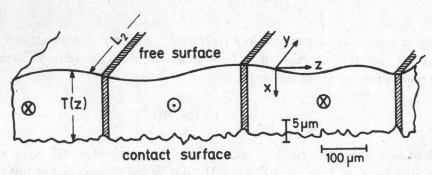

Fig. 7.10. Domain walls in a ribbon with fluctuating thickness $T(y, z)$, showing a smooth free surface and a rough contact surface.

have a natural surface roughness due to the uneven roller as well as due to the inhomogeneous solidification process. In general, the contact surface reveals a larger surface roughness, $\Delta T / T \simeq 0.05 - 0.1$, than the free surface with $\Delta T / T \sim 0.01$ [7.5]. In Fig. 7.10 it is schematically shown that, in addition to the different surface roughnesses, the wavelengths of the surface irregularities are different from each other [7.10]. On the free surface wavelengths of $\lambda \approx 100$ μm and on the contact side wavelengths of $\lambda \approx 10$ μm are observed. Due to the varying ribbon thickness $T(y, z)$ the dw moving through the ribbon changes its area as a function of z. A planar dw extending a length L_2, parallel to the y-axis, is described by a total wall energy

$$\phi_\gamma(z) = \gamma_B \int\limits_0^{L_2} T(y, z)\, dy = \gamma_B L_2 \langle T(z) \rangle, \tag{7.34}$$

where $\langle T(z) \rangle$ denotes the statistical average of the ribbon's thickness in the y-direction. The force acting on the dw is now given by

$$P_{\text{surf}}(z) = -\gamma_B L_2 \frac{d\langle T(z) \rangle}{dz}. \tag{7.35}$$

For $\langle T(z) \rangle$ a sinusoidal ansatz,

$$\langle T(z) \rangle = \langle T \rangle + \Delta T \sin(2\pi z/\lambda), \tag{7.36}$$

is used, where ΔT corresponds to the amplitude of the surface irregularities. For the calculation of the coercive field eqs. (7.5)–(7.7) are used, modified for the case of a two-dimensional distribution of pinning centres of area density ρ_s, giving [7.5]

$$\mu_0 H_c^{\text{surf}} = \frac{\pi}{M_s} \frac{\Delta T}{\langle T \rangle} \rho_s^{1/2} \gamma_B \left(\frac{L_2}{2\lambda}\right)^{1/2} \left(\ln \frac{L_3}{2L_0}\right)^{1/2}. \tag{7.37}$$

Assuming that the upper bound of the density of pinning centres just corresponds to $\rho_s = 1/\lambda^2$ we obtain

$$\mu_0 H_c^{\text{surf}} = \frac{\pi}{M_s} \frac{\Delta T}{\langle T \rangle \lambda_s} \gamma_B \left(\frac{L_2}{2\lambda}\right)^{1/2} \left(\ln \frac{L_3}{2L_0}\right)^{1/2}. \tag{7.38}$$

Inserting into eq. (7.38) the material parameters of $Fe_{40}Ni_{40}P_{14}B_6$ ($J_s = 1$ T, $\Delta T/\lambda = 0.1$, $\Delta T/T = 0.05$, $\gamma_B = 10^{-4}$ J/m^2, $\lambda \sim \langle T \rangle = 50$ μm, $L_2 = L_3 = 100$ μm) we find $\mu_0 H_c^{\text{surf}} \sim 7 \cdot 10^{-6}$ T. Changing the topological parameters of the surface by special treatments, eq. (7.38) may be tested experimentally. Figure 7.11 shows the influence of thinning a ribbon by special preparations. In all cases H_c^{surf} is found to depend linearly on $1/\langle T \rangle$ as predicted by eq. (7.38). The increase of H_c with decreasing $\langle T \rangle$ is largest after grinding by glass-paper. Polishing by diamond paste leads to an intermediate increase of H_c, whereas electropolishing leads only to a small increase of H_c. This behaviour of H_c with respect to different surface treatments is compatible with the different values of ΔT and λ expected for the above mentioned treatments.

If volume as well as surface pinning takes place simultaneously, both terms contribute to the coercive field. Here we have to distinguish between two limiting cases with respect to the wavelength of the surface roughness and the average wavelength of the volume pinning potential. In the case of statistically

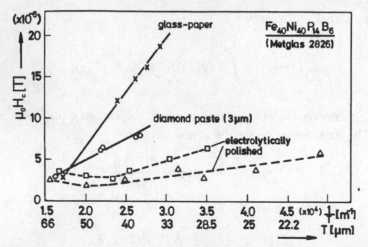

Fig. 7.11. The coercive field as a function of the inverse ribbon thickness after different polishing treatments.

distributed volume pinning centres the average wavelength of the dw pinning potential is of the order of $2\delta_B$. For $\lambda \gg 2\delta_B$ the two contributions to H_c have to be added,

$$H_c = H_c^{surf} + H_c^V, \qquad (7.39)$$

and in the case where $\lambda \approx 2\delta_B$,

$$H_c = \sqrt{\left(H_c^{surf}\right)^2 + \left(H_c^V\right)^2} \qquad (7.40)$$

holds. In general the condition $\lambda \gg 2\delta_B$ applies, i.e., the total coercive field is just the additive of both individual coercive fields.

7.2.5 Nanocrystalline alloys

The conventional crystalline soft magnetic materials are permalloy, sendust, FeSi-transformer steel and amorphous alloys. These materials are characterized by a low magnetocrystalline anisotropy and a low magnetostriction. In order to achieve large susceptibilities in these materials defect structures should be reduced as much as possible. However, some of these materials have a low critical shear stress and consequently plastic deformations are serious sources of deterioration of soft magnetic properties. Only a few crystalline alloys obey the condition of vanishing magnetocrystalline anisotropy and magnetostriction. Therefore, amorphous alloys have become of importance because these alloys a priori are characterized by vanishing anisotropy. In the case of the amorphous alloy $Co_{71}Fe_1Mo_1Mn_4Si_{14}B_9$, with a low magnetostriction of $\lambda_s < 3 \cdot 10^{-7}$, the second condition is also fulfilled.

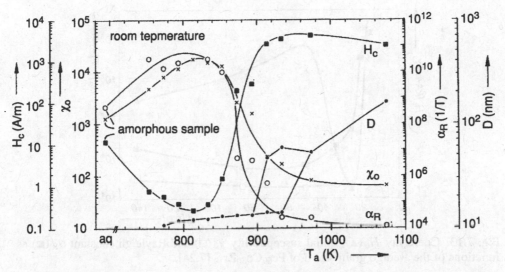

Fig. 7.12. Dependence of H_c, χ_0, and α_R on the annealing temperature T_a and the grain diameter of $Fe_{73.5}Si_{13.5}Nb_3Cu_1B_9$. The measured points at aq correspond to the as-quenched amorphous state [7.23].

Disadvantages of the amorphous alloys are, however, their thermal instability, a low Curie temperature, a quite considerable electrical conductivity, which results in magnetic losses, and a low remanence.

Recently, promising iron-based nanocrystalline alloys have been developed which are more stable above room temperature and have a larger remanence. Most investigations were performed on the typical $Fe_{73.5}Cu_1Nb_3Si_{13.5}B_9$ alloy with a nanocrystalline grain structure produced by the crystallization of amorphous melt-spun ribbons [7.19–7.23]. Another alloy which has been investigated is the $Fe_{60}Co_{30}Zr_{10}$ alloy [7.24]. As an example, Fig. 7.12 shows the dependence of χ_0, H_c and α_R on the annealing temperature, T_a, of the FeCuNbSiB alloy. In addition, the grain size has been determined, and it varies between 10 nm at $T_a = 750$ K and 160 nm at $T_a = 1070$ K. A drastic increase of H_c is observed at $T_a \sim 830$ K connected with a decrease of χ_0 and α_R. A similar behaviour is observed for $Fe_{60}Co_{30}Zr_{10}$, where Fig. 7.13 shows the dependence of χ_0, H_c and α_R on the grain size, d. The corresponding annealing dependence of d is shown in Fig. 7.14.

For an interpretation of the grain size dependence as proposed by Herzer [7.20] the random anisotropy model as developed by Alben *et al.* [7.25] may be applied. Here we consider an ensemble of exchange coupled grains with their easy directions distributed randomly as shown in Fig. 7.15. Due to the exchange coupling between the grains the magnetization cannot abruptly follow the fluctuations of the easy direction. Within a range of the exchange length of K_1, $l_K = \sqrt{A/K_1}$, the

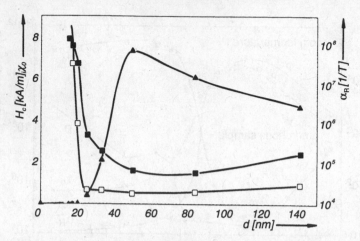

Fig. 7.13. Coercivity H_c (▲), initial susceptibility χ_0 (□) and Rayleigh constant α_R (■) as functions of the average grain size d of $Fe_{60}Co_{30}Zr_{10}$ [7.24].

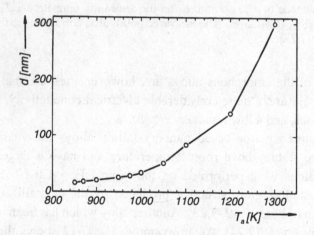

Fig. 7.14. Grain size d as a function of the annealing temperature T_a for annealing times of 10 min at each temperature [7.24].

magnetization rotates into the direction of the easy axis. In the case where $l_K > d$ holds, an averaging over the grains lying within the volume $(l_K)^3$ takes place as demonstrated by Fig. 7.16. The number of grains within this volume is given by $N = (l_K/d)^3$. The resulting effective anisotropy is now given by

$$\langle K \rangle = \frac{K_1}{\sqrt{N}} = K_1 \left(\frac{d}{l_K}\right)^{3/2}. \qquad (7.41)$$

Taking into account that the exchange length l_K is also related self-consistently to $\langle K \rangle$ leads to an effective exchange length

$$l_K^{\mathrm{eff}} = \sqrt{A/\langle K \rangle}. \qquad (7.42)$$

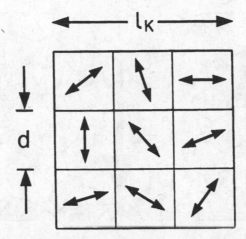

Fig. 7.15. Model of exchange coupled nanocrystalline grains. The arrows present the statistically distributed easy directions. The number of grains within the exchange length l_K is given by $n = (l_K/d)^3$.

Replacing l_K by l_K^{eff} in eq. (7.41) gives finally

$$\langle K \rangle = \frac{K_1^4}{A^3} d^6 = K_1 \left(\frac{d}{\delta_B} \right)^6 \pi^6. \tag{7.43}$$

Equation (7.43) should hold not only for uniaxial crystals but also for cubic crystals because only statistical and scaling arguments have been used. $\langle K \rangle$ may now be used to determine χ_0 and H_c. In the case of an ensemble of randomly distributed single domain grains in a polycrystalline material, the magnetization occurs predominantly by spontaneous nucleation processes. The average coercive field has been determined by Stoner and Wohlfarth for uniaxial grains. Applying similar concepts for the average susceptibility and taking into account the combined magnetization process for misaligned grains – reversible rotation and reversion by nucleation – we obtain for grains with diameters smaller than the wall width

$$H_c = P_c \frac{\langle K \rangle}{J_s} \propto d^6, \qquad \chi_0 = P_\chi \frac{J_s^2}{\mu_0 \langle K \rangle} \propto d^{-6}, \tag{7.44}$$

with the parameters $P_c = 0.96$ and $P_\chi = 0.33$ for uniaxial grains and $P_c = 0.64$ and $P_\chi = 0.33$ for cubic grains with positive K_1. The parameters P_χ were obtained from the initial susceptibilities of individual grains of uniaxial and cubic symmetry, $\chi_0 = (J_s^2/2K_1) \sin^2 \theta$, and averaging over $\sin^2 \theta$, where θ denotes the angle between the easy axis and the applied field ($\langle \sin^2 \theta \rangle = 2/3$). The interesting features of eqs. (7.44) are the fact that they contain only the intrinsic material constants J_s, A and K_1 and the d^6-law for H_c and a d^{-6}-law for χ_0. For the Rayleigh constant a

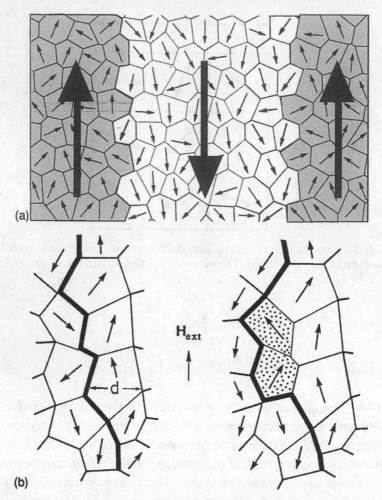

Fig. 7.16. (a) Schematic presentation of a dw in a nanocrystalline matrix. (b) Dw displacements occur either by rotations in individual grains or by the reversal of a cluster of grains which simulates a dw displacement.

d^{-12}-law holds making use of the self-consistency relation (7.8). It is furthermore of interest that the relation

$$\frac{\chi_0 \mu_0 H_c}{J_s} = P_c \cdot P_\chi \simeq 0.32 \tag{7.45}$$

holds for uniaxial and cubic symmetry. A test of the above mentioned power laws for H_c, χ_0 and α_R has been performed for $Fe_{60}Co_{30}Zr_{10}$, which in the nanocrystalline state is composed of the Laves phase $(Fe_{1-x}Co_x)_2Zr$ and the binary alloy α-Fe (Co). Figure 7.17a,b shows logarithmic plots of H_c, χ_0 and α_R as a function d. The experimental results are in nice agreement with the theoretical predictions.

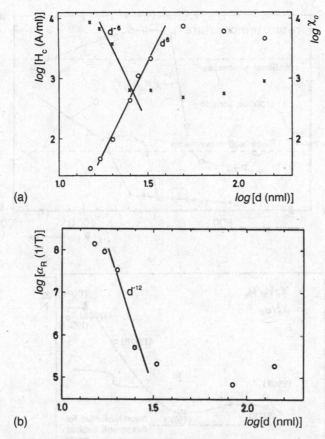

Fig. 7.17. (a) Coercivity H_c (○) and initial susceptibility χ_0 (×) of FeCoZr as a function of grain size, d, showing the predicted d^6 and d^{-6} dependences. (b) Rayleigh constant α_R of FeCoZr as a function of grain size, d, showing the predicted d^{-12} law [7.24].

In order to analyse the type of magnetization processes, the quantity $\chi_0 H_c / M_s$ has been represented in Fig. 7.18a,b for FeCuNbSiB and FeCoZr as a function of the annealing temperature and of the coercive field in the latter case. This quantity changes from $4 \cdot 10^{-3}$ below 840 K to a value of 0.6 for 870 K $< T_a < 950$ K and then decreases to a value 0.2 at higher annealing temperatures. According to eq. (7.45) the larger values are compatible with rotation processes, whereas the lower values below 840 K are compatible with dw displacements. According to eqs. (7.8) the corresponding self-consistency condition for dw displacements is written

$$\frac{\chi_0 H_c}{M_s} = \frac{\sqrt{\pi}}{2} v \frac{\delta_B}{L_3} \left(\ln \frac{L_3}{2\delta_B} \right)^{1/2}, \tag{7.46}$$

where v corresponds to the relative volume and the dws are oriented parallel to the applied field. With the material parameters for Fe_3Si, $K_1 = 800$ J/m^3,

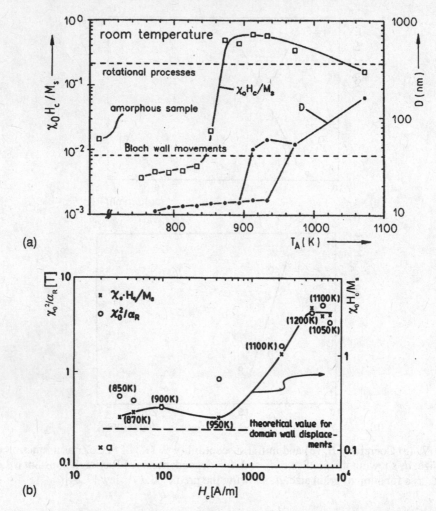

Fig. 7.18. The quantity $\chi_0 H_c/M_s$ as a function of the annealing temperature (grain size) or of the coercive field and comparison with predictions for dw displacements and rotational processes. (a) $Fe_{73.5}Si_{13.5}Nb_3Cu_1B_9$ (T_a). (b) $Fe_{60}Co_{30}Zr_{10}$ (H_c) [7.23, 7.24]

$A = 10^{-11}$ J/m, we obtain for $d = 12$ nm, $\langle K \rangle = 13$ J/m^3 and $\delta_B = 2.76$ μm. The numerical value for $\chi_0 H_c/M_s$ then gives for $L_3 = 200$ μm, $v = 0.4$ a value of of $0.9 \cdot 10^{-2}$, which is of the order of the experimental values for the temperature range $T_a < 850$ K. The corresponding values, $\chi_0 H_c/M_s \sim 0.32$, for rotational magnetization processes are achieved for $T_a > 870$ K. According to these results we may conclude that the magnetization process in the nanocrystalline state of Finemet with $d < l_K^{\text{eff}}$ takes place by dw displacements. A transition to rotational processes takes place for $d > l_K^{\text{eff}}$. At very large grains, $d \gg l_K^{\text{eff}}$, finally the magnetization process is best described by the bowing of dws at grain boundaries,

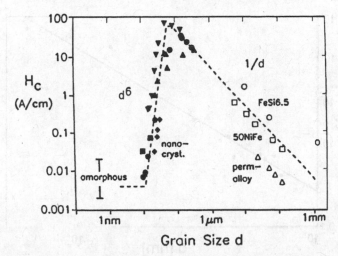

Fig. 7.19. Coercivity, H_c, versus grain size, d, for several soft magnetic alloys showing the amorphous, nanocrystalline and microcrystalline regions according to Herzer [7.27].

where H_c and χ_0 are given by [6.72, 7.26]

$$H_c = \frac{2\gamma_B}{J_s \cos\varphi_0} \cdot \frac{1}{d}, \qquad \chi_0 = \frac{2}{3}\frac{J_s^2 \cos\varphi_0}{4\pi\mu_0\gamma_B}\frac{d^2}{L_3}, \qquad (7.47)$$

with φ_0 the angle of the applied field with respect to the positive magnetization within the domains. The $1/d$-law is observed for grain sizes $d > 100$ nm. In Fig. 7.19, where the results of a large number of nanocrystalline alloys are summarized [7.27], the $1/d$-law for large grain sizes is clearly demonstrated.

If we perform a similar analysis for the FeCoZr alloy, according to Fig. 7.18b for small grain sizes, i.e., $T_a < 950$ K ($d < 25$ nm), $\chi_0 H_c/M_s \sim 0.25$ holds. Here obviously rotational processes govern the magnetization process. Above 950 K the quantity $\chi_0 H_c/M_s$ increases up to a value of 3 at $T_a = 1050$ K ($d > 50$ nm). According to the current theories no conventional magnetization process exists that is compatible with this large value, which is due to the large values of H_c and χ_0 in the upper annealing range. Here it should be noted that in Finemet H_c changes from 0.5 A/m to $2 \cdot 10^3$ A/m, whereas in FeCoZr H_c changes from 20 A/m to $\sim 10^4$ A/m. Similarly χ_0 changes by a factor of 100 in Finemet but only by a factor of 15 in the case of FeCoZr. This indicates that the microstructure of the latter alloy is more complex than in Finemet. A possible explanation would be a change from single domain grains to multidomain grains which leads to a drastic increase of the susceptibility.

So far it has been assumed that in the nanocrystalline alloys no macroscopic induced anisotropy, K_u, exists, which may result from shape anisotropy, induced anisotropy or magnetoelastic anisotropy (see Chapter 10). Applying the same

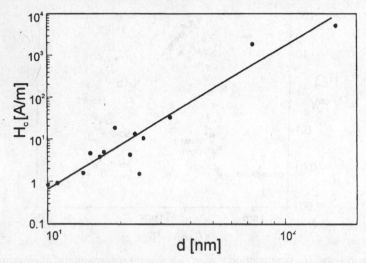

Fig. 7.20. Dependence of the coercivity field on grain diameter d for $Fe_{73.5}Si_{13.5}Nb_3Cu_1B_9$ and $Fe_{73.5}Si_{13.5}Ta_3Cu_1B_9$ with uniaxial induced anisotropy ($H_c = 2.9 \cdot 10^{-4}d^{3.35}$ A/m) according to [7.29].

procedure as in the preceding case by calculating the mean square deviation, Herzer and Varga [7.28] obtained for the total averaged anisotropy

$$\langle K \rangle = \sqrt{K_u^2 + v K_1^2 \left(\frac{d}{l_K^{\text{eff}}}\right)^3}. \tag{7.48}$$

From eq. (7.48) we derive for $K_u < \langle K \rangle$,

$$\langle K \rangle = v^2 K_1 \left(\frac{d}{l_K}\right)^6, \tag{7.49}$$

and for $K_u > \langle K \rangle$,

$$\langle K \rangle = K_u + \frac{1}{2}v\sqrt{K_u K_1}\left(\frac{d}{l_K}\right)^3. \tag{7.50}$$

This means that, in the case of an appreciable macroscopic induced anisotropy, H_c according to eq. (7.50) scales with a d^3-law. Experiments performed by Murillo and González [7.29] in fact point to the validity of eq. (7.50) as demonstrated by Fig. 7.20, where H_s vs. d leads to a $d^{3.35}$-law. Deviations from the d^6-law were also reported by Suzuki *et al.* [7.30] and ascribed to the role of an induced anisotropy.

References

[7.1] Rayleigh, Lord, 1887, *Philos. Mag.* **23**, 225.
[7.2] Barkhausen, H., 1919, *Phys. Z.* **20**, 401.

[7.3] Pfeffer, K.-H., 1967, *Phys. Stat. Sol.* **21**, 857; *ibid.* **19**, 735; **20**, 395; **21**, 837.

[7.4] Kronmüller, H., 1972, *Int. J. Nondestr. Test.* **3**, 315.

[7.5] Kronmüller, H., 1981, *J. Magn. Magn. Mater.* **24**, 159.

[7.6] Kronmüller, H., Fähnle, M., Domann, M., Grimm, R., and Gröger, B., 1979, *J. Magn. Magn. Mater.* **13**, 53.

[7.7] Kronmüller, H., 1965, 'Theorie der plastischen Verformung'. In *Moderne Probleme der Metallphysik*, Vol. 1 (Ed. A. Seeger, Springer-Verlag, Berlin–Heidelberg–New York) p. 126.

[7.8] Köster, E., 1967, *Phys. Stat. Sol.* **19**, 655.

[7.9] Kronmüller, H., 1970, *Z. Angew. Physik* **30**, 9.

[7.10] Kronmüller, H., 1988, *J. Phys. (Paris)* **42**, 1285.

[7.11] Diehl, J., 1965, 'Atomare Fehlstellen und Strahlenschädigung'. In *Moderne Probleme der Metallphysik*, Vol. 1 (Ed. A. Seeger, Springer-Verlag, Berlin–Heidelberg–New York) p. 227.

[7.12] Cohen, M.H., and Turnbull, D., 1959, *J. Chem. Phys.* **31**, 1164.

[7.13] Turnbull, D., 1964, *Physics of Non-Crystalline Solids* (Ed. J.W. Prins, North-Holland).

[7.14] Egami, T., Maeda, K., and Vitek, V., 1980, *Philos. Mag.* **A 41**, 883.

[7.15] Kronmüller, H., 1981, *J. Appl. Phys.* **52**, 1859.

[7.16] Kronmüller, H., Lenge, L., and Habermeier, H.-U., 1984, *Phys. Lett.* **101 A**, 439.

[7.17] Gröger, B., and Kronmüller, H., 1981, *Appl. Phys.* **24**, 287.

[7.18] Gröger, B., and Kronmüller, H., 1980, *J. Magn. Magn. Mater.* **19**, 161.

[7.19] Yoshizawa, Y., Oguma, S., and Yamauchi, K., 1988, *J. Appl. Phys.* **64**, 6044.

[7.20] Herzer, G., 1989, *IEEE Trans. Magn.* **25**, 3327; *ibid.*, 1990, **26**, 1397.

[7.21] Yoshizawa, Y., and Yamauchi, K., 1989, *IEEE Trans. Magn.* **25**, 3324.

[7.22] Reininger, Th., Hofmann, B., and Kronmüller, H., 1992, *J. Magn. Magn. Mater.* **111**, L 220.

[7.23] Hofmann, B., Reininger, Th., and Kronmüller, H., 1992, *Phys. Stat. Sol. (A)* **134**, 247.

[7.24] Guo, H.-Q., Reininger, Th., Kronmüller, H., Rapp, M., and Skumreyev, V.K., 1991, *Phys. Stat. Sol. (A)* **127**, 519; *ibid.*, 1992, *J. Magn. Magn. Mater.* **112**, 287.

[7.25] Alben, R., Becker, J.J., and Chi, M.C., 1978, *J. Appl. Phys.* **49**, 1653.

[7.26] Mayer, A., 1952, *Ann. Phys. (Leipzig)* **11**, 15.

[7.27] Herzer, G., 1995, *Scripta Met.* **33**, 1741.

[7.28] Herzer, G., and Varga, L.K., 2000, *J. Magn. Magn. Mater.* **215-216**, 506.

[7.29] Murillo, N., and González, J., 2000, *J. Magn. Magn. Mater.* **218**, 53.

[7.30] Suzuki, K., Herzer, G., and Cadogan, J.M., 1998, *J. Magn. Magn. Mater.* **177**, 949.

8

Law of approach to ferromagnetic saturation and high-field susceptibility

8.1 Introduction

In 1940–1941 W.F. Brown [2.1, 2.2] published his pioneering work on the theory of micromagnetism of ferromagnetic materials. This continuum theoretical approach originally was developed by Brown to describe the law of approach to saturation (LAFS). Experimentally, the magnetization at high magnetic fields thus far had been described by [2.4]

$$J(H) = J_s - \frac{a_1}{H} - \frac{a_2}{H^2}, \tag{8.1}$$

and it was the aim of Brown [2.2] to find an explanation for the a_1/H term by considering the effects of different types of defects. A more generalized relation for the high-field magnetization was derived later on [8.1], being given by

$$J(H) = J_s - \frac{a_{1/2}}{H^{1/2}} + \frac{a_1}{H} + \frac{a_{3/2}}{H^{3/2}} + \frac{a_2}{H^2} + \frac{a_3}{H^3} + \alpha T\sqrt{H} + \chi_P \cdot \mu_0 H. \tag{8.2}$$

Here the terms $a_{n/2}$ result from intrinsic as well as from extrinsic properties. Intrinsic properties are magnetocrystalline anisotropies, spin waves and the Pauli paramagnetic susceptibility χ_P. Extrinsic properties are stress centres such as point defects, dislocations, grain boundaries and nonmagnetic precipitations. Each of these properties gives rise to a characteristic field dependence:

1. Intrinsic properties
   (a) The magnetocrystalline anisotropy gives rise to a $1/H^2$ dependence. Local fluctuations of the magnetocrystalline anisotropy as in amorphous alloys result in a $1/H^{1/2}$-term at low fields and a $1/H^2$-term at large fields.
   (b) According to Holstein and Primakoff [8.2] the energy gap in the spin wave spectrum due to the Zeeman energy gives rise to the $\sqrt{H}$-term.
   (c) The linear term, $\chi_P \cdot H$, is due to the enhanced Pauli paramagnetism of the band structure.

2. Extrinsic properties

Due to magnetoelastic interactions stress sources produce inhomogeneous spin states. The corresponding deviations from saturation depend on the geometry of the defect, the spatial variation of the stresses and the correlation between the defects. The deviations, $\Delta J_{\text{Def}} = J_s - J(H)$, from saturation are due to the following defects:

(a) Point defects of radius r:

$$\sigma \propto \frac{1}{r^3}; \qquad r_0 < l_H, \, \Delta J_{\text{Def}} = a_{1/2}/H^{1/2} \qquad r_0 > l_H, \, \Delta J_{\text{Def}} = a_2/H^2$$

(b) Straight dislocation dipoles of width D:

$$\sigma \propto \frac{1}{r^2}; \qquad D < l_H, \, \Delta J_{\text{Def}} = a_1/H \qquad D > l_H, \, \Delta J_{\text{Def}} = a_2/H^2$$

(c) Circular dislocation dipoles of radius R:

$$R < l_H; \qquad \Delta J_{\text{Def}} = a_{1/2}/H^{1/2}$$
$$R > l_H; \qquad \Delta J_{\text{Def}} = a_2/H^2$$

(d) Individual straight dislocations:

$$\sigma \propto \frac{1}{r}; \qquad \Delta J_{\text{Def}} = a_2/H^2$$

(e) Nonmagnetic spherical precipitations of radius r_0:

$$r_0 < l_H; \qquad \Delta J_{\text{Def}} = a_{1/2}/H^{1/2}$$
$$r_0 > l_H; \qquad \Delta J_{\text{Def}} = a_{3/2}/H^{3/2}$$

These field dependences are usually derived from the linearized micromagnetic equations as given in Section 3.4. In the following we consider characteristic examples of the LAFS; however, in order to simplify the calculations we use the magnetoelastic coupling energy of eq. (2.77) for isotropic magnetostriction. As a suitable coordinate system we choose, as in Section 3.4, the y-coordinate parallel to the magnetic field direction. In this case the magnetic polarization in field direction can be expressed by the volume averages, $\langle \gamma_i^2 \rangle$, of the squares of the direction cosines,

$$J = J_s \langle \gamma_2 \rangle = J_s \left\{ 1 - \frac{1}{2} (\langle \gamma_1^2 \rangle + \langle \gamma_3^2 \rangle) \right\}, \tag{8.3}$$

where we have made use of the approximation of eq. (3.30). In the special case where we deal with a uniaxial crystal we describe the orientation of the magnetization with respect to the easy axis by the angle φ (see Section 6.2.1) and the orientation of the applied magnetic field by ψ_0. The polarization parallel to H is then given by

$$J = J_s \cos(\psi_0 - \varphi) = J_s \cos\theta \simeq J_s \left(1 - \frac{1}{2} (\psi_0 - \varphi)^2 \right), \tag{8.4}$$

where $\theta = \psi_0 - \varphi$ denotes the angle between J_s and H.

8.2 Approach to saturation in uniaxial crystals

The angle θ between J_s and H for homogeneous magnetization is obtained from the equilibrium condition of eq. (6.34) by replacing $\varphi = \psi_0 - \theta$, which gives for large fields $H > 2K_1/J_s$

$$\theta = \frac{K_1 \sin 2\psi_0}{2K_1 \cos 2\psi_0 + HJ_s}. \tag{8.5}$$

It should be noted that for $\psi_0 = 0$ and $\psi_0 = \pi/2$ the crystal is completely saturated for $H = 0$ and $H = 2K_1/J_s$, respectively. For all intermediate orientations saturation is only possible for $H \to \infty$, and the LAFS for $H > 2K_1/J_s$ is written

$$J(H) = J_s\left(1 - \frac{1}{2}\left(\frac{K_1 \sin 2\psi_0}{2K_1 \cos 2\psi_0 + HJ_s}\right)^2\right). \tag{8.6}$$

Averaging over all possible easy directions in the case of a polycrystalline material, a Taylor series of eq. (8.6) for $H \to \infty$ leads to the well-known result of Danan [8.3]

$$J(H) = J_s\left(1 - \frac{4}{15}\frac{K_1^2}{J_s^2}\frac{1}{H^2}\right), \tag{8.7}$$

i.e. $a_2 = (4/15)(K_1/J_s)^2$.

Taking into account the anisotropy constant K_2 we obtain [8.4]

$$J(H) = J_s\left(1 - \frac{4}{15}\frac{K_1^2}{J_s^2}\frac{1}{H^2} - \frac{64}{105}\frac{K_1 K_2}{J_s^2}\frac{1}{H^2} - \frac{128}{315}\frac{K_2^2}{J_s^2}\frac{1}{H^2}\right). \tag{8.8}$$

If we relax the condition $H \to \infty$ and consider the case $2K_1/J_s < H < \infty$ by averaging eq. (8.6) we find

$$J(H) = J_s\left[1 - \frac{1}{8}\left(-\frac{3}{8} + \frac{3HJ_s - 2K_1}{8K_1}\sqrt{\frac{4K_1}{HJ_s - 2K_1}}\right.\right.$$

$$\left.\left. \times \arctan\sqrt{\frac{4K_1}{HJ_s - 2K_1}}\right)\right], \tag{8.9}$$

which for $H \to \infty$ gives

$$J(H) = J_s\left(1 - \frac{4}{15}\frac{K_1^2}{J_s^2}\frac{1}{H^2} - \frac{16}{105}\frac{K_1^3}{J_s^3}\frac{1}{H^3}\right). \tag{8.10}$$

The additional term in $1/H^3$ was also mentioned by Néel [8.4]. Néel also considered a $1/H$ term due to nonmagnetic spherical inclusions in α-Fe [8.5, 8.6].

8.3 Approach to saturation in cubic crystals

The LAFS for cubic crystals was originally treated by Akulov [8.7], Gans [8.8] and Becker and Döring [2.4]. These authors obtained from the minimization of the Gibbs free energy at large fields, $H > 2K_1/J_s$, using the magnetocrystalline anisotropy given by eq. (2.31):

$$J(H) = J_s\left(1 - \frac{a_2}{H^2} - \frac{a_3}{H^3} \cdots\right),$$

(8.11)

with

$$a_2 = \frac{2K_1^2}{J_s^2}\left[\sum_{i=1}^{3}\beta_i^6 - \left(\sum_{i=1}^{3}\beta_i^4\right)^2\right]$$

$$a_3 = \frac{8K_1^3}{J_s^3}\left[3\sum_{i=1}^{3}\beta_i^8 - 7\sum_{i=1}^{3}\beta_i^4 \cdot \sum_{i=1}^{3}\beta_i^6 + 4\sum_{i=1}^{3}(\beta_i^4)^3\right],$$

(8.12)

where the β_i denote the direction cosines of H with respect to the cubic axes. Averaging over all easy directions in the upper half sphere gives for the isotropic polycrystal

$$J(H) = J_s\left(1 - \frac{8}{105}\frac{K_1^2}{J_s^2}\frac{1}{H^2} - \frac{192}{5005}\frac{K_1^3}{J_s^3}\frac{1}{H^3}\right).$$

(8.13)

Here it should be noted that the results obtained in Sections 8.2 and 8.3 are of an intrinsic nature and may be used to determine the anisotropy constants from high-field magnetization measurements using either single or polycrystals.

8.4 Approach to saturation in the presence of stress sources

8.4.1 Introduction

In this section we consider the LAFS for inhomogeneous spin states which are induced by stress sources such as point defects or dislocations. The micromagnetic background for these calculations is the linearized micromagnetic equations, which for a first-order approximation, where we neglect the terms g_i^k, g_{ij}^k and g_{ij}^{el} in eq. (3.37) as well as the stray field terms $H_{s,i}$, are given by

$$2A\Delta\gamma_i - J_s H_{ext}\gamma_i = g_i^{el}(r).$$

(8.14)

The spatially dependent function $g_i^{el}(r)$ has to be derived from the magnetoelastic coupling energy as given by eq. (2.76) for cubic symmetry or by eq. (2.77) for the isotropic magnetostriction. In general the calculation of $g_i^{el}(r)$ is a bothersome problem because the following three different coordinate systems have to be taken into account:

1. Brown's coordinate system (x, y, z), where the linearization holds with the y-axis parallel to the magnetic field and the x- and z-axes chosen suitably.
2. The coordinate system (x', y', z') adopted to the crystal symmetry, e.g., the magnetoelastic coupling energy of cubic crystals is given in the coordinate system of the cubic axes.
3. The coordinate system (x'', y'', z'') suitable for the stress centres. In the case of dislocations the stress tensor has its simplest form in the coordinate system with the x-axis parallel to the Burgers vector, b, the y-axis parallel to the glide plane normal, n, and the z-axis parallel to the dislocation line, l.

For the following investigation we have to consider the quantity g_i^{el} which acts as a source of inhomogeneities in the direction of J_s. According to eq. (3.32) g_i^{el} is defined as $d\phi'_{el}/d\gamma_i$ where ϕ'_{el} denotes the magnetoelastic coupling energy as given by eq. (2.76) for cubic symmetry. In the first step we have to transform the expression of eq. (2.76) on the coordinate system x'' of the dislocation. Since the magnetoelastic coupling energy as an invariant may be written as (neglecting the term $\varepsilon^Q \cdot\cdot\ c \cdot\cdot\ \varepsilon^Q$)

$$\phi'_{el} = -\sum_{i,j} \sigma''_{ij} \cdot \varepsilon_{ij}^{Q''}, \qquad (8.15)$$

where the components σ''_{ij} and $\varepsilon_{ij}^{Q''}$ refer to the coordinate system x'' of the dislocation. $\varepsilon_{ij}^{Q''}$ is obtained by a transformation of the tensor $\varepsilon^{Q'}$ of eq. (2.53) given by

$$\varepsilon_{ik}^{Q''} = \sum_{l,m} \beta'_{il} \beta'_{km} \varepsilon_{lm}^{Q'}. \qquad (8.16)$$

Here β'_{il} denotes the direction cosines of the x''-system with respect to the cubic system x'. As a further transformation we have to replace the direction cosines γ'_i by the direction cosines γ_i of Brown's coordinate system x according to

$$\gamma'_i = \sum_k \beta_{ik} \gamma_k, \qquad (8.17)$$

where the β_{ik} correspond to the direction cosines between the x-coordinates and the x'-coordinates of the crystal. Inserting eq. (8.17) into $\varepsilon_{ij}^{Q'}$ and eq. (8.16) into eq. (8.15) we obtain

$$g_i^{el} = \left. \frac{\partial \phi'_{el}}{\partial \gamma_i} \right|_{\gamma_{1,3} \ll 1;\ \gamma_2 = 1} = \sum_{k,l} c_{kl}^{(i)} \sigma''_{kl}, \qquad (8.18)$$

with the transformed magnetostriction tensor

$$c_{kl}^{(i)} = \sum_{m,n} \beta_{km}' \beta_{ln}' c_{mn} (\beta_{2m}\beta_{in} + \beta_{im}\beta_{2n}), \qquad (8.19)$$

and

$$c_{mn} = \begin{cases} 3\lambda_{100} & \text{for} \quad m = n \\ 3\lambda_{111} & \text{for} \quad m \neq n. \end{cases}$$

The solution of eq. (8.14) is performed by Fourier transforms $\tilde{\gamma}_i(\mathbf{k})$ and $\tilde{g}_i(\mathbf{k})$ defined as

$$\tilde{\gamma}_i(\mathbf{k}) = \frac{1}{(2\pi)^{3/2}} \int_V \gamma_i(\mathbf{r}) \exp(-i\mathbf{k} \cdot \mathbf{r}) d^3 r,$$

$$\gamma_i(\mathbf{r}) = \frac{1}{(2\pi)^{3/2}} \int_{V_k} \tilde{\gamma}_i(\mathbf{k}) \exp(i\mathbf{k} \cdot \mathbf{r}) d^3 k, \qquad (8.20)$$

with similar expressions for $g_i^{\text{el}}(\mathbf{r})$ and $\tilde{g}_i^{\text{el}}(\mathbf{k})$.

Inserting $\gamma_i(\mathbf{r})$ and $g_i(\mathbf{r})$ as Fourier integrals into eq. (8.14) gives the solution (see also Section 3.4)

$$\tilde{\gamma}_i(\mathbf{k}) = -\frac{1}{2A} \frac{\tilde{g}_i^{\text{el}}(\mathbf{k})}{k^2 + \kappa_H^2} ; \qquad i = 1, 3, \qquad (8.21)$$

with

$$\kappa_H^2 = l_H^{-2} = \frac{\mu_0 M_s H_{\text{ext}}}{2A}. \qquad (8.22)$$

The LAFS is obtained from eq. (8.3) by means of Parseval's theorem [8.9]

$$\langle \gamma^2(\mathbf{r}) \rangle = \frac{1}{V} \int |\tilde{\gamma}^2(\mathbf{k})| d^3 k, \qquad (8.23)$$

which gives for the deviation ΔJ from saturation

$$\Delta J(H) = J_s - J(H) = \frac{J_s}{8VA^2} \int \frac{|g_1^{\text{el}}(\mathbf{k})|^2 + |g_3^{\text{el}}(\mathbf{k})|^2}{(k^2 + \kappa_H^2)^2} d^3 k. \qquad (8.24)$$

In the following we apply eq. (8.24) in order to determine ΔJ for point defects, dislocation loops, straight dislocations and dislocation dipoles.

8.4.2 Isotropic spherical defects

The elastic stress fields of point defects are usually described as elastic dipoles [5.9] which are characterized by a local volume dilatation, ΔV. For a determination of the stress field tensor $\boldsymbol{\sigma}$ we may use Beltrami's equations, eqs. (2.65). Introducing the Fourier transforms of the stress tensor components, $\tilde{\sigma}_{ij}(\boldsymbol{k})$ and those of the quasiplastic incompatibility $\tilde{\eta}_{ij}^Q(\boldsymbol{k})$ we obtain from eq. (2.65) for elastic isotropy

$$\tilde{\sigma}_{ij}(\boldsymbol{k}) = \frac{2G}{k^2}\left\{\tilde{\eta}_{ij}^Q + \frac{1}{1-\nu}\,\tilde{\eta}_{\mathrm{I}}^Q\left(\frac{k_i k_j}{k^2} - \delta_{ij}\right)\right\}, \tag{8.25}$$

where $\tilde{\eta}_{\mathrm{I}}^Q = \sum_{i=1}^{3}\tilde{\eta}_{ii}^Q$, G is the shear modulus, ν is Poisson's ratio and δ_{ij} is the Kronecker symbol. The incompatibility tensor according to eq. (2.66) follows from the local quasiplastic deformation characterizing the point defects. Assuming a spherically symmetric local deformation with a volume expansion ΔV we may write [8.10]

$$\varepsilon^Q = \boldsymbol{I}\,\frac{\Delta V}{\pi^{3/2}\,r_0^3}\,\exp\!\left(-\frac{r^2}{r_0^2}\right), \tag{8.26}$$

where r_0 denotes the effective radius of the point defect and $\boldsymbol{I}$ the unit tensor. With ε^Q we obtain from eq. (2.66)

$$\tilde{\eta}^Q(\boldsymbol{k}) = -\frac{\Delta V}{(2\pi)^{3/2}}(\boldsymbol{k}\boldsymbol{k} - k^2\cdot\boldsymbol{I})\,\exp\!\left(-\frac{k^2 r_0^2}{4}\right). \tag{8.27}$$

Inserting eq. (8.27) into eq. (8.25) now gives

$$\tilde{\sigma}(\boldsymbol{k}) = \frac{2G\Delta V}{(2\pi)^{3/2}k^2}\frac{1+\nu}{1-\nu}(\boldsymbol{k}\boldsymbol{k} - k^2\boldsymbol{I})\,\exp\!\left(-\frac{k^2 r_0^2}{4}\right). \tag{8.28}$$

The distribution of the spontaneous magnetization around a spherical defect is shown schematically in Fig. 8.1. Inserting $\tilde{\sigma}(\boldsymbol{k})$ into $\tilde{g}_i(\boldsymbol{k})$ and using Parseval's theorem of eq. (8.23), we obtain from eq. (8.24)

$$\Delta J(H) = \frac{1}{2}J_{\mathrm{s}}(\langle\gamma_1^2\rangle + \langle\gamma_3^2\rangle) = J_{\mathrm{s}}\frac{G^2(\Delta V)^2}{2(2\pi)^3 A^2 V}\left(\frac{1+\nu}{1-\nu}\right)^2$$

$$\times \int_0^{\infty}\frac{\sum_{ij}^{3}\varepsilon_{ij}\big[k_i k_j - k^2\delta_{ij}\big]^2\Big[\big(c_{ij}^1\big)^2 + \big(c_{ij}^3\big)^2\Big]\cdot\exp(-k^2 r_0^2/2)\,\mathrm{d}^3 k}{k^4(k^2 + \kappa_H^2)^2},$$

$$\tag{8.29}$$

where $\varepsilon_{ij} = 1$ for $i = j$ and $\varepsilon_{ij} = 2$ for $i \neq j$.

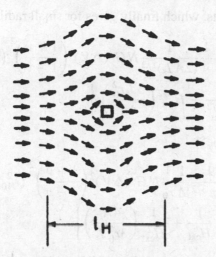

Fig. 8.1. Spin arrangement around a point-like defect. Integral spin deviation $\propto 1/H^{1/2}$.

Performing the integration over the polar coordinates of k-space we find

$$\Delta J(H) = J_s \frac{1}{30\pi^2} \frac{G^2(\Delta V)^2}{A^2 V} \left(\frac{1+\nu}{1-\nu}\right)^2 \sum_{i,j=1}^{3} \left[\left(c_{ij}^1\right)^2 + \left(c_{ij}^3\right)^2\right]\varepsilon'_{ij}$$

$$\times \int_0^\infty \frac{k^2 \exp\left[-k^2 r_0^2/2\right]}{(k^2 + \kappa_H^2)^2} \, dk, \tag{8.30}$$

with $\varepsilon'_{ij} = 4$ for $i = j$ and $\varepsilon'_{ij} = 1$ for $i \neq j$.

The integral in eq. (8.30) gives

$$I = \int_0^\infty \frac{k^2 \exp\left[-k^2 r_0^2/2\right]}{(k^2 + \kappa_H^2)^2} \, dk$$

$$= \frac{\pi}{4\kappa_H}\left(1 + \kappa_H^2 r_0^2\right) \cdot \left(1 - \mathrm{erf}\left(\frac{\kappa_H r_0}{\sqrt{r_2}}\right)\right) \exp\left(\frac{\kappa_H^2 r_0^2}{2}\right) - \frac{\sqrt{2\pi}}{4} r_0. \tag{8.31}$$

The integral which describes the field dependence of the LAFS may be considered for small and large dilatation centres, i.e., $\kappa_H r_0 \gtrless 1$ by Taylor series expansion of $\exp\left(\kappa_H^2 r_0^2/2\right)$ or the asymptotic expansion of $\mathrm{erfc}(x) = 1 - \mathrm{erf}(x)$. Whereas in eq. (8.30) the angular dependence of ΔJ is determined by the quantities $\left(c_{ij}^1\right)^2$ and $\left(c_{ij}^3\right)^2$, the result for a polycrystal is obtained by averaging over the $\left(c_{ij}^{1,3}\right)^2$

orientational components, which finally gives for small radii, $\kappa_H r_0 < 1$,

$$\Delta J = J_s \frac{3}{25\pi} \frac{1}{A} \frac{1}{(2AJ_s)^{1/2}} NG^2(\Delta V)^2 \left(\frac{1+\nu}{1-\nu}\right)^2 (\lambda_{100}^2 + \lambda_{111}^2)$$

$$\times \left\{ \frac{1}{H_{ext}^{1/2}} + H_{ext}^{1/2} \left(\frac{3J_s r_0^2}{4A}\right) \right\}, \tag{8.32}$$

and for large radii, $\kappa_H r_0 \gg 1$,

$$\Delta J = \frac{9}{25\pi^{3/2}} \frac{1}{A} \frac{1}{r_0} NG^2(\Delta V)^2 \left(\frac{1+\nu}{1-\nu}\right)^2 (\lambda_{100}^2 + \lambda_{111}^2)$$

$$\times \left\{ \frac{1}{H_{ext}} + \frac{1}{H_{ext}^2} \left(\frac{4A}{3r_0^2 J_s}\right) \right\}. \tag{8.33}$$

Here we have introduced the defect density $N = 1/V$.

According to eq. (8.32) and eq. (8.33) the field dependences of ΔJ follow an $H^{-1/2}$- or H^{-1}-law depending whether we deal with small or large stress centres. In the case of vacancies or interstitial atoms, the density of these defects and their volume expansion is too small to give a measurable effect in the LAFS. However, in thin amorphous films of composition $Fe_{80}B_{20}$ produced by the sputtering technique on quartz substrates, the LAFS throughout the temperature range from $10\,K$ to $240\,K$ follows an $H^{-1/2}$-law in the field range from $20\,mT$ to $1\,T$ as demonstrated in Fig. 8.2 [8.11]. The deviations from the $H^{-1/2}$-law at large fields, $>1\,T$, is due to the parasusceptibility $\alpha T \cdot \sqrt{H}$ of Holstein and Primakof [8.2].

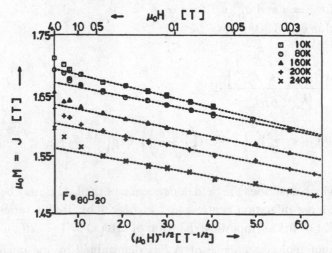

Fig. 8.2. Magnetic polarization $J = \mu_0 M$ versus $1/(\mu_0 H)^{1/2}$ of a sputtered amorphous film of $Fe_{80}B_{20}$. Magnetic field applied parallel to film plane. Film thickness 1500 nm [8.11].

From Fig. 8.2 we may also learn how to determine the spontaneous magnetization in the case of a measurable high-field susceptibility. Extrapolating the linear $H^{-1/2}$-plot from lower fields to larger fields, thus eliminating the paraeffect, gives $J_s(T)$ for $H^{-1/2} \to 0$. A quantitative analysis further shows that the minimum radii of the stress centres are of the order of 1 nm if it is assumed that the maximum defect density is given by $N^{\text{max}} = (3/4\pi) l_H^{-3}$. This condition guarantees that the defects have distances larger than l_H, which is a requirement in order that the $1/H^{1/2}$-law holds. Here it should be noted that the exchange length l_H is of the order of 3 nm for $\mu_0 H = 1$ T.

8.4.3 Dislocation loops

In many cases the incompatibility tensor η is only defined on dislocation lines. In this case the Fourier transform of η is given by [8.10]

$$\tilde{\eta}(k) = i(2\pi)^{-3/2} \int \exp(-i k \cdot r) \ \text{Sym} \{ dl \, [b \times k] \}, \tag{8.34}$$

where b corresponds to the Burgers vector and dl to the line element of the dislocation loop. The symmetric part of the wavy bracket of the dyadic product has to be considered. Dislocation loops are produced by the agglomeration of vacancies or interstitial atoms as produced by high-energy particle irradiation or by cyclic deformation [7.11, 5.8]. For N_L dislocation loops per unit volume distributed equally on the four $\{111\}$ planes of a cubic crystal with radius ρ_0 and Burgers vector b_z perpendicular to the loop's plane we obtain for $\kappa_H \rho_0 < 1$

$$\Delta J = \frac{\pi N_L G^2 b_z^2 \rho_0^4 J_s^{1/2}}{60(2A)^{3/2}} (21.6 \lambda_{100}^2 + 33.6 \lambda_{111}^2) \frac{1}{H_{\text{ext}}^{1/2}}, \tag{8.35}$$

and for large loops, $\kappa_H \rho_0 > 1$,

$$\Delta J = \frac{N_L G^2 b_z^2 \rho_0}{8 J_s} (19.9 \lambda_{100}^2 + 30.6 \lambda_{111}^2) \ln \left(\frac{2\kappa_H \rho_0}{2.2} \right) \frac{1}{H_{\text{ext}}^2}. \tag{8.36}$$

The result for $\kappa_H \rho_0 < 1$ is very similar to the result for the spherical defect of eq. (8.32). However, the result for large fields differs from that of the spherical defect where a $1/H_{\text{ext}}$ law holds, whereas for dislocation loops we find a $1/H_{\text{ext}}^2$-law. As will be shown in the following section, straight dislocation lines also lead to a $1/H^2$-law, i.e., circular dislocation loops magnetically act as straight dislocation lines if the radius ρ_0 is larger than the exchange length l_H.

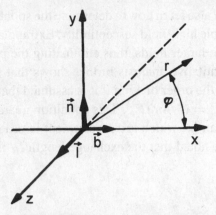

Fig. 8.3. Definition of coordinate systems of an edge dislocation. The polar coordinates in the (x, y)-plane are the radius r and the azimuthal angle φ.

8.4.4 Straight dislocation lines

In the preceding sections the LAFS has been determined by means of Fourier transforms since only knowledge of the quadratic averages of the γ_is is required. In order to obtain insight into the spin arrangement around a dislocation, in this section we determine an explicit solution of the linearized micromagnetic equations, eqs. (3.37), for edge and screw dislocations. A fully explicit solution is only possible by neglecting the stray field terms, i.e., we start with eq. (8.14) and insert eq. (8.18) for the stress components of the dislocations. Choosing a coordinate system as shown in Fig. 8.3, the components of the stress tensor in polar coordinates are given by the following expressions:

1. Edge dislocation (Burgers vector $b \parallel x$-axis)

$$
\sigma_{11}'' = -B_\perp \frac{3 \sin \varphi + \sin 3\varphi}{r}; \qquad \sigma_{22}'' = B_\perp \frac{\sin 3\varphi + \sin \varphi}{r};
$$

$$
\sigma_{33}'' = \nu(\sigma_{11} + \sigma_{22}) = -B_\perp \nu \frac{4 \sin \varphi}{r}; \tag{8.37}
$$

$$
\sigma_{12}'' = B_\perp \frac{\cos \varphi + \cos 3\varphi}{r}; \qquad B_\perp = \frac{Gb}{4\pi(1 - \nu)}.
$$

2. Screw dislocation (Burgers vector $b \parallel z$-axis)

$$
\sigma_{13}'' = B_\odot \frac{\sin \varphi}{r}; \qquad \sigma_{23}'' = B_\odot \frac{\cos \varphi}{r}; \qquad B_\odot = \frac{Gb}{4\pi}. \tag{8.38}
$$

The functions g_i^{el} as defined by eq. (8.18) may now be written as

$$
g_i^{\mathrm{el}}(r, \varphi) = \frac{1}{r} \sum_{n=1, 3} \left(c_{n,\mathrm{s}}^{(i)} \sin n\varphi + c_{n,\mathrm{c}}^{(i)} \cos n\varphi \right), \tag{8.39}
$$

where the constants $c_{n,\mathrm{s,c}}^{(i)}$ in terms of the parameters $c_{kl}^{(i)}$ of eq. (8.19) are given by

$$
\begin{aligned}
c_{1,\mathrm{s}}^{(i)} &= -B_\perp\left(3c_{11}^{(i)} + c_{33}^{(i)} + 4vc_{22}^{(i)}\right); & c_{3,\mathrm{s}}^{(i)} &= B_\perp\left(c_{33}^{(i)} - c_{11}^{(i)}\right); \\
c_{1,\mathrm{c}}^{(i)} &= B_\perp c_{13}^{(i)}; & c_{3,\mathrm{c}}^{(i)} &= B_\perp c_{13}^{(i)}; \qquad (8.40)\\
c_{1,\mathrm{s}}^{(i)} &= B_\odot c_{12}^{(i)}; & c_{3,\mathrm{s}}^{(i)} &= 0; \\
c_{1,\mathrm{c}}^{(i)} &= B_\odot c_{22}^{(i)}; & c_{3,\mathrm{c}}^{(i)} &= 0.
\end{aligned}
$$

In cylindrical coordinates, r, φ, z, with the z-axis parallel to the dislocation line, the micromagnetic equations are written [2.7]

$$
\frac{\partial^2 \gamma_i}{\partial r^2} + \frac{1}{r}\frac{\partial \gamma_i}{\partial r} + \frac{\gamma_i}{r^2} + \frac{1}{r^2}\frac{\partial^2 \gamma_i}{\partial \varphi^2} - \kappa_H^2 \gamma_i = \frac{1}{2A}\, g_i(r, \varphi), \qquad i = 1, 3. \quad (8.41)
$$

The solution of the differential eq. (8.41) is composed of a solution of the homogeneous equation and a particular solution of the inhomogeneous differential equation, giving for the edge dislocation

$$
\begin{aligned}
\gamma_i^\perp &= \frac{1}{2A\kappa_H}\left(K_1(\kappa_H r) - \frac{1}{\kappa_H r}\right)\left(c_{1,\mathrm{s}}^{(i)} \sin\varphi + c_{1,\mathrm{c}}^{(i)} \cos\varphi\right); \\
&\quad + \frac{1}{2A\kappa_H}\left(K_3(\kappa_H r) - \frac{1}{\kappa_H r} + \frac{8}{\kappa_H^3 r^3}\right)\left(c_{3,\mathrm{s}}^{(i)} \sin 3\varphi + c_{3,\mathrm{c}}^{(i)} \cos 3\varphi\right),
\end{aligned}
\tag{8.42}
$$

and for the screw dislocation

$$
\gamma_i^\odot = \frac{1}{2A\kappa_H}\left(K_1(\kappa_H r) - \frac{1}{\kappa_H r}\right)\left(c_{1,\mathrm{s}}^{(i)} \sin\varphi + c_{1,\mathrm{c}}^{(i)} \cos\varphi\right). \tag{8.43}
$$

Here $K_n(\kappa_H r)$ denotes the modified Bessel functions of the second kind. It is of interest to note that for $\kappa_H r \to 0$ the singularities $1/r^n$ vanish and $\gamma_i \to 0$ holds. This is due to the symmetry of the stress components which have singularities at $r \to 0$; however, change their sign and the magnetoelastic coupling energy also changes sign, which results in a zero deviation γ_i. The radial function $K_1(\kappa_H r) - 1/(\kappa_H r)$ and $K_3(\kappa_H r) + 1/(\kappa_H r) + 8/(\kappa_H^3 r^3)$ are presented in Fig. 8.4 showing maxima at $\kappa_H r = 1$ and $\kappa_H r = 3$ for the above functions. The distribution of the spontaneous magnetization around an edge dislocation is shown schematically in Fig. 8.5. The maximum deviation of J_s from the field direction parallel to the x-axis for $\varphi = 0$, $\kappa_H r = 1$ and a field of $\mu_0 H = 0.2$ T in the case of Ni is found to be $\sim 11°$, in agreement with Brown's approximation.

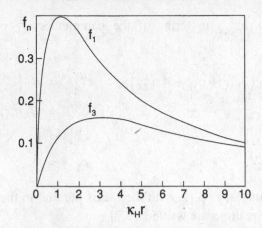

Fig. 8.4. The functions $f_1 = K_1(\kappa_H r) - 1/(\kappa_H r)$ and $f_3 = K_3(\kappa_H r) - 1/(\kappa_H r) + 8/(\kappa_H^3 r^3)$ as a function of $\kappa_H r$.

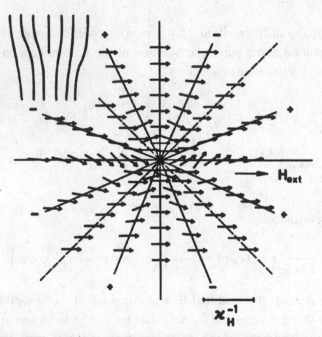

Fig. 8.5. Spin arrangement in the stress field of an edge dislocation for $\mu_0 H = 0.1$ T applied parallel to the Burgers vector. The signs $(+)$ and $(-)$ refer to the sign of the shear stress σ_{12}. Integral spin deviation $\propto 1/H^2$.

At large fields, $\kappa_H r > 1$, the modified Bessel functions, $K_n(\kappa r_0)$, vanish exponentially and the deviations γ_i are determined by the term $(\kappa^2 r)^{-1}$, i.e., the deviations decrease according to the long-range stress field of the dislocation with a $1/r$-law, and deviations from magnetic saturation, which are given by the squares $\langle \gamma_i^2 \rangle$, follow a $1/H^2$-law.

8.4.5 Dislocation groups

In plastically deformed materials in general the dislocations are arranged in groups, networks or in strings, as in the case of cyclically deformed crystals. Figure 8.6 shows some characteristic dislocation arrangements in plastically deformed Ni crystals. Since the stress fields of dislocations have long-range features their special arrangement influences the LAFS sensitively. For a quantitative treatment of ensembles of dislocations, we have to consider instead of $\tilde{g}_i(k)$ the Fourier transform

$$\tilde{g}_i(k) = \sum_{j=1}^{n} \tilde{g}_i^0(k) \, \exp\left(-i k \cdot R_j\right), \tag{8.44}$$

where g_i^0 refers to the individual dislocation and R_j denotes the positions of n other dislocations. In the following we assume groups of n parallel dislocation lines surrounded by other groups at distances of $2R_0$. The mutual interaction between these groups is allowed for by assuming that the surface of a cylinder of radius R_0 is stress free, independent of the special arrangement of dislocations within the groups. As an example we give the Fourier transform of the stress component $\tilde{\sigma}_{xx}$ of a group of dislocations of length L, arranged within a cylinder of radius R_0:

$$\tilde{\sigma}_{xx}(k, R_j) = \frac{iGb}{1-\nu} \frac{1}{(2\pi)^{3/2}} \frac{\sin\frac{1}{2}k_{0z}L}{k_{0z}k_0} (3\sin\varphi - \sin 3\varphi) \tag{8.45}$$

$$\times \sum_{j=1}^{n} \left[\exp(i k \cdot R_j) - J_0(k_0 \cdot R_0)\right],$$

where the k-vector is given in cylindrical coordinates

$$k = \begin{pmatrix} k_0 \cos\phi_0 \\ k_0 \sin\phi_0 \\ k_{0z} \end{pmatrix}, \tag{8.46}$$

and J_0 denotes the Bessel function of zero order. For the other stress components we just have to introduce in eq. (8.45) the angular functions of eqs. (8.37). Inserting the Fourier transforms for σ_{ij} into eq. (8.18) and eq. (8.24) gives, after a lengthy calculation [8.12] for the leading terms of the LAFS, assuming magnetostrictive isotropy [7.4],

$$\Delta J(\kappa_H, R_{ij}) = \frac{9}{8} F(\beta_i) \frac{G^2 b^2 \lambda_s^2 N_0}{J_s} \frac{1}{H^2} \tag{8.47}$$

$$\times \left\{ n \ln(\kappa_H R_0) - 2 + 2 \cdot \left[\sum_{i \neq j}^{n} \ln(R_0/R_{ij}) + K_0(\kappa_H R_{ij})\right]\right\}.$$

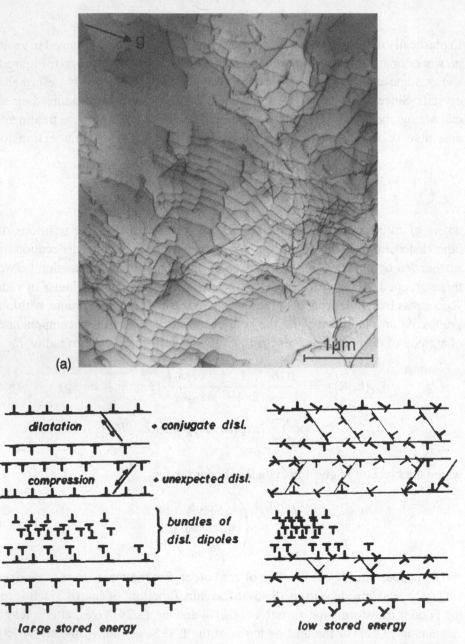

Fig. 8.6. (a) Dislocation arrangement within the primary glide plane of a Ni single crystal plastically deformed up to 50 MPa at 78 K and annealed for 1 h at 660 °C in order to annihilate dipole structures [8.16]. (b) Different types of dislocation arrangements within the primary glide plane due to reactions with secondary glide systems.

(N_0 = number of dislocation groups per unit area, K_0 = modified Bessel function of second kind and order zero), $F(\beta_i)$ describes the anisotropy of the LAFS which for the case of a linearly arranged dislocation group is written [7.6, 8.13, 8.14]

$$F(\beta_i) = 10\beta_1^2(1 - \beta_1^2) + 8\nu\beta_2^2(1 - \beta_2^2)(2\nu - 1) \tag{8.48}$$
$$+ 2\beta_3^2(1 - \beta_3^2) + 2(1 - \beta_2^2 - 6\beta_1^2\beta_3^2) - 16\nu\beta_1^2\beta_2^2,$$

with $\beta_1 \sphericalangle (H, b)$; $\beta_2 \sphericalangle (H, l)$; $\beta_3 \sphericalangle (H, n)$. From eq. (8.47) we may derive two limiting cases of the LAFS. We obtain the result for isolated dislocations by putting $|R_{ij}| = |R_j - R_i| = R_0$, giving for $\kappa_H R_0 \gg 1$

$$\Delta J = \frac{9}{8} F(\beta_i) \frac{G^2 b^2 \lambda_s^2 N_0 n}{J_s \cdot H^2} \{\ln \kappa_H R_0 - 2\}. \tag{8.49}$$

In the case where the dislocation distances are small, $\kappa_H R_{ij} < 1$, we find

$$\Delta J = \frac{9}{8} F(\beta_i) \frac{G^2 (nb)^2 \lambda_s^2 N_0}{J_s \cdot H^2} \{\ln \kappa_H R_0 - 2\}. \tag{8.50}$$

According to this result, the dislocation group acts magnetically as a superdislocation with Burgers vector $B = nb$. The deviation ΔJ is enhanced by a factor n, the number of group dislocations. The above two limiting cases are represented schematically in Fig. 8.7. For $\kappa_H R_{ij} > 1$ the magnetization follows each individual dislocation, whereas for low fields, $\kappa_H R_{ij} < 1$, the dislocation group acts magnetically as a superdislocation. In the general case as described by eq. (8.47) both aspects are included. The short-range behaviour is described by the first term in eq. (8.47), and the long-range stresses are taken into account by the two terms in the bracket.

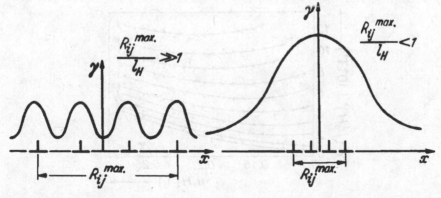

Fig. 8.7. Schematic representation of distribution of magnetization around dislocations for the two limiting cases $\kappa_H R_{ij} > 1$ and $\kappa_H R_{ij} < 1$.

The theoretical results obtained so far have been used widely to analyse dislocation structures of plastically deformed crystals [7.4, 7.6, 8.12, 8.15–8.18]. A comparison with experimental results of the LAFS, often obtained as measurements of the high-field susceptibility,

$$\chi = \mathrm{d}\Delta J/\mathrm{d}(\mu_0 H), \tag{8.51}$$

gives information about three characteristic properties of dislocation structures:

1. The absolute value of χ is related to the dislocation density.
2. The field dependence gives information on the long-range stress fields with $\chi \propto 1/H^3$ for dislocation groups and $\chi = 1/H^2$ for dislocation dipoles (see Section 8.4.6).
3. The anisotropy of χ as given by the function $F(\beta_i)$ gives information on the types of dislocations involved and on their long-range arrangements.

For the analysis of experimental results, either the quantity $\chi \cdot H^3$ is plotted vs. H or the magnetization is plotted vs. $1/H$ or $1/H^2$. The plot $\chi \cdot H^3$ leads to a linear relation

$$\chi H^3 = 2a_2 + a_1 H, \tag{8.52}$$

if eq. (8.1) holds. The plots vs. $1/H$ or $1/H^2$ allow a direct test of the existence of long-range stress fields varying as $1/r$ or of dipole fields varying as $1/H$ (see Section 8.4.6). As an example, Fig. 8.8 presents χH^3 of a Ni single crystal for different degrees of deformation. Here it is evident that the flat curves indicate a predominant $1/H^3$ term to be attributed to long-range stress fields of dislocations. The positive slope of the χH^3 curves at larger fields may be attributed to the so-called paraprocess, which according to eq. (8.2) corresponds to a $1/H^{1/2}$-law for χ.

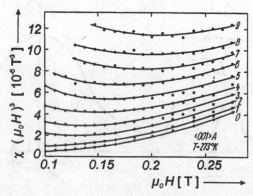

Fig. 8.8. χH^3 of a plastically deformed [100]-Ni single crystal (273 K) for deformations from zero to 115 MPa [8.15].

8.4.6 Dislocation dipoles

In cyclically deformed metals and in amorphous materials the stress fields are predominantly those of dislocation dipoles. This fact is due to the reduced elastic energy of these dipole configurations that the materials adopt if the dislocations are cyclically displaced or if during a rapid quenching process free volumes agglomerate in amorphous alloys. We consider in the following two edge dislocations of opposite Burgers vectors b and $-b$, of length L and a distance D between the dislocations, which are arranged at an angle of $45°$ with respect to the positive Burgers vector. The spin arrangement of such a configuration is shown schematically in Fig. 8.9. Choosing the same coordinate system as in the case of individual edge dislocations with the origin in the middle between the two dipole dislocations, we find for the Fourier transform of the incompatibility tensor only the component $\tilde{\eta}_{33}$, which in a cylindrical coordinate system (k_ρ, k_z, φ) is written

$$\tilde{\eta}_{33} = \tilde{\eta}_{\mathrm{I}} = -\frac{4ibk_y}{(2\pi)^{3/2}} \sin\left[k_\rho D \cos\left(\varphi - \frac{\pi}{4}\right)\right] \frac{\sin\frac{1}{2}k_z L}{k_z}. \tag{8.53}$$

The Fourier transforms of the stress tensor components are given by

$$\tilde{\sigma}_{ik} = \frac{2G}{1-\nu}\frac{1}{k^4}\left(k_i k_k - k^2 \delta_{ij}\right)\tilde{\eta}_{\mathrm{I}}; \qquad i,\, k \neq 3,$$
$$\tilde{\sigma}_{33} = \nu\left(\tilde{\sigma}_{11} + \tilde{\sigma}_{22}\right). \tag{8.54}$$

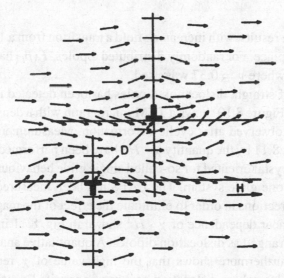

Fig. 8.9. Spin arrangement in the stress field of an edge dislocation dipole. Integral spin deviation $\propto 1/H$.

Inserting eq. (8.54) into eq. (8.18) and calculating ΔJ according to eq. (8.24) gives for isotropic magnetostriction [8.14]

$$\Delta J = \frac{9}{16\pi} F(\beta_i) \frac{G^2 b^2 \lambda_s^2 N_{\text{dip}}}{(1-v)^2 J_s H^2} \left\{ \ln\left(\frac{1}{2}\kappa_H D\right) \right.$$

$$\left. + K_0(\kappa_H D) + \frac{\kappa_H D}{2} K_1(\kappa_H D) + \gamma_0 - 0.5 \right\} \qquad (\gamma_0 = 0.5772). \quad (8.55)$$

From eq. (8.55) we may derive field dependences for two limiting curves. For large exchange lengths, $\kappa_H D < 1$, the dislocation dipole also behaves magnetically as a dipole giving

$$\Delta J = \frac{9}{256\pi} F(\beta_i) \frac{G^2 b^2 D^2 \lambda_s^2 N_{\text{dip}}}{(1-v)^2 A} \cdot \frac{1}{H}. \quad (8.56)$$

In the case of small exchange lengths, $\kappa_H D > 1$, the magnetic power of resolution, $l_H = \kappa_H^{-1}$, is smaller than D and each dislocation is indicated magnetically as an individual dislocation leading to

$$\Delta J = \frac{9}{16\pi} F(\beta_i) \frac{G^2 b^2 \lambda_s^2 N_{\text{dip}}}{(1-v)^2 J_s} \cdot \frac{1}{H^2} \left\{ \ln \frac{1}{2}\kappa_H D + 0.0772 \right\}. \quad (8.57)$$

According to these results, with increasing field a transition from a $1/H$- to a $1/H^2$-dependence takes place. For randomly distributed dipoles, $F(\beta_i)$ has to be replaced by $F(\beta_i) = 1.67$, where $v = 0.37$ was used.

The existence of straight dislocation dipoles has been detected in cyclically deformed materials. Figure 8.10 shows a TEM micrograph with a dense configuration of dislocations as observed after cyclic deformation. Measurements of the LAFS represented in Fig. 8.11 by the quantity $\chi \cdot H^3$ have been performed for a cyclically deformed single crystal oriented for so-called single glide behaviour, i.e., activation of predominantly one glide system. The LAFS has been measured parallel to the four $\langle 111 \rangle$ easy directions in order to eliminate the effect of the magnetocrystalline anisotropy. The linear dependence of $\chi \cdot H^3$ shown in Fig. 8.11 indicates that the dislocations are arranged as dislocation dipoles. A quantitative analysis performed by Diehl [8.19] furthermore shows that the high value of χ requires the existence of superdipoles, where dislocation groups of opposite Burgers vector of 5 to 10 dislocations are arranged within distances smaller than 20 nm.

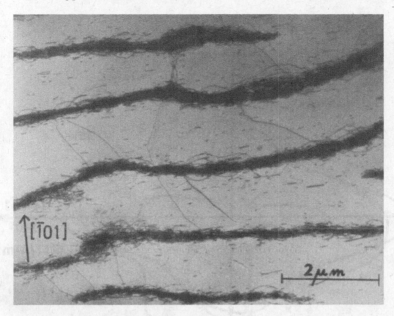

Fig. 8.10. TEM of the primary glide plane of a cyclically deformed Ni single crystal showing dense dipole bundles. Accumulated total plastic strain $a_{pl} = 50$, amplitude of cycle 0.055, flow stress 50 MPa, local dislocation density $N > 10^{11}/\text{cm}^2$.

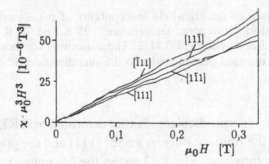

Fig. 8.11. $\chi \cdot H^3$ for a cyclically deformed Ni single crystal measured parallel to the four $\langle 111 \rangle$ directions. Accumulated plastic strain $a_{pl} = 50$, amplitude of cycle 0.075, flow stress 56.6 MPa.

8.4.7 Anisotropy of the high-field susceptibility

Measurements of the anisotropy of the high-field susceptibility, $\chi = d\Delta J/dH$, have been performed for plastically deformed iron single crystals in the $(\bar{1}01)$-, $(1\bar{2}1)$- and the (111)-planes. From an analysis of χ on the basis of the anisotropy function $F_k(\beta_i)$ (eq. (8.48)), the type and the arrangement of dislocations can be deduced. As an example, Fig. 8.12 shows the polar diagram of $\chi(\beta_i)$ as measured

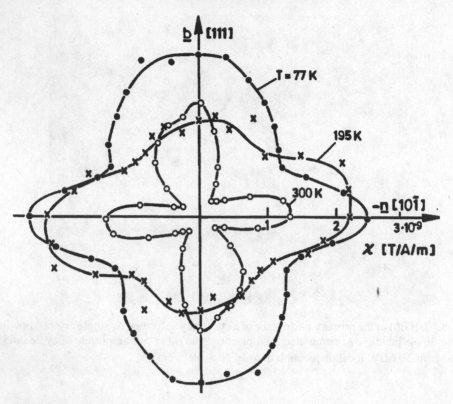

Fig. 8.12. Polar diagram of the high-field susceptibility of middle oriented α-Fe single crystals plastically deformed at room temperature, 195 K and 77 K for a shear strain of $a_{pl} = 0.18$. Applied magnetic field 0.11 T. The anisotropy has been measured in the $(1\bar{2}1)$ plane which is oriented perpendicular to the line direction of the edge disloction [8.17].

within the $(1\bar{2}1)$-plane (perpendicular to l) after deformation at RT, 195 K and 77 K. This plane contains the primary Burgers vector [111] and the primary glide plane normal $[\bar{1}01]$. For comparison, Fig. 8.13 shows the calculated polar diagrams for primary edge and screw dislocations together with the screw dislocations of the conjugate glide system (101) $[\bar{1}11]$. From these results the following conclusions can be drawn:

1. At room temperature mainly dislocations of the primary glide system together with some conjugate screw dislocations are formed.
2. At 195 K mainly screw dislocations of the primary and conjugate glide systems are produced. The ratio of the density of conjugate to primary dislocations is of the order of 0.6.
3. For deformation at 77 K the density of conjugate dislocations increases corresponding to a ratio of the density of conjugate to primary dislocations of 1.6.

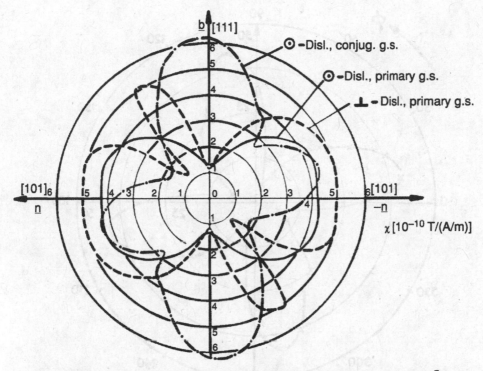

Fig. 8.13. Calculated polar diagram of the high-field susceptibility in the $(1\bar{2}1)$-plane of α-Fe for primary edge dislocations $(\bar{1}01)[\bar{1}11]$, primary screw dislocations and dislocations of the conjugate glide system $(101)\,[\bar{1}11]$. Dislocation density $N = 10^9/\mathrm{cm}^2$, $\mu_0 H = 0.11$ T.

4. The analysis of the amplitude of χ shows that the dislocation are arranged as dislocation groups with $n = 16$–18 dislocations.

The anisotropy of χ has also been investigated by magnetic small angle neutron scattering for plastically deformed α-Fe single crystals [8.20]. For this experiment the glide plane to be investigated was aligned parallel to the plane subtended by the neutron beam and the applied field. By rotating the sample by an angle ϕ around the glide plane normal, the azimuthal dependence of the scattering cross-section for a fixed orientation of the applied field has been measured as a function of the orientation of the dislocation lines with respect to the field direction. The results shown in Fig. 8.14 cannot be explained exclusively by primary dislocations. A quantitative fit of the angular dependence is possible by a dislocation structure composed of 80% primaries and 20% secondaries of Burgers vector $1/2[\bar{1}11]$, as demonstrated by the theoretical result shown in Fig. 8.15. It is of interest to note that the primary and secondary dislocations in this case lead to a Burgers vector $b = [010]$, which in fact has been observed by TEM investigations [8.20].

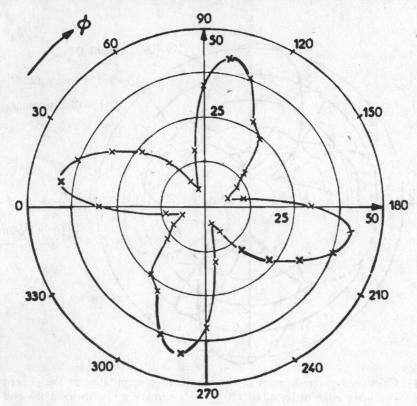

Fig. 8.14. Polar diagram of the neutron scattering cross-section in the $(\bar{1}01)$ plane $((l, b)$-plane) of a middle oriented α-Fe single crystal, plastically deformed at RT up to 28 MPa. For $\phi = 0$ the [111]-direction (glide plane normal) is parallel to the neutron beam [8.20]. Magnetic field 0.11 T [8.20].

8.4.8 Amorphous alloys

8.4.8.1 General remarks

In amorphous alloys deviations of the magnetization from saturation may result from three contributions

$$\Delta J = \Delta J_{sw} + \Delta J_{intr} + \Delta J_{def}, \qquad (8.58)$$

which are attributed to the following origins:

1. ΔJ_{sw} corresponds to the effect of thermally excited spin waves. Each magnon reduces the spontaneous magnetization by $2\mu_B$.
2. ΔJ_{intr} denotes the effect of intrinsic fluctuations of the materials parameters, e.g., local spin density, local magnetic anisotropy energy, local magnetic stray fields resulting from the local fluctuations of the spin density.
3. ΔJ_{def} results from the spin inhomogeneities induced by the magnetoelastic coupling energy around stress sources.

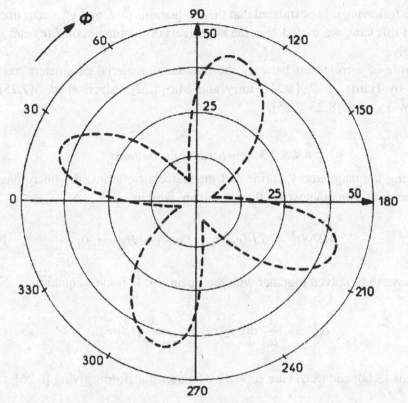

Fig. 8.15. Calculated polar diagram of the neutron scattering cross-section for primary dislocations ($N_{pr} = 2.8 \cdot 10^9/\text{cm}^2$) and secondary dislocations ($N_s = 0.7 \cdot 10^9/\text{cm}^2$) of Burgers vector $b - \frac{1}{2}[\bar{1}1\bar{1}]$. For a quantitative fitting a number of $n = 13$ dislocations per dislocation group has to be assumed. Magnetic field 0.11 T [8.20].

ΔJ_{sw} and ΔJ_{def} may be considered similarly as discussed in the preceding sections. For example, the magnetoelastic coupling energy is described by an isotropic magnetostriction and we may use the results for spherical defects and dislocation dipoles. Accordingly, a very new contribution to ΔJ results from the intrinsic material properties fluctuations. The basic micromagnetic equations for a treatment of these fluctuations are derived from the energy functional of eq. (3.33) with spatially varying material parameters. In particular we consider the following fluctuations:

$$J_2(r) = \langle J_2(r) \rangle + \delta J_2(r),$$
$$g_{ij}^K(r) = \langle g_{ij}^K(r) \rangle + \delta g_{ij}^K(r), \tag{8.59}$$

where $\langle J_2(r) \rangle = J_s$, $\langle \delta J_2(r) \rangle = 0$ and $\langle g_{ij}^K(r) \rangle = 0$. The stray field components $H_{s,i}$ are obtained from Poisson's equation (3.39) by inserting $J_2(r)$ of eq. (8.59).

In the following it is postulated that the fluctuations of J_s and g_{ij}^K occur uncorrelated. In this case, we may determine the effects of fluctuation of $J_s(r)$ and $g_{ij}^K(r)$ separately.

The role of correlations between the fluctuating material parameters has been studied by Harris *et al.* [8.21], Imry and Ma [8.22], Alben *et al.* [7.25] and Chudnovsky *et al.* [8.23–8.25].

8.4.8.2 Magnetostatic fluctuations

Neglecting the magnetocrystalline and magnetoelastic terms, the micromagnetic equations for the magnetostatic fluctuation are written

$$2A(\nabla\gamma_i)^2 + \langle J_2(r)\rangle H_{s,i}(r) - \langle J_2\rangle H\gamma_i = 0, \tag{8.60}$$

which have to be solved together with the linearized Poisson's equation

$$\Delta U = \frac{1}{\mu_0}\, \text{div}\, J_2(r) = \frac{1}{\mu_0}\frac{\partial}{\partial y}(\delta J_2(r)). \tag{8.61}$$

Equations (8.60) and (8.61) are solved by Fourier transforms giving [8.26]

$$\tilde{\gamma}_i(k) = \frac{\kappa_s^2 k_i k^2}{k^4 + k^2(\kappa_H^2 + \kappa_s^2) - k_2^2\kappa_s^2}\frac{\delta\tilde{J}_2(k)}{J_s}. \tag{8.62}$$

Using Parseval's theorem, the volume averages $\langle\gamma_i^2(r)\rangle$ may be determined and we then find

$$\Delta J_{intr} = J_s - \langle J_2(H_{ext})\rangle = \frac{2\pi}{\langle J_s\rangle V}\int_0^\infty |\delta\tilde{J}_2(k)|^2 \cdot G(k, \kappa_H, \kappa_s) \cdot k^2\, dk, \tag{8.63}$$

with

$$G(k, \kappa_H, \kappa_s) = -\frac{3}{2} + \frac{3k^2 + 3\kappa_H^2 + 2\kappa_s^2}{2\kappa_s\sqrt{k^2 + \kappa_H^2 + \kappa_s^2}}\, \text{arctanh}\frac{\kappa_s}{\sqrt{k^2 + \kappa_H^2 + \kappa_s^2}}. \tag{8.64}$$

The function G is shown in Fig. 8.16 as a function of κ_H^2/κ_s^2 for different parameters k_0^2/κ_s^2. The square of the Fourier transform $|\delta\tilde{J}_2(k)|^2$ may be replaced by the

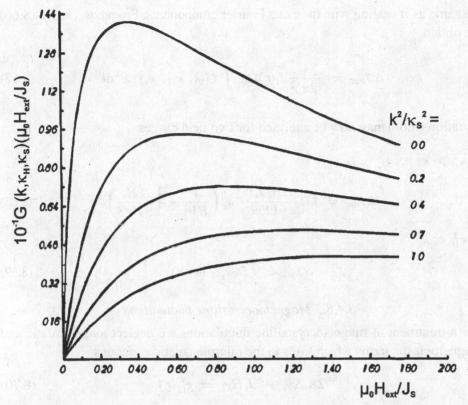

Fig. 8.16. The function $G(k, \kappa_H, \chi_s)$ as a function of κ_H^2/κ_s^2 for different parameters k^2/κ_s^2.

ensemble averaged Fourier components given by

$$\langle |\delta \tilde{J}_2(k)|^2 \rangle = \frac{1}{(2\pi)^3} \int_V \int d^3r \, d^3r' \exp\left(-ik \cdot (r - r')\right) \langle \delta J_2(r) \cdot \delta J_2(r') \rangle.$$

$$(8.65)$$

For a spatially random distribution of magnetic ions, which describes a system with one type of magnetic ions, we have, with $\langle J_2 \rangle = J_s$,

$$\langle \delta J_2(r) \cdot \delta J_2(r') \rangle = \Omega_0 J_s(r) \cdot \delta(r - r'),$$

$$(8.66)$$

where Ω_0 denotes the atomic volume. Naturally, in reality there are corrections to eq. (8.66) because of the finite size of the magnetic ions. Since, however, the assumed correlation length of 3 Å is much smaller than the magnetic correlation lengths ($\kappa_H^{-1} \sim 100 \, \text{Å}$, $\kappa_s^{-1} \sim 25 \, \text{Å}$) the main results of our calculation should be .

the same as if dealing with the exact Fourier components. From eqs. (8.63)–(8.66) we obtain

$$\Delta J_{\text{intr}} = \frac{1}{(2\pi)^2} J_s(\boldsymbol{r})\Omega_0 \int\limits_0^\infty G(k, \kappa_H, \kappa_s) \cdot k^2 \, \mathrm{d}k. \qquad (8.67)$$

Equation (8.67) may now be analysed for two field ranges

1. $\kappa_H^2 \gg \kappa_s^2$

$$\Delta J_{\text{intr}} = J_s\Omega_0 \frac{(2A/J_s)^{1/2}}{120\pi}\kappa_s^4 \left(\frac{1}{H^{1/2}} - \frac{1}{2} \cdot \frac{M_s}{H^{3/2}}\right). \qquad (8.68)$$

2. $\kappa_H^2 \ll \kappa_s^2$

$$\Delta J_{\text{intr}} \simeq J_s\Omega_0\kappa_s^3 \cdot 14 \cdot 10^3. \qquad (8.69)$$

8.4.8.3 Magnetocrystalline fluctuations

For a treatment of magnetocrystalline fluctuations we neglect magnetostatic and magnetoelastic terms which leads to the micromagnetic equations

$$2A \, \Delta\gamma_i - J_s H\gamma_i = g_{i2}^K(\boldsymbol{r}). \qquad (8.70)$$

Solving eq. (8.70) by Fourier transforms leads to

$$\tilde{\gamma}_i(\boldsymbol{k}) = -\frac{1}{2A}\frac{\tilde{g}_{i2}^K(\boldsymbol{k})}{k^2 + \kappa_H^2}, \qquad (8.71)$$

and by applying Parseval's theorem according to eq. (8.23) gives

$$\Delta J_{\text{intr}} = J_s - J(H) = \frac{J_s}{8A^2V} \int \frac{\left|\tilde{g}_{12}^K(\boldsymbol{k})\right|^2 + \left|\tilde{g}_{32}^K(\boldsymbol{k})\right|^2}{\left(k^2 + \kappa_H^2\right)^2} \, \mathrm{d}^3k. \qquad (8.72)$$

The Fourier components $\left|g_{i2}^K\right|^2$ are now replaced by the ensemble averaged components

$$\langle\left|\tilde{g}_{i2}^K(\boldsymbol{k})\right|^2\rangle = \frac{1}{(2\pi)^3} \iint \mathrm{d}^3r \, \mathrm{d}^3r' \, \exp(-\boldsymbol{k} \cdot (\boldsymbol{r} - \boldsymbol{r}'))\langle g_{i2}^K(\boldsymbol{r}) \, g_{i2}^K(\boldsymbol{r}')\rangle. \qquad (8.73)$$

For spatially random fluctuations of $g_{i2}^K(\boldsymbol{r})$ and a long-range anisotropy $g_{i2}(\boldsymbol{r})$ we obtain

$$\langle g_{i2}^K(\boldsymbol{r}) \cdot g_{i2}^K(\boldsymbol{r}')\rangle = \langle\left(\delta g_{i2}^K(\boldsymbol{r})\right)^2\rangle \delta(\boldsymbol{r} - \boldsymbol{r}') + \langle g_{i2}(\boldsymbol{r})\rangle^2. \qquad (8.74)$$

Inserting eq. (8.74) into eq. (8.73) and further into eq. (8.72) gives finally [8.27]

$$
\begin{aligned}
\Delta J_{\text{intr}} &= J_s \Omega_0 \frac{\langle (\delta \tilde{g}_{12}^K)^2 \rangle + \langle (\delta \tilde{g}_{32}^K)^2 \rangle}{16 \pi^2 A^2} \int_0^\infty \frac{k^2 \, dk}{(k^2 + \kappa_H^2)^2} \\
&\quad + J_s \frac{(\langle g_{12} \rangle^2 + \langle g_{32} \rangle^2)}{8 A^2} \frac{1}{\kappa_H^4}, \\
&= J_s \Omega_0 \frac{\langle (\delta \tilde{g}_{12}^K)^2 \rangle + \langle (\delta \tilde{g}_{32}^K)^2 \rangle}{64 \pi^2 A^2} \left(\frac{2A}{J_s} \right)^{1/2} \left(\frac{1}{H^{1/2}} \right) \\
&\quad + \frac{1}{2 J_s} (\langle g_{12} \rangle^2 + \langle g_{32} \rangle^2) \left(\frac{1}{H^2} \right).
\end{aligned}
\tag{8.75}
$$

According to these results the intrinsic deviation ΔJ_{intr} due to magnetocrystalline fluctuations may be written as

$$
\Delta J_{\text{intr}} = \frac{a_{1/2}}{H^{1/2}} + \frac{a_2}{H^2},
\tag{8.76}
$$

where the $1/H^2$-term exists only if long-range anisotropies $\langle k_{i2} \rangle \neq 0$ exist.

8.4.8.4 Magnetoelastic fluctuations

If elastic stresses exist in the amorphous material, the magnetocrystalline coefficients $g_{ij}^K(r)$ become dependent on the stress components $\sigma_{ij}(r)$. In a first approximation we consider the first two terms of a Taylor series of $g_{ij}^K(r)$ in powers of $\sigma_{ij}(r)$, giving

$$
g_{ij}^K(r) = g_{ij}^{K,0}(r) + g_{ij}^{K,1}(r) \, \sigma_{ij}(r),
\tag{8.77}
$$

with

$$
g_{ij}^{K,1}(r) = \frac{\partial g_{ij}^K(r)}{\partial \sigma_{ij}},
\tag{8.78}
$$

and $g_{ij}^{K,0}(r)$ corresponding to spatially varying magnetocrystalline constants used in our previous sections. σ_{ij} either corresponds to external stresses or may result from deviations from an ideal amorphous Bernal structure [8.28]. At present, there exists no unique theory of defects in amorphous alloys. One way to define such defects is based on the type of stress fields induced by inhomogeneities of the atomic structure. Here we may differentiate between three types of stress fields.

1. Short-range stresses varying according to a $1/r^3$-law resulting from point-like spherical defects. These defects are characterized by the volume expansion, ΔV, they produce in the amorphous matrix. Vacancy-type defects are characterized by a volume constriction

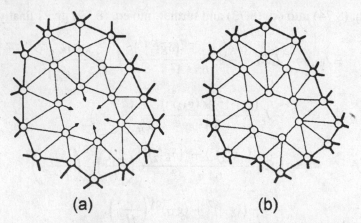

Fig. 8.17. Vacancy-type defects in the amorphous state. (a) unrelaxed network model; (b) relaxed network model.

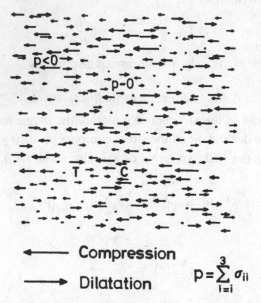

◄——— Compression

———► Dilatation $p = \sum\limits_{i=i}^{3} \sigma_{ii}$

Fig. 8.18. Distribution of the hydrostatic pressure for three successive atomic layers. T, centre of tensile stress; C, centre of compressive stress (according to Egami *et al.* [7.14]).

($\Delta V < 0$) and interstitial-type defects by a volume expansion ($\Delta V > 0$). Figure 8.17 gives a schematic example of a vacancy-type defect in the unrelaxed and the relaxed state. Point-like defects in amorphous structures have been simulated by Egami *et al.* [7.14]. Figure 8.18 represents the hydrostatic pressure for three successive atomic layers of the amorphous structure. Regions of high hydrostatic pressure often appear in pairs as indicated in Fig. 8.18 by T for the tensile region and by C for the compressive region.

2. Short-range stresses varying according to a $1/r^2$-law are generated by defects extending over regions larger than the nearest neighbour distance and extending in a preferred direction. Defects of this kind may be due to an agglomeration of vacancy-type defects described under 1. The agglomerate as the vacancy-type defect corresponds to a region of reduced mass density. Small agglomerates probably form a more or less spherical hole up to 5–10 vacancies, whereas larger agglomerates become unstable and collapse into a kind of 'dislocation loop' similar to vacancy agglomerates in crystals. In the case of circular shapes of the collapsed region the stress field varies according to a $1/r^3$-law. If the collapsed region is linearly extended in one direction we deal with a straight dislocation dipole with a stress field varying as $1/r^2$. Regions of this kind, as shown in Fig. 7.6, are characterized by their width, D, their extension, L, and the width, b_{eff}, of the collapsing region corresponding to an effective Burgers vector of the order of the average nearest-neighbour distance. We may describe the collapsed region by the concept of quasidislocations [2.59] and therefore in the following these defects are denoted as 'quasidislocation dipoles'. Figure 8.19 shows a schematic model of the formation of loops of quasidislocations by the agglomeration of free volumes [8.11] in sputtered amorphous films.

3. Long-range stresses are usually defined as stresses varying according to a $1/r$-law or even varying slowly over distances much larger than the nearest neighbour distance. In crystalline materials long-range stresses are due to dislocations or arrangements of

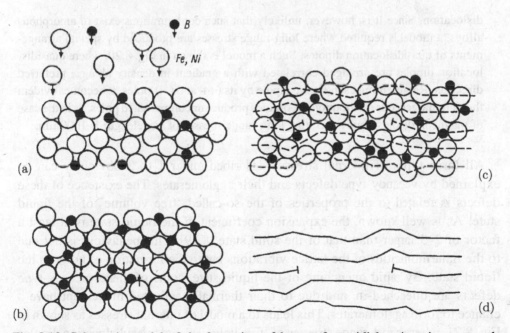

Fig. 8.19. Schematic model of the formation of loops of quasidislocations in sputtered amorphous films. (a) Formation of free volumes during the sputtering process, (b) agglomeration of free volumes, (c) collapse of vacancy-type agglomerates, (d) collapsed region forming a quasidislocation loop showing the missing part of a quasinet plane [8.11].

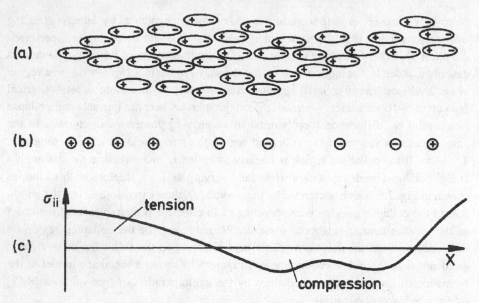

Fig. 8.20. Long-range stresses due to an inhomogeneous arrangement of dislocation dipoles $+/-$. (a) Schematic arrangements of dipoles. (b) Distribution of excess free volumes and of antifree volumes. (c) Spatial variation of the long-range stress component.

dislocation. Since it is, however, unlikely that such configurations exist in amorphous alloys, a model is required where long-range stresses are produced by special arrangements of quasidislocation dipoles. Such a model is shown in Fig. 8.20, where quasidislocation dipoles are arranged correlated with a gradient in density along a preferred direction. Representing each elastic dipole by its $(+)$- and $(-)$-poles, it becomes evident that inhomogeneous distributions of dipoles produce uncompensated poles, as in the case of magnetic dipoles, thus generating long-range stresses of wavelength $1–1000 \, \mu m$.

All three types of internal stresses described under 1 to 3 obviously can be explained by vacancy-type defects and their agglomerates. The existence of these defects is related to the properties of the so-called 'free volume' of the liquid state. As is well known, the expansion coefficient of the liquid state is at least a factor of 2–3 larger than that of the solid state [8.29]. This behaviour is not due to the inharmonicities of the lattice vibrations but to the formation of holes in the liquid state. By rapid quenching of the liquid state, some of these vacancy-type defects are quenched-in, and due to their thermally activated movement have a chance to form agglomerates. This leads to a model of internal stresses as shown in Fig. 8.21, where short-range stresses of 'free volumes' and quasidislocation dipoles are superimposed on the long-range stresses due to gradients of the elastic dipolar defects.

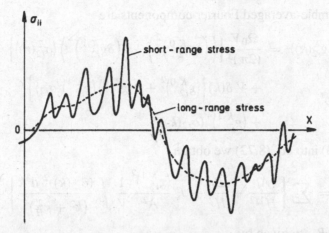

Fig. 8.21. Superposition of short-range stresses due to free volumes and of long-range stresses due to an inhomogeneous arrangement of quasidislocation dipoles.

For a quantitative treatment we consider first the case of an ideally amorphous material with constant stress components σ_{i2}. Only these components have to be considered because in Brown's coordinate system $\boldsymbol{H}$ is applied parallel to the x_2-direction. We then obtain

$$g_{i2}^{K,0}(\boldsymbol{r}) = \delta g_{i2}^{K,0} \quad \text{and} \quad g_{i2}^{K,1}(\boldsymbol{r}) = \delta g_{i2}^{K,1}. \tag{8.79}$$

The ensemble-averaged Fourier components are

$$\langle \tilde{g}_{i2}^2(\boldsymbol{k}) \rangle = \frac{\Omega_0 V}{(2\pi)^3} \left\{ \left\langle \left(\delta g_{i2}^{K,0} \right)^2 \right\rangle + \left\langle \left(\delta g_{i2}^{K,1} \right)^2 \right\rangle \sigma_{i2}^2 \right\}. \tag{8.80}$$

Inserting eq. (8.80) into eq. (8.72) then gives

$$\Delta J_{\text{Def}} = J_s \Omega_0 \sum_{i=1,3} \frac{\left\langle \left(\delta g_{i2}^{K,0} \right)^2 \right\rangle + \left\langle \left(\delta g_{i2}^{K,1} \right)^2 \right\rangle \sigma_{i2}^2}{64\pi A^2} \left(\frac{2A}{J_s} \right)^{1/2} \frac{1}{H^{1/2}}. \tag{8.81}$$

In an ideally amorphous material with vanishing averages, $\langle g_{i2}^{K,0} \rangle = 0$ and long-range stresses, we cannot distinguish between different stress-generating configurations because the fluctuations $\delta g_{i2}^{K,0}$ as well as $\delta g_{i2}^{K,1}$ give rise to a $1/H^{1/2}$-law.

In a second step we consider the case of nonvanishing volume averages of $g_{i2}^{K,0}$ and of $g_{i2}^{K,1}$, i.e., we have

$$g_{i2}^{K,0}(\boldsymbol{r}) = \langle g_{i2}^{K,0} \rangle + \delta g_{i2}^{K,0}(\boldsymbol{r}), \tag{8.82}$$

$$g_{i2}^{K,1}(\boldsymbol{r}) = \langle g_{i2}^{K,1} \rangle + \delta g_{i2}^{K,1}(\boldsymbol{r}).$$

Then the ensemble-averaged Fourier components are

$$\langle \tilde{g}_{i2}^K(\boldsymbol{k}) \rangle = \frac{\Omega_0 V}{(2\pi)^3} \left\{ \left\langle \left(\delta g_{i2}^{K,0} \right)^2 \right\rangle + \left\langle \left(\delta g_{i2}^{K,1} \right)^2 \right\rangle \langle \sigma_{i2}^2(\boldsymbol{r}) \rangle \right\}$$
$$+ V \delta(\boldsymbol{k}) \left\{ \langle g_{i2}^{K,0} \rangle^2 + 2 \langle g_{i2}^{K,0} \rangle \langle g_{i2}^{K,1} \rangle \langle \sigma_{i2} \rangle \right\}$$
$$+ \langle g_{i2}^{K,1} \rangle^2 (\sigma_{i2}(\boldsymbol{k}))^2. \tag{8.83}$$

With eq. (8.83) into eq. (8.72) we obtain

$$\Delta J_{\mathrm{def}} = \sum_i \left\{ \frac{A_i}{H^{1/2}} + \frac{B_i}{H^2} + J_s \frac{\langle g_{i2}^{K,1} \rangle^2}{A^2} \frac{1}{V} \int \frac{(\tilde{\sigma}_{i2}(\boldsymbol{k}))^2 \, d^3k}{(k^2 + \kappa_H^2)^2} \right\}, \tag{8.84}$$

where A_i and B_i are given by

$$A_i = J_s \Omega_0^{-1} \frac{\left\langle \left(\delta g_{i2}^{K,0} \right)^2 \right\rangle + \left\langle \left(\delta g_{i2}^{K,1} \right)^2 \right\rangle \langle \sigma_{i2}^2 \rangle}{64\pi \, A^2} \left(\frac{2A}{J_s} \right)^{1/2}, \tag{8.85}$$

$$B_i = \frac{2}{J_s} \left\{ \langle g_{i2}^{K,0} \rangle^2 + 2 \langle g_{i2}^{K,0} \rangle \langle g_{i2}^{K,1} \rangle \langle \sigma_{i2} \rangle \right\}.$$

In eq. (8.84) the quantity $\langle g_{i2}^{K,1} \rangle^2$ may be identified with $[(3/2)\lambda_s]^2$ which follows from the comparison of eq. (8.77) with eq. (2.77). In eq. (8.84) the first two terms correspond to those of eq. (8.75) modified by the stress-induced terms. The third term in eq. (8.84) corresponds to the conventional effect of stress sources considered in sections 8.4.1 to 8.4.6. We therefore may insert for this term the results obtained in these sections, i.e., a $1/H^{1/2}$ term for small spherical defects or small quasidislocation loops, a $1/H$ term for quasidislocation loops and a $1/H^2$ term for large quasidislocation loops or long-range stresses.

Taking care of all possible contributions, the LAFS of amorphous alloys may be written as

$$\Delta J = \frac{a_{1/2}}{H^{1/2}} + \frac{a_1}{H} + \frac{a_2}{H^2} + C(T) \cdot f_{\mathrm{HP}}(H), \tag{8.86}$$

where the last term corresponds to the spin wave paraeffect. According to the results of Holstein and Primakoff [8.2] the functions $C(T)$ and $f_{\mathrm{HP}}(H)$ are given by

$$C(T) = \left(\mu_B^{3/2} k_B / (2\pi) \right) \left[\mu_0 / (2D_{\mathrm{sp}})^{3/2} \right] \cdot T, \tag{8.87}$$

$$f_{\mathrm{HP}}(H) = 3\sqrt{\mu_0 H} + \sqrt{J_{s0}} \frac{\mu_0 H + J_{s0}}{J_{s0}} \sin^{-1} \left(\frac{J_{s0}}{\mu_0 H + J_{s0}} \right)^{1/2}, \tag{8.88}$$

with k_B = Boltzmann's constant, D_{sp} the spin wave stiffness constant of the spin wave dispersion law $\varepsilon_k = D_{\mathrm{sp}} \cdot k^2$ and J_{s0} the spontaneous magnetization at $T \to 0$.

8.4.8.5 Analysis of experimental results

In rapidly quenched amorphous alloys or sputtered thin films, magnetic saturation in general can be achieved only at very high fields of more than 10 T. This behaviour is different from perfect single crystals where in the main crystallographic directions $\langle 100 \rangle$, $\langle 110 \rangle$ and $\langle 111 \rangle$ saturation is achieved at finite fields, e.g., in Ni at $H_{\text{sat}} = 2K_1/J_s$ in the $\langle 100 \rangle$-direction or for α-Fe at $H_{\text{sat}} = 2K_1/J_s$ in the $\langle 110 \rangle$-direction. As outlined in the preceding Section 8.4.8.4 amorphous alloys contain internal stress sources and intrinsic fluctuations of the material parameters which give rise to deviations from saturation. In addition the paraeffect due to spin wave excitation gives rise to a $\sqrt{H}$ term in the LAFS. According to our above results these different contributions to ΔJ may be determined from an analysis of the field dependence of ΔJ. From the analysis of the temperature dependence of ΔJ and of the paraeffect interesting properties of the thermally excited spin waves can be derived. In the following we discuss the LAFS for some FeNiB-type alloys where a complete analysis has been performed. The magnetic field will be presented as $\mu_0 H$, i.e., in teslas.

$Fe_{40}Ni_{40}P_{14}B_6$ A number of measurements of $J(T, H)$ curves are shown in Fig. 8.22 and Fig. 8.23 for $Fe_{40}Ni_{40}P_{14}B_6$ [8.30, 8.31]. According to Fig. 8.22, $J(T, H)$ in the temperature range from 9 K to 250 K is a linear function of $1/(\mu_0 H)$ for fields $\mu_0 H < 0.3$ T. This field dependence clearly indicates the presence of quasidislocation dipoles treated in detail in Section 8.4.6. A quantitative analysis using the material parameters used previously by Kronmüller *et al.* [7.6] $\left(\kappa_H = 1.44 \cdot 10^6 \text{ cm}^{-1}, \ A = 3.1 \cdot 10^{-12} \text{ J/m}, \ J_s = 0.8 \text{ T}, \ H = 1.6 \cdot 10^5 \text{ A/m}, \ \lambda_s = 11 \cdot 10^{-6} \right)$ gives an upper limit for the dipole density, $N_{\text{dip}} = 6 \cdot 10^{-16} \text{ m}^{-2}$, a width $D = 4.1$ nm and an effective Burgers vector of $b_{\text{eff}} = 0.25$ nm. Whereas in Fig. 8.22 the LAFS follows a $1/H$-law with a steep increase at $H > 0.3$ T, which results from the paraeffect, in Fig. 8.23 the LAFS is characterized by three different ranges. Figure 8.23 presents $J(T, H)$ as a function of $1/H$ for the as-quenched state, an annealed state and for the plastically deformed material for temperatures between 4.2 K and 190 K. At small magnetic fields, $\mu_0 H < 0.04$ T, a $1/H$-law is observed; in an intermediate range, $0.04 \text{ T} < \mu_0 H < 0.3$ T, a $1/H^2$-law dominates; and above 0.3 T a further increase of the polarization takes place. First we consider the $1/H$ and the $1/H^2$ terms. For a more detailed analysis we plot the quantity

$$\chi \cdot (\mu_0 H)^3 = a_1(\mu_0 H) + 2a_2, \tag{8.89}$$

in Fig. 8.24 and Fig. 8.25. For small fields, $\mu_0 H < 0.04$ T, the LAFS is determined by the $1/H$-law and above this field the constancy of χH^3 demonstrates the presence of the $1/H^2$-term. This type of field dependence can be interpreted in terms of

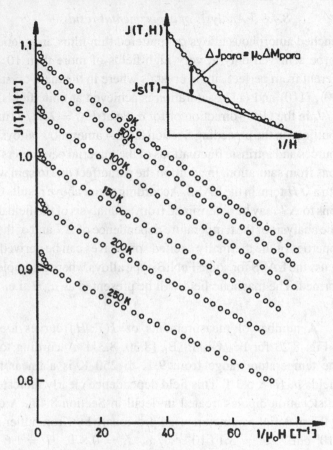

Fig. 8.22. Field dependence of the magnetic polarization $J(T, H)$ of amorphous $Fe_{40}Ni_{40}P_{14}B_6$ as a function of $1/(\mu_0 H)$ for various temperatures [7.6].

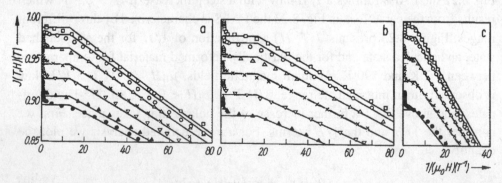

Fig. 8.23. Magnetic polarization $J(T, H)$ of amorphous $Fe_{40}Ni_{40}P_{14}B_6$ as a function of $1/(\mu_0 H)$ as measured for different temperatures and different pre-treatments. (a) As-quenched state. □ 4.2, ○ 40, △ 80, × 116, ▽ 145, ▲ 176, ● 90 K. (b) Annealed 1 h at 643 K. □ 4.2, ○ 40, △ 81, × 120, ▽ 160, ▲ 196 K. (c) Plastically deformed by rolling $\triangle d/d = 0.35\%$. □ 4.2, ○ 52, △ 105, × 151, ● 199 K [8.31].

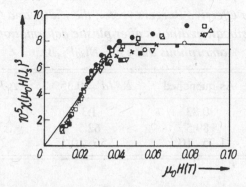

Fig. 8.24. The high-field quantity χH^3 of as-quenched $Fe_{40}Ni_{40}P_{14}B_6$ showing the $1/(\mu_0 H)$-law at low fields and the transition to a $1/(\mu_0 H)^2$-law at larger fields. □ 4.2, ● 40, △ 80, ⨯ 116, ○ 145, ■ 176, ▽ 190 K [8.31].

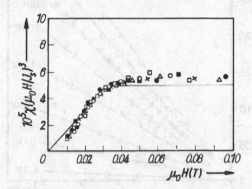

Fig. 8.25. The high-field quantity χH^3 of annealed (1 h at 643 K) $Fe_{40}Ni_{40}B_{20}$ showing the $1/(\mu_0 H)$-law at small fields and the transition to a $1/(\mu_0 H)^2$-law at larger fields. □ 4.2, ● 40, △ 81, ⨯ 120, ■ 160, ○ 196 K [8.31].

quasidislocation dipoles where the exchange length, l_H, is smaller than the dipole width, D_{dip}, for $H < 0.0\,\mathrm{T}$ and as a consequence the $1/H$-law holds. For larger fields $l_H < D_{dip}$ holds and a $1/H^2$-law is observed. A similar behaviour is observed for the plastically deformed specimens. The results of a quantitative analysis are presented in Table 8.1 with a dipole density of $N_{dip} = 1.25 \cdot 10^{12}$ cm^{-2} for the as-quenched specimen and $N_{dip} = 2.66 \cdot 10^{12}$ cm^{-2} for the specimen deformed by 0.6%.

In a second step we consider the paraeffect. Above the field of 0.3 T a systematic deviation from the linear $1/(\mu_0 H)$-plot is observed in Fig. 8.22. Subtracting the linearly extrapolated $1/H$-curves from the measured $J(T, H)$ curves we obtain the contribution ΔJ_{para}. For its analysis we present in Fig. 8.26 the quantity $\Delta J_{para} = J(T, H) - a_1/(\mu_0 H)$ as a function of f_{HP} (eq. (8.88)). From the slopes of the straight lines of this representation we obtain the quantity $C(T)$, which according to Fig. 8.27 is found to be linear in T as predicted by the Holstein–Primakoff theory.

Table 8.1. *The effective Burgers vector, the width, and the density of quasidislocation dipoles after plastic deformation in the amorphous alloy* $Fe_{40}Ni_{40}P_{14}B_6$.

	As-quenched	$\Delta d/d = 0.35\%$	$\Delta d/d = 0.6\%$
b_{eff} (Å)	0.83	1.3	1.4
D (Å)	89.5	62.5	61.5
N_{dip} (cm^{-2})	$1.25 \cdot 10^{12}$	$2.56 \cdot 10^{12}$	$2.66 \cdot 10^{12}$

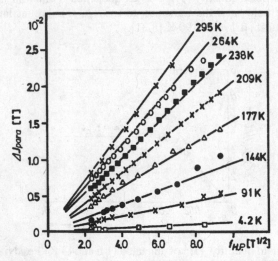

Fig. 8.26. The paraeffect, $\triangle J_{\mathrm{para}} = \triangle J - a_1/H$, for $Fe_{40}Ni_{40}P_{14}B_6$ represented for different temperatures as a function of the Holstein–Primakoff function f_{HP} [7.6].

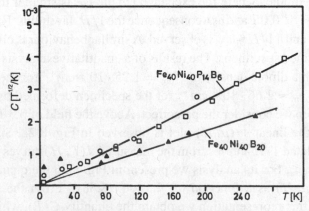

Fig. 8.27. The temperature function $C(T)$ of $Fe_{40}Ni_{40}P_{14}B_6$ and $Fe_{40}Ni_{40}B_{20}$ [7.6].

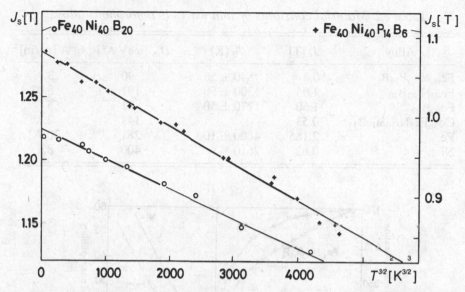

Fig. 8.28. Test of Bloch's $T^{3/2}$-law for amorphous $Fe_{40}Ni_{40}P_{14}B_6$ and $Fe_{40}Ni_{40}B_{20}$ alloys [7.6].

An interesting feature of the analysis of the field dependence is that it opens the possibility to determine the temperature dependence of the spontaneous magnetization, $J_s(T, 0) = J_{s0}$, at zero magnetic field. J_{s0} is obtained by extrapolating the $1/H$- or the $1/H^2$-plot to larger fields, i.e., $1/H^P = 0$. By this procedure the high-field paraprocess is eliminated and we obtain the spontaneous magnetization $J(T, 0)$ at $1/H^P \to 0$. The temperature dependences of $J_s(T)$ are represented in Fig. 8.28 as a function of $T^{3/2}$ which leads to a linear relation

$$J_s(T, 0) = J_{s0}\left(1 - (T/T_0)^{3/2}\right), \tag{8.90}$$

corresponding to Bloch's $T^{3/2}$-law for thermally excited spin waves. The $T^{3/2}$-law holds if the energy dispersion relation of spin waves is a quadratic function of the wave number k,

$$\varepsilon_k = D_{sp}k^2. \tag{8.91}$$

Here D_{sp} denotes the spin wave stiffness constant which is related to the characteristic temperature T_0 of Bloch's $T^{3/2}$-law by

$$T_0 = \left(\frac{M_{s0}}{0.117\,\mu_B}\right)^{2/3}\frac{D_{sp}}{k_B}. \tag{8.92}$$

From the stiffness constant, D_{sp}, the exchange constant, A, is obtained from the relation

$$A(T) = M_s(T, 0)D_{sp}/(2g\mu_B), \tag{8.93}$$

Table 8.2. *Material constants of spin waves in amorphous alloys.*

Alloy	J_s [T]	T_0 [K]	D_{sp} [meV Å^2]	A [pJ/m]
$Fe_{40}Ni_{40}P_{14}B_6$	0.8	840 ± 50	90	3.1
$Fe_{40}Ni_{40}B_{20}$	1.05	1300 ± 50	191	8.07
$Fe_{80}B_{20}$	1.60	1270 ± 50	91	5
$Co_{58}Fe_5Ni_{10}Si_{16}B_{11}$	0.53		143	2.7
Fe	2.185	4600 ± 100	281	20.7
Ni	0.62	2610	400	8.5

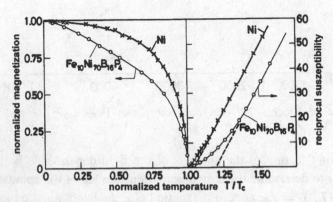

Fig. 8.29. Comparison of the temperature dependence of the spontaneous magnetization and of the parasusceptibility of crystalline Ni and amorphous $Fe_{40}Ni_{40}P_{14}B_6$.

(g = Landé factor). The material parameters derived from the analysis of the paraeffect are summarized in Table 8.2. From the comparison with the material parameters of crystalline metals such as Ni or α-Fe it becomes evident that in amorphous metals the excitation of spin waves requires lower energies which gives rise to a much stronger temperature dependence of the spontaneous magnetization as demonstrated in Fig. 8.29.

$Fe_{40}Ni_{40}B_{20}$ In Fig. 8.30 magnetization curves of $Fe_{40}Ni_{40}B_{20}$ are shown as a function of $1/H$ for as-quenched ribbons [8.32]. For fields below 0.05 T a linear behaviour is observed, whereas at larger fields a flattening takes place indicating a crossover from a $1/H$- to a $1/H^2$-law. As in the case of $Fe_{40}Ni_{40}P_{14}B_6$, at large fields the magnetization curves show a steep increase due to the paraeffect. In order to analyse the crossover phenomenon, in Fig. 8.31 the quantity $\Delta J(\mu_0 H)^2 = a_1\mu_0 H + a_2$ is plotted as a function of H. At low fields a linear behaviour in H is observed, i.e., $a_2 = 0$, whereas at larger fields a constant value a_2 is observed with $a_1 = 0$. As in the case of $Fe_{40}Ni_{40}P_{14}B_6$, this crossover behaviour is attributed

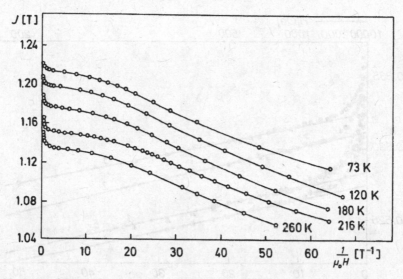

Fig. 8.30. Field dependence of the magnetic polarization $J(T, H)$ of amorphous $Fe_{40}Ni_{40}B_{20}$ as a function of $1/(\mu_0 H)$ for various temperatures [8.32].

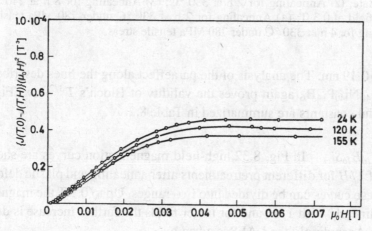

Fig. 8.31. The quantity $\triangle J(\mu_0 H)^2$ as a function of $\mu_0 H$ for $Fe_{40}Ni_{40}B_{20}$ [8.32].

to the decrease of the exchange length with increasing field. Usually one would expect that at low fields the $1/H^2$-term prevails. The contrary behaviour is due to the spin inhomogeneities around quasidislocation dipoles. At small fields, i.e., at a large exchange length, $\kappa_H^{-1} > D_{\mathrm{dip}}$, the dipole acts magnetically as a dipole, whereas at large fields with small exchange lengths, $\kappa_H^{-1} < D_{\mathrm{dip}}$, the dipole acts magnetically as two individual dislocations. A quantitative analysis of the LAFS of $Fe_{40}Ni_{40}B_{20}$ leads to an average dipole width of 24.5 nm, $N_{\mathrm{dip}} = 1.8 \cdot 10^{15}$ m^{-2}

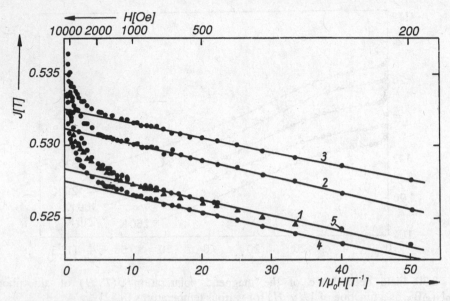

Fig. 8.32. The magnetic polarization of the nearly nonmagnetostrictive amorphous $Co_{58}Fe_5Ni_{10}B_{16}Si_{11}$ as a function of $1/(\mu_0 H)$ for different pretreatments [8.33]: (1) As-quenched state; (2) Annealing for 2 h at 330 °C; (3) Annealing for 8 h at 240 °C under an applied field of 0.3 T; (4) Annealing for 2 h at 330 °C under 780 MPa tensile stress; (5) Annealing for 4 h at 330 °C under 780 MPa tensile stress.

and $b_{eff} = 0.19$ nm. The analysis of the paraeffect along the lines described in the case of $Fe_{40}Ni_{40}P_{14}B_6$ again proves the validity of Bloch's $T^{3/2}$-law (Fig. 8.28). The material constants are summarized in Table 8.2.

$Co_{58}Fe_5Ni_{10}B_{16}Si_{11}$ In Fig. 8.32 high-field magnetization curves are shown as a function of $1/H$ for different pretreatments after annealing and plastic deformation [8.33]. These curves can be divided into two ranges. Up to 0.1 T the magnetization increases linearly with $1/H$ and for larger fields the further increase is due to the paraeffect. Accordingly the LAFS is given by

$$J(T, H) = J_s(T, 0)(1 - a_1/H) + C(T) f_{HP}(H). \qquad (8.94)$$

According to Fig. 8.32 the absolute values of the polarization change slightly from curve to curve, whereas the slopes of the $1/H$-plot remain almost constant. The field dependence of the polarization at large fields above 0.1 T is presented in Fig. 8.33 as a function of $\sqrt{H}$, which is a rather good approximation for f_{HP}. From the linear slopes we may conclude that Bloch's $T^{3/2}$-law holds. The values for D_{sp} and the exchange constant are given in Table 8.2.

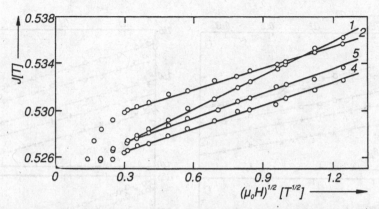

Fig. 8.33. High-field polarization of $Co_{58}Fe_5Ni_{10}B_{16}Si_{11}$ as a function of $(\mu_0H)^{1/2}$ for the as-quenched state and different pretreatments (see Fig. 8.32) [8.33].

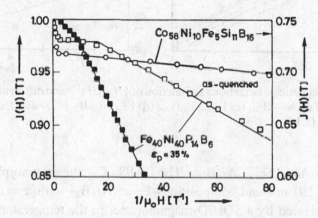

Fig. 8.34. The magnetic polarization as a function of $1/(\mu_0H)$ for different amorphous alloys [8.31].

The parameters a_1 of $Co_{58}Fe_5Ni_{10}B_{16}Si_{11}$ are found to be much smaller than in the case of $Fe_{40}Ni_{40}P_{14}B_6$ and $Fe_{40}Ni_{40}B_{20}$, which has to be attributed to the low magnetostriction of $\sim10^{-7}$ as compared to 10^{-5} for the FeNiB alloys. The existing $1/H$-term due to quasidislocation dipoles may then be attributed to long-range fluctuation of λ_s within the bulk of the sample which may result from long-range chemical fluctuations. The significant difference between magnetostrictive and non-magnetostrictive alloys is demonstrated in Fig. 8.34 where the $1/H$-plots show a large slope for plastically deformed $Fe_{40}Ni_{40}P_{14}B_6$ and only a small slope for $Co_{58}Fe_5Ni_{10}B_{16}Si_{11}$. This behaviour may be taken as clear evidence for the magnetoelastic origin of the $1/H$-term.

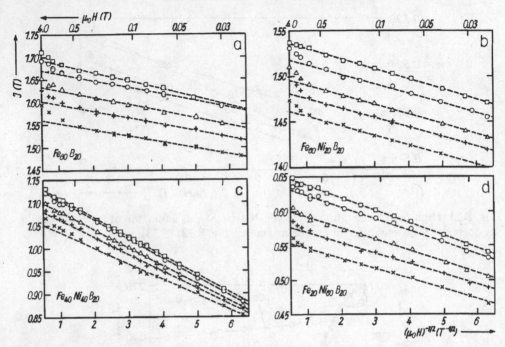

Fig. 8.35. The magnetic polarization as a function of $1/(\mu_0 H)^{1/2}$ and different temperatures. (a) $Fe_{80}B_{20}$; (b) $Fe_{60}Ni_{20}B_{20}$; (c) $Fe_{40}Ni_{40}B_{20}$; (d) $Fe_{20}Ni_{60}B_{20}$; □ $T = 20$, ○ 80, △ = 160, + 200, × 240 K [8.11].

Sputtered amorphous $Fe_{80-x}Ni_x B_{20}$ The LAFS of sputtered amorphous films of thickness 70–1750 nm and compositions $Fe_{80-x}Ni_x B_{20}$ with $x = 0, 20, 40, 60$ has been investigated by a SQUID-magnetometer in the temperature range from 5 to 240 K. According to Fig. 8.35a–d, over a wide range of magnetic fields and temperatures the field dependence may be represented as [8.11]

$$J(H, T) = J_s(0, T)\left(1 - \frac{a_{1/2}}{(\mu_0 H^{1/2})}\right) + C(T) f_{HP}(H). \qquad (8.95)$$

As in the preceding sections, from an extrapolation of the linear $1/(\mu_0 H)$-plot to $\mu_0 H \to \infty$ the temperature dependence of the spontaneous magnetization has been determined. The parameter $a_{1/2}$, the spin wave stiffness constant and the exchange constant A as determined from eqs. (8.92) and (8.93) are represented in Table 8.3.

In the case of amorphous ribbons produced by the melt-spin technique, the LAFS is described by $1/H$- and $1/H^2$-terms indicating the presence of so-called

Table 8.3. *Spin wave and microstructural parameters of sputtered*
$Fe_{80-x}Ni_xB_{20}$ *films [8.11].*

Magnet	J_s [T]	D_{sp} [meV Å^2]	A [pJ/m]	$a_{1/2}$ [$10^{-2}T^{3/2}$], 240 K
$Fe_{80}B_{20}$	1.56	100	5.4	1.34
$Fe_{60}Ni_{20}B_{20}$	1.46	157	7.9	1.05
$Fe_{40}Ni_{40}B_{20}$	1.05	151	5.5	3.46
$Fe_{20}Ni_{60}B_{20}$	0.56	109	2.1	1.47

quasidislocation dipoles. In the present case of sputtered films exclusively a $1/H^{1/2}$-term has been found which may have different origins:

1. Intrinsic fluctuations of the spin moments, eq. (8.68), or intrinsic fluctuations of the magnetocrystalline anisotropy, eq. (8.84).
2. Stress sources of spherical symmetry (eq. (8.32)) or circular quasidislocation loops, eq. (8.35) of extensions $d < l_H$. Average densities of $N \approx 5 \cdot 10^{19}$ cm^{-3} have been determined [8.11] with diameters $d < 4$ nm. The formation of such dipoles during the sputtering process is shown schematically in Fig. 8.19.

Here it should be noted that the $H^{1/2}$-law is only valid if the radii of the defects or the correlation lengths of intrinsic fluctuations are smaller than the exchange length, l_H, and if in addition the distance between the defects is larger than l_H. A quantitative estimation of the LAFS for spin moment fluctuations shows that these give rise to much smaller effects than determined experimentally. Magnetocrystalline fluctuations and internal stress centres of the type mentioned under 1 and 2 in principle are also able to describe the $1/H^{1/2}$-term quantitatively. For a final decision therefore, other criteria have to be considered. One of these criteria is the temperature dependence of the $a_{1/2}$-parameters which is represented in Fig. 8.36 for $Fe_{80}B_{20}$ and $Fe_{40}Ni_{40}B_{20}$ as a function of λ_s^2, thus testing whether we deal with magnetoelastic or magnetocrystalline interactions. From the smooth course of the $a_{1/2}$ vs. λ_s^2 plots in Fig. 8.36 it is obvious that these curves do not extrapolate to $a_{1/2} \to 0$ for $\lambda_s^2 \to 0$. It is therefore evident that $a_{1/2}$ contains contributions from both magnetoelastic and magnetocrystalline interactions.

8.4.9 Nonmagnetic holes and nonferromagnetic precipitations

In a homogeneously magnetized material a nonmagnetic hole acts as a magnetic dipole. If the preferred direction of the magnetization is the z-direction, the magnetic

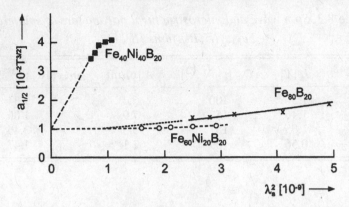

Fig. 8.36. The dependence of the parameter $a_{1/2}$ as a function of λ_s^2 for $Fe_{80}B_{20}$, $Fe_{60}Ni_{20}B_{20}$, $Fe_{40}Ni_{40}B_{20}$.

field within a spherical hole of radius r_0 corresponds to the Lorentz field

$$H_L^{(i)} = \frac{J_s}{3\mu_0} = \frac{1}{3}M_s. \tag{8.96}$$

For the coordinate system shown in Fig. 8.37, the magnetic field components outside the sphere are given by [8.34]

$$H_z^{(o)}(r, \vartheta) = -\frac{M_s}{3}\left(\frac{r_0}{r}\right)^3 (3\cos^2\vartheta - 1),$$

$$H_x^{(o)}(r, \vartheta) = -M_s\left(\frac{r_0}{r}\right)^3 \sin 2\vartheta \cos\varphi, \tag{8.97}$$

$$H_y^{(o)}(r, \vartheta) = -M_s\left(\frac{r_0}{r}\right)^3 \sin 2\vartheta \sin\varphi.$$

It is of interest that at the pole site the z-component of the stray field is given by $H_z^{(o)}(0, 0) = -2H_L^{(i)}$. The inhomogeneous magnetic stray field exerts a magnetic torque $[M_s \times H_s]$ on the magnetization, rotating it into the direction of the surface tangent of the sphere as shown in Fig. 8.38. Accordingly, the magnetic surface charges are reduced and as a consequence the field inside the hole is also reduced. Due to the inhomogeneous magnetization, the exchange energy also becomes important and the whole problem may only be solved on the basis of the micromagnetic equations. A quantitative treatment starts from the surface charges of spherical symmetry, i.e., $J_s(\vartheta, \varphi) = J_s(\pi + \vartheta, \varphi)$. The normal component of J_s at the hole surface may be represented by the following series

$$J_n(\vartheta) = J_s \sum_{n=0}^{\infty} a_n P_{2n+1}(\cos\vartheta), \tag{8.98}$$

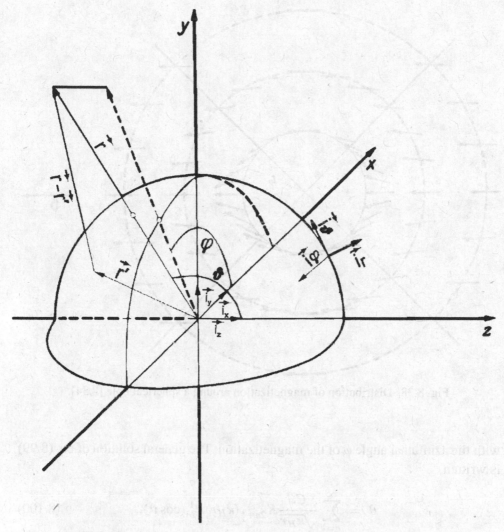

Fig. 8.37. Definition of the coordinate systems for magnetization around a sphere.

where P_{2n+1} denotes the spherical function of first kind and a_n are developing coefficients for which the condition $\sum_{n=0}^{\infty} a_n = 1$ holds, since for $\vartheta = 0$, $J_n = J_s$ and $P_{2n+1}(1) = 1$. The coefficients a_n have to be determined from a minimization of the total Gibbs free energy. For the determination of the distribution of J_s we have to determine the angle θ between J_s and the field direction, which is chosen to be the z-axis. Neglecting the stray field energy terms, the micromagnetic equation for the angle θ in spherical coordinates is given by

$$\Delta\theta - \left(\kappa_H^2 + \frac{1}{r^2 \sin^2 \vartheta}\right)\theta = 0, \tag{8.99}$$

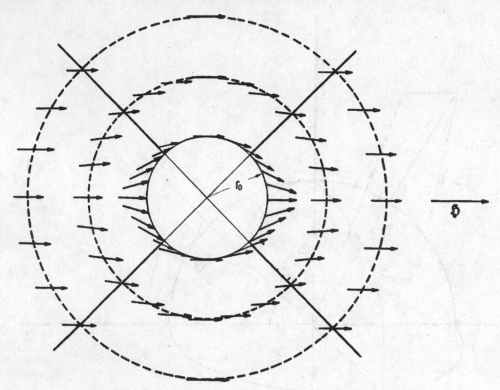

Fig. 8.38. Distribution of magnetization around a spherical hole [8.34].

with the azimuthal angle φ of the magnetization. The general solution of eq. (8.99) is written

$$\theta = \sum_{n=1}^{\infty} \frac{C_n}{\sqrt{\kappa_H r}} K_{2n+\frac{1}{2}}(\kappa_H r) P_{2n}^1(\cos \vartheta). \qquad (8.100)$$

Here $K_{2n+\frac{1}{2}}$ corresponds to the modified Hankel function of order $2n + \frac{1}{2}$ which is given by

$$K_{2n+\frac{1}{2}}(\kappa_H r) = \left(\frac{\pi}{2\kappa_H r}\right)^{1/2} \exp(-\kappa_H r) \sum_{\nu=0}^{2n} \frac{(2n+\nu)!}{\nu!(2n-\nu)!(2\kappa_H r)^\nu}. \qquad (8.101)$$

The coefficients C_n are related to the parameter a_n by the condition for the normal component J_n,

$$J_n = J_s \sum_{n=0}^{\infty} a_n P_{2n+1}(\cos \vartheta) = J_s \cos(\vartheta - \theta). \qquad (8.102)$$

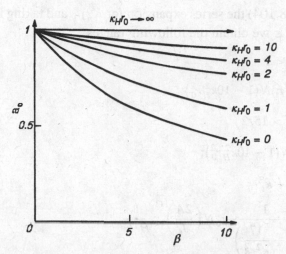

Fig. 8.39. The reduction parameter a_0 as a function of $\beta = \kappa_s^2 r_0^2$ for different parameters $\kappa_H r_0$.

In the approach to saturation θ remains small and the right side of eq. (8.102) can be linearized in θ which is then replaced by the solution (eq. (8.100)). From eq. (8.102) the relations between the a_n and the C_n are obtained and then the parameters a_n may be determined by a minimization of the total magnetic Gibbs free energy. The main results are the following:

1. The field in the centre of the hole corresponds to a reduced Lorentz field

$$H_z^{(i)}(0, 0) = a_0 H_L, \qquad (8.103)$$

where $\varepsilon = 1 - a_0$ is a function of $\beta = \kappa_s^2 r_0^2$ represented in Fig. 8.39 with $\kappa_H r_0$ as a parameter.

2. In the LAFS we have to distinguish between small and large holes $\kappa_H r_0 \gtrless 1$. In addition, the size of the hole in comparison to the exchange length, $l_s = \kappa_s^{-1}$, plays a role. According to previous calculations the deviation, ΔJ, from saturation is given approximately by

$$\Delta J = \frac{5}{6}\pi J_s \varepsilon^2 r_0^3 N \left(1 - \frac{K_{3/2}(\kappa_H r_0) K_{7/2}(\kappa_H r_0)}{K_{5/2}^2(\kappa_H r_0)}\right), \qquad (8.104)$$

with

$$\varepsilon = \frac{2}{15}\beta \qquad \text{for} \quad \kappa_H r_0 < 1, \ \kappa_s r_0 < 1$$

$$\varepsilon = 1 \qquad \text{for} \quad \kappa_H r_0 < 1, \ \kappa_s^2 r_0^2 > 15/2$$

$$\varepsilon = \frac{\kappa_s^2 r_0^2}{17 + 8\kappa_H r_0 + \kappa_s^2 r_0^2} \qquad \text{for} \quad \kappa_H r_0 \gg 1.$$

Inserting into eq. (8.104) the series expansion for $K_{n+\frac{1}{2}}$ and taking into account the different ranges of ε we obtain the following ranges:

1. $\kappa_H r_0 < 1$, $\kappa_s r_0 < 1$

$$\Delta J = \frac{2}{27}\pi J_s \kappa_s^2 r_0^5 N \left(1 - 10\kappa_H^2 r_0^2\right);$$

2. $\kappa_H r_0 < 1$, $\kappa_s^2 r_0^2 > 15/2$

$$\Delta J = \frac{5}{9}\pi J_s r_0^3 N \left(1 - 10\kappa_H^2 r_0^2\right);$$

3. $1 < 8\kappa_H r_0 < 17 + \kappa_s^2 r_0^2$

$$\Delta J = \frac{5}{6}\pi J_s \frac{1}{\left(1 + \dfrac{17}{\kappa_s^2 r_0^2}\right)^2} r_0^2 N \left(\frac{2A}{J_s}\right)^{1/2} \frac{1}{H^{1/2}};$$

4. $8\kappa_H r_0 \gg 17 + \kappa_s^2 r_0^2$

$$\Delta J = \frac{5\pi}{192} J_s r_0^4 N \left(\frac{J_s M_s}{2A}\right)^{1/2} \left(\frac{M_s}{H}\right)^{3/2}.$$

There is a continuous change of the LAFS with increasing magnetic field. In terms of the susceptibility, $\chi = dJ/d(\mu_0 H)$, this means that χ changes from a constant value to a smoothly decreasing value according to a $1/H^{3/2}$-law which finally changes to a $1/H^{5/2}$-dependence. It is of interest that within the framework of micromagnetism no $1/H$- or $1/H^2$-term appears, which is in contrast to Néel's results [8.5, 8.6] which predicted $1/H$- and $1/H^2$-terms within the framework of a purely magnetostatic model. Obviously, the neglect of the role of the exchange energy leads to a stronger decrease of the ΔJ-effect. Measurements of the LAFS have been performed for porous α-Fe which contained carbon inclusions [8.5, 8.6] and for EuS single crystals containing paramagnetic Eu grains [8.35, 8.36]. In the case of the iron–carbon system for fields below 5 T a $1/H$-term has been detected. In EuS (Eu) single crystals the LAFS is written

$$J(T, H) = J_s(T, 0)\left(1 - \frac{a_{1/2}}{(\mu_0 H)^{1/2}}\right) + C(T) f_{HP}(H). \tag{8.105}$$

Since $f_{HP}(H)$ approximately corresponds to $f_{HP} \simeq 3 \cdot \sqrt{H}$, the susceptibility is given by

$$\chi = \frac{dJ}{d(\mu_0 H)} = \frac{J_s(T, 0) a_{1/2}}{2(\mu_0 H)^{3/2}} + \frac{3C(T)}{2\sqrt{\mu_0 H}}. \tag{8.106}$$

The relation

$$\chi \cdot (\mu_0 H)^{3/2} = \frac{J_s(T, 0)}{2}a_{1/2} + \frac{3}{2}C(T)\mu_0 H, \tag{8.107}$$

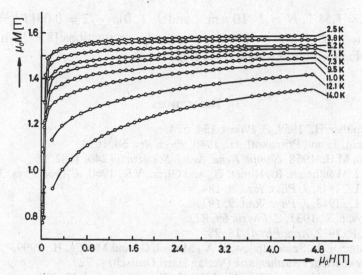

Fig. 8.40. Magnetization curves of an EuS single crystal [8.37].

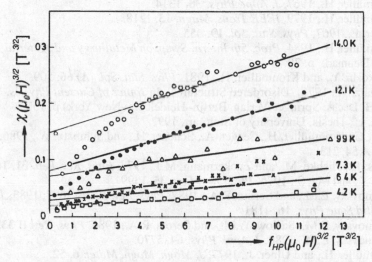

Fig. 8.41. Test of the parameters of EuS by plotting $\chi(\mu_0 H)^{3/2}$ for different temperatures versus the Holstein–Primakoff function $f_{\text{HP}}(\mu_0 H)^{3/2}$ of eq. (8.88) [8.37].

then should lead to a linear plot in $\mu_0 H$. Figure 8.40 shows measurements of the magnetization curves of EuS (Eu) single crystals performed in the temperature range from 2.5 K to 14 K, and Fig. 8.41 presents the plots of $\chi \cdot (\mu_0 H)^{3/2}$ showing over a wide range of fields the linear behaviour predicted by eq. (8.107) and revealing a finite intersection with the ordinate axis, which corresponds to the quantity $J_s(T, 0) a_{1/2}/2$. With the material parameters of EuS, $A = 10^{-13}$ J/m,

$J_s = \mu_0 M_s = 1.54$ T, $N = 2 \cdot 10^{12}$ m^{-3} and $J_s(T, 0)a_{1/2}/2 = 0.05(T^{3/2})$, we find $r_0 = 4.5$ μm in agreement with microstructural investigations [8.37] which show paramagnetic precipitations within the EuS matrix.

References

[8.1] Kronmüller, H., 1959, *Z. Physik* **154**, 574.

[8.2] Holstein, T., and Primakoff, H., 1940, *Phys. Rev.* **58**, 1098.

[8.3] Danan, M.H., 1958, *Compt. Rend. Acad. Sci. (Paris)* **246**, 1182.

[8.4] Néel, L., Pauthenet, R., Rimet, G., and Giron, V.S., 1960, *J. Appl. Phys.* **31**, 27 S.

[8.5] Néel, L., 1948, *J. Phys. Rad.* **9**, 184.

[8.6] Néel, L., 1948, *J. Phys. Rad.* **9**, 193.

[8.7] Akulov, N.S., 1931, *Z. Physik* **69**, 822.

[8.8] Gans, R.,1932, *Ann. Physik* **15**, 28.

[8.9] Bronstein, I.N., Semendjajew, K.A., Musiol, G., and Mühlig, H., 1999, *Taschenbuch der Mathematik* (Verlag Harri Deutsch) p. 727.

[8.10] Seeger, A., and Kronmüller, H., 1960, *J. Phys. Chem. Sol.* **12**, 298.

[8.11] Lenge, L., and Kronmüller, H., 1986, *Phys. Stat. Sol. (A)* **95**, 621.

[8.12] Kronmüller, H., and Seeger, A., 1961, *J. Phys. Chem. Sol.* **18**, 93.

[8.13] Kronmüller, H., 1967, *J. Appl. Phys.* **38**, 1314.

[8.14] Kronmüller, H., 1979, *IEEE Trans. Magn.* **15**, 1218.

[8.15] Köster, E., 1967, *Phys. Stat. Sol.* **19**, 655.

[8.16] Kronmüller, H., 1984, *Proc. 5th Intern. Symp. on Metallurgy and Material Science*, Risø, Denmark, p. 79.

[8.17] Umakoshi, Y., and Kronmüller, H., 1981, *Phys. Stat. Sol. (A)* **66**, 509.

[8.18] Schmatz, W., 1978, 'Disordered Structures'. In *Topics of Current Physics*, Vol. 6, (Ed. H. Dachs, Springer-Verlag, Berlin–Heidelberg–New York) p. 150.

[8.19] Diehl, L., Thesis, University of Stuttgart, 1992.

[8.20] Göltz, G., Kronmüller, H., Seeger, A., Scheuer, H., and Schmatz, W., 1986, *Philos. Mag. A* **54**, 213.

[8.21] Harris, R., Plishke, M., and Zuckermann, M.J., 1973, *Phys. Rev. Lett.* **31**, 160.

[8.22] Imry, Y., and Ma, S., 1975, *Phys. Rev. Lett.* **35**, 1399.

[8.23] Chudnovsky, E.M., and Serota, R.A., 1982, *Phys. Rev. B.* **26**, 2697; 1983, *J. Phys. C: Solid State Phys.* **16**, 4181.

[8.24] Chudnovsky, E.M., Saslow, W.M., and Serota, R.A., 1986, *Phys. Rev.* B **33**, 251.

[8.25] Chudnovsky, E.M., 1988, *J. Appl. Phys.* **64**, 5770.

[8.26] Kronmüller, H., and Ulner, J., 1977, *J. Magn. Magn. Mater.* **6**, 52.

[8.27] Fähnle, M., and Kronmüller, H., 1978, *J. Magn. Magn. Mater.* **8**, 149.

[8.28] Bernal, J.D., 1960, *Nature* **185**, 68; 1964, *Proc. R. Soc., London* **280**, 299.

[8.29] Cohen, M.H., and Turnbull, D., 1959, *J. Chem. Phys.* **31**, 1164.

[8.30] Grimm, H., and Kronmüller, H., 1980, *J. Magn. Magn. Mater.* **15–18**, 1411.

[8.31] Grimm, H., and Kronmüller, H., 1983, *Phys. Stat. Sol. (B)* **117**, 663.

[8.32] Domann, M., Grimm, H., and Kronmüller, H., 1979, *J. Magn. Magn. Mater.* **13**, 81.

[8.33] Vazquez, M., Fernengel, W., and Kronmüller, H., 1989, *Phys. Stat. Sol. (A)* **115**, 547.

[8.34] Kronmüller, H., 1962, *Z. Physik*, **168**, 478.

[8.35] Herz, R., and Kronmüller, H., 1980, *J. Magn. Magn. Mater.* **15–18**, 1299.

[8.36] Chang, H.P.. Herz, R., and Kronmüller, H., 1980, *Appl. Phys.* **22**, 155.

[8.37] Chang, H.P., Herz, R., and Strunk, H.-P., 1979, *Appl. Phys.* **18**, 29.

9

Microstructure and domain patterns

9.1 Origin of domain patterns

The first attempt to calculate the size of Weiss domains was undertaken by Landau and Lifshitz [2.3] and by Kittel and Galt [2.5]. Extended representations of magnetic domains have been given by several authors [2.8, 2.11, 2.14, 3.2, 9.1, 9.2]. As is well known, ferro- and ferrimagnetic materials in the global demagnetized state are subdivided into domains within which the spontaneous magnetization lies parallel to an easy direction. This subdivision into domains is due to the magnetostatic stray field energy which amounts to $\frac{1}{2}\mu_0 N M_s^2$ for a homogeneously magnetized sample, if N is the demagnetization factor for the sample parallel to M_s. In the case of a platelet extending infinitely in two dimensions and of finite thickness, T, $N = 1$ holds and the magnetostatic energy per unit surface area is given by $\phi^{s'} = \frac{1}{2}\mu_0 M_s^2 T = J_s^2 T/2\mu_0$. By means of a division into individual domains, the magnetostatic energy may be drastically reduced. However, due to the formation of domains the total domain wall energy increases. Therefore the domains cannot become arbitrarily small and an equilibrium domain width exists. In general it is possible to reduce the magnetostatic stray field energy by the formation of so-called closure domains at the surfaces of the specimen. Within these closure structures the magnetization M_s orients more or less parallel to the surface depending on the parameter

$$Q = \frac{2K_1}{\mu_0 M_s^2} = \frac{2\mu_0 K_1}{J_s^2}.$$
(9.1)

For $Q > 1$ the closure structure is dominated by the stray field energy, i.e., the demagnetization is less inclined to the surface, whereas for $Q < 1$ a strong inclination takes place and K_1 determines the closure domain energy. Thus far, the domain width D is determined by the total wall energy, ϕ_γ, and the closure

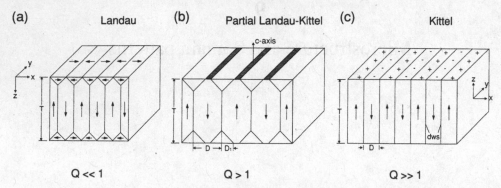

Fig. 9.1. Review of characteristic domain patterns in platelets of varying Q-parameters. (a) Landau structure in $\langle 100 \rangle$-Fe platelets, and in uniaxial platelets with small Q values. (b) Partial Landau–Kittel structure for intermediate Q values. (c) Open Kittel structure.

domain energy, ϕ_{cl}. In the model of laminar domain patterns – examples of which are shown in Fig. 9.1 – the total energy per unit area is composed of two terms

$$\phi_{tot} = \phi_\gamma + \phi_{cl} = \gamma_B \frac{T}{D} + \phi_{cl} \cdot D, \tag{9.2}$$

where the first term ϕ_γ denotes the wall energy per unit area, T the thickness of the specimen, D the domain width, and ϕ_{cl} the energy per unit area of the closure domains. The equilibrium domain width, D, is obtained by minimizing eq. (9.2) with respect to D, giving

$$D = \sqrt{\frac{\gamma_B T}{\phi_{cl}}}. \tag{9.3}$$

Here we have assumed that T is large with respect to the extension of the closure domains. In all domain calculations performed so far the micromagnetic equations (3.1) are fulfilled only approximately, because in general the geometry of the domain pattern is assumed to be known, and then by calculating ϕ_{cl} and using the known values of γ_B, the domain width, D, is obtained from eq. (9.3). This means that the micromagnetic equations are fulfilled for the domain walls; however, for the closure domains only the global energy minimum is determined and not necessarily the torque equation (3.1). This method has been applied to calculations of the three laminar domain patterns shown in Fig. 9.1.

9.2 Laminar domain patterns

9.2.1 Landau structure

The Landau structure according to Fig. 9.1a is characterized by a vanishing mag-netostatic energy because no free poles exist if the surfaces of the sample of cubic symmetry are (100) planes and if $K_1 > 0$ holds. In this case the magnetization in the closure domains is oriented parallel to the surface. In the case of a [100]-platelet of α-Fe with $\langle 100 \rangle$-easy directions, the relevant energy which then determines D is the magnetoelastic energy resulting from the incompatible magnetostrictions of the closure and the bulk domains. The magnetostriction in the closure domain is given by $(3/2)\lambda_{100}$ leading to a magnetoelastic energy per unit area of the crystal surface, which in a first approximation (neglecting the elastic reaction of the bulk) is given by

$$\phi_{cl} = \frac{c_{11}}{2}\left(\frac{3}{2}\lambda_{100}\right)^2 \cdot \frac{D}{2} = \frac{9}{16}c_{11}\lambda_{100}^2 D. \tag{9.4}$$

The equilibrium distance follows from eq. (9.3)

$$D = \frac{4}{3\lambda_{100}}\sqrt{\frac{\gamma_B^{180°}T}{c_{11}}}. \tag{9.5}$$

It is of interest to note that the 90°-walls of the closure domains do not affect D.

If we deal with a uniaxial crystal with the easy direction perpendicular to the platelet, the closure domains of Fig. 9.1a are characterized by the crystal energy and ϕ_{cl} is given by $\phi_{cl} = \frac{K_1}{2}D$ and

$$D = \sqrt{\frac{2\gamma_B^{180°}T}{K_1}}. \tag{9.6}$$

Naturally the assumption of a stray field free closure structure in the case of a uniax-ial crystal is an artificial assumption. As we shall see in Section 9.2.3 actually only a partial reduction of the surface charges takes place. This may also be described by the so-called μ^*-effect, as shown in Section 9.2.4. If the anisotropy energy in the closure domains is determined by internal stresses, K_1 has to be replaced by the magnetoelastic coupling energy, $K_\sigma = (3/2)\lambda_s\sigma$, where σ corresponds to com-pressive stresses in the bulk of the material. From eq. (9.6) the following relations for relevant magnetic parameters may be derived:

$$K_1 = 64AT/D^4, \qquad \gamma_B = 32AT/D^2, \qquad \delta_B = \frac{\pi}{8}D^2/T, \tag{9.7}$$

where the exchange constant A has to be determined from the spin wave stiffness constant according to eq. (8.93).

9.2.2 Kittel structure

The Kittel structure according to Fig. 9.1c is characterized by free poles at the surface alternating in sign. The magnetostatic energy is given by the surface integral (2.48)

$$\phi_s = \frac{1}{2} \int_F U^s \cdot J_s \cdot df, \tag{9.8}$$

where the integral extends over the total surface of the sample and U^s corresponds to the magnetostatic potential at the surface which has to be determined from the Laplace eq. (3.19) under consideration of the boundary conditions of eq. (3.20) and eq. (3.21) [9.3]. Representing the surface charges by a Fourier series,

$$J_n = - \sum_{n=1}^{\infty} \frac{4 J_s}{n\pi} \sin\frac{n\pi x}{D}; \qquad n = 1, 3, 5\ldots, \tag{9.9}$$

the potentials U_i^s and U_o^s inside and outside the specimen are given by

$$U_i^s(x, z) = - \sum_{i=1,3\ldots}^{\infty} \frac{16 J_s D}{n^2 \pi} \exp\left[- \frac{n\pi T}{2D}\right] \sinh\frac{n\pi z}{D} \cdot \sin\frac{n\pi x}{D}, \tag{9.10}$$

$$U_o^s(x, z) = \mp \sum_{i=1,3\ldots} \frac{16 J_s D}{n^2 \pi} \sinh\left(- \frac{n\pi T}{2D}\right) \exp\left[\mp\frac{n\pi z}{D}\right] \cdot \sin\frac{n\pi x}{D},$$

where the upper sign holds for $z > T/2$ and the lower one for $z < T/2$. The magnetostatic energy per unit area of the sample surface is given by

$$\phi_{cl} = \frac{1}{2D} \int_0^{2D} U_i^s\left(x, \frac{L}{2}\right) \cdot J_s(x)\, dx \tag{9.11}$$

$$= \frac{16 D J_s^2}{4\pi^3 \mu_0} \sum_{n=1,3\ldots} \frac{1}{n^3} \{1 - \exp(-n\pi T/D)\}.$$

For $\pi L > 1$, $\phi_{cl} = \frac{1.7}{4\pi \mu_0} J_s^2 D$ holds and the equilibrium distance is given by

$$D = 2 \sqrt{\frac{\pi \mu_0 \gamma_B T}{1.7 J_s^2}} \tag{9.12}$$

9.2.3 Partial Landau–Kittel structure

For a quantitative understanding of the role of the magnetostatic and the magnetocrystalline energy we now consider the case where the Landau and the Kittel closure structures are only developed partially as shown in Fig. 9.1b. As in the preceding case, the domain width is described by D and the width of the closure domain by D_1. The Fourier transform of the surface charges is then given by

$$J_n = - \sum_{n=1,3...} \frac{4J_s}{n\pi} \cos\left(\frac{n\pi D_1}{D}\right) \sin\frac{n\pi x}{D}. \tag{9.13}$$

The solution of the Laplace equation with the boundary conditions leads to the internal potential

$$U_i^s(x,z) = - \sum_{n=1,3...}^{\infty} \frac{16 J_s D}{n^2 \pi \mu_0} \exp\left[-\frac{n\pi T}{2D}\right] \cos\frac{n\pi D_1}{D} \tag{9.14}$$

$$\sinh\frac{n\pi z}{D} \cdot \sin\frac{n\pi x}{D}.$$

The domain parameters D and D_1 are obtained by minimizing the total energy per unit surface, which is given by

$$\phi_t = \gamma_B^{180°} \frac{T}{D} + \frac{2K_1 D_1^2}{D} + \frac{1}{2D} \int U_i^s M_s \cdot df. \tag{9.15}$$

With the parameter $\alpha = D_1/D$ the equilibirium conditions are written

$$D = \left[\frac{\gamma_B^{180°} T}{\sum_{k=0}^{\infty} \frac{4J_s M_s}{\mu_0 \pi^3 (2k+1)^3} \cos^2[(2k+1)\pi\alpha] + 2\alpha^2 K_1}\right]^{1/2}, \tag{9.16}$$

$$\alpha = \sum_{k=0}^{\infty} \frac{J_s^2}{\pi^2 K_1 \mu_0 (2k+1)^3} \sin 2(2k+1)\pi\alpha.$$

The parameter α can be determined as a function of the parameter $\beta = J_s^2/(\mu_0 \pi^2 K_1)$, i.e., $\beta = \frac{2}{\pi^2}\frac{1}{Q}$ from eq. (9.15). As β varies from 0 to ∞, α varies monotonically from 0 to 1/2 as represented in Fig. 9.2. For these two extreme cases we obtain the well-known results of eq. (9.6) and eq. (9.11). The corresponding energies per unit surface area are

$$\phi_{\alpha=0}^s = [1.7 \gamma_B T M_s J_s/\pi]^{1/2}, \tag{9.17}$$

$$\phi_{\alpha=1/2}^s = [2\gamma_B T K_1]^{1/2}.$$

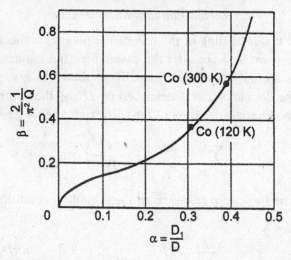

Fig. 9.2. Relative extension D_1/D_0 of the closure domains as a function of the parameter $\frac{2}{\pi^2}\frac{1}{Q}$.

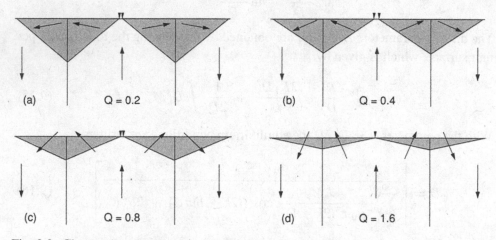

Fig. 9.3. Closure domain patterns with tilted magnetization for different values of Q according to Hubert and Schäfer [2.14].

From the behaviour of the function $\alpha(\beta)$ we may conclude that domain structures with surface charges are more and more favoured with increasing crystal energy, whereas for small crystal energies domain structures free from stray fields become more stable.

Hubert [2.14] has considered a refinement of the partial Landau–Kittel structure by allowing a tilting of the magnetization in the closure domains. With this additional parameter he obtained closure domain structures shown in Fig. 9.3. Here the surface charges are reduced by the tilting of the magnetization, thus avoiding

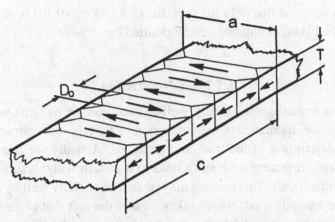

Fig. 9.4. Schematic model of transverse laminar domains with in-plane easy direction for a soft magnetic material with closure domains.

the hard magnetic direction, which leads to an energetically lower configuration as in the case obtained by eq. (9.17) for $\alpha = 1/2$.

9.2.4 Kittel-type structure for in-plane easy direction

The model domain pattern for the in-plane uniaxial anisotropy in a thin platelet is shown in Fig. 9.4. The calculation of the domain width is different from that of the Kittel structure because the magnetic surface charges are limited by D and T. For the model shown in Fig. 9.4 the wall energy per unit length of the platelet with width a and thickness T is given by

$$\Phi_\gamma = \frac{T \cdot a\gamma_B}{D}. \tag{9.18}$$

The magnetostatic stray field energy is that of a periodic arrangement of two rows of alternating surface charges, J_s, within rectangular areas of wavelength $2D$ and a width T. For thin platelets $T < D$ the magnetostatic stray field energy of the closure structure per unit length has been determined as [9.4]

$$\phi_{cl} = \frac{4J_s^2 T^2}{\pi^3 \mu_0} \left\{ \sum_{n=1,3\ldots} \left[\frac{1}{n^2} \ln \frac{4D}{n\pi T} - K_0\left(\frac{n\pi a}{D}\right) - 6 - \gamma \right] \right\}, \tag{9.19}$$

with $\gamma = 0.577$ and K_0 the modified Hankel function of first kind and order zero. The equilibrium width D follows from the minimum $\phi_t = \phi_\gamma + \phi_{cl}$ with respect to D, giving

$$D = \frac{\pi^3 \mu_0 a\gamma_B}{4J_s^2 T}. \tag{9.20}$$

Here it has to be noted that only the term for $n = 1$ in eq. (9.19) is of importance and the modified Hankel function can be omitted for $a \gg D$.

9.2.5 The μ^*-correction

So far we have treated geometrically well-defined domain patterns with discrete orientations of the spontaneous magnetization. This type of treatment gives an approximate description of the true configurations. Actually the transition from the laminar domain pattern with stray field to a pattern without surface charges takes place continuously. This process may be described fairly well by introducing an effective permeability, μ^*, which takes care of the fact that at the surface the magnetization inclines parallel to the surface as discussed by Williams *et al.* and others [9.5, 2.8, 2.14]. This effect is denoted as the μ^* correction, where μ^* is given by

$$\mu^* = 1 + J_s^2/2\mu_0 K_1 = 1 + Q^{-1}, \tag{9.21}$$

if the magnetization rotates by 90° out of the easy axis, and is given by

$$\mu^* = 1 + J_s^2/4\mu_0 K_1, \tag{9.22}$$

if the rotation angle corresponds to 45°. The μ^* correction is introduced into eq. (9.12) by replacing μ_0 by $\mu_0 \cdot \mu^*$ giving for the rotation angle 45°

$$D = 2\sqrt{\frac{\pi\mu_0(1 + J_s^2/4\mu_0 K_1)\gamma_B T}{1.7 J_s^2}}. \tag{9.23}$$

Equation (9.23) describes the transition from the Kittel structure to the Landau structure rather well. For soft magnetic materials, i.e., $J_s^2/4\mu_0 K_1 \gg 1$ and $\mu^* \gg 1$, eq. (9.6) is obtained with a slightly different numerical value (instead of 2, the value $(\pi/1.7)$), and for hard magnetic materials, i.e., $\mu^* = 1$, eq. (9.12) is obtained. The inclination of the magnetization is important for soft magnetic materials with $Q \ll 1$. Naturally the μ^* correction also has to be applied to the in-plane Kittel structure of Fig. 9.4 of soft magnetic materials. Replacing μ_0 by $\mu_0 \cdot \mu^*$ in eq. (9.20) according to eq. (9.22), we find for $Q \ll 1$

$$D = \frac{\pi^3 a \gamma_B}{16 T K_1}. \tag{9.24}$$

The inclination of the magnetization in the surface regions has been numerically treated by Hubert [2.14] leading to configurations shown in Fig. 9.4 for different values of $Q(\mu^*)$. The application of eq. (9.24) will be discussed in Section 9.3.

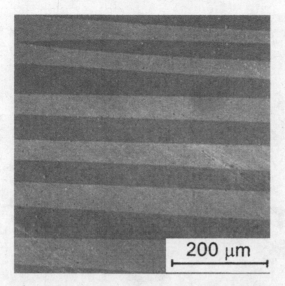

Fig. 9.5. Laminar domain patterns in a platelet of $Nd_2Fe_{14}B$ of thickness 2 mm and width 3 mm with easy direction inclined by an angle of 4.9° with respect to the surface [9.11].

9.2.6 Branching of domains in hard magnetic materials

Whereas for soft magnetic materials the μ^*-correction gives an excellent description of the domain patterns, in hard magnetic materials the so-called branching effect of the domains at surfaces becomes important. For a test of the relations derived for the Kittel-type domain patterns suitable materials are hard magnetic intermetallic compounds $RE_2Fe_{14}B$. Investigations of this type have been performed by several authors [9.6–9.11]. Here we present some characteristic results obtained for single crystals and sintered magnets. Figures 9.5–9.8 show laminar domain patterns as observed in macroscopic specimens of $Nd_2Fe_{14}B$. In the case of a sample with $a = 3$ mm and $T = 2$ mm a domain width of 50 μm is observed [9.11] and for a bulk sample of $T = 0.5$ mm a domain width of $D = 10$ μm is found [9.16]. Using the value of $\gamma_B = 24\,mJ/m^2$ of Table 6.1, we obtain for the theoretical D-values from eq. (9.20) and eq. (9.12), $D^{theor} = 16$ μm and 3.0 μm, to be compared with the experimental values of 50 and 10 μm. Accordingly there exists a discrepancy of a factor of 3. If we consider grains with a thickness in the range of 10 μm the situation improves somewhat [9.12]. Within the grain of thickness 10 μm of the sintered magnet shown in Fig. 9.8, we observe an experimental domain width of 1.7 μm, whereas the theoretical value gives $D^{theor} = 1.2$ μm if we consider only the domains in the centre of the grain. Actually the domain pattern does not correspond precisely to the Kittel model because on the upper side we observe typical domain branching with a domain width of 0.75 μm. The average domain width therefore is 1.23 μm, which agrees fairly well with the theoretical predictions. There exist

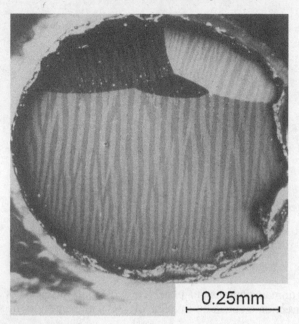

0.25mm

Fig. 9.6. Laminar domain pattern in a Nd$_2$Fe$_{14}$B single crystal with in-plane easy direction and thickness 0.5 mm. [6.46].

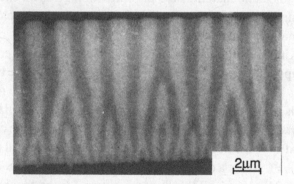

2μm

Fig. 9.7. Laminar domain pattern in a platelet of Nd$_2$Fe$_{14}$B with perpendicular easy direction and thickness 10 μm [9.12].

only a few grains without the branching phenomenon. In a rather small grain with $T = 5$ μm, a domain width of 0.5 μm is observed. The corresponding theoretical value using eq. (9.12) gives $D^{theor} = 0.65$ μm. This result is quite satisfying if we take into account that the grain is embedded into a magnetic environment.

The branching of domains at the surface of grains or macroscopic specimens is schematically shown in Fig. 9.9. It has been shown by Hubert [9.13], Bodenberger and Hubert [9.14] and Plusa *et al.* [9.15] that in this case the wall width in the centre

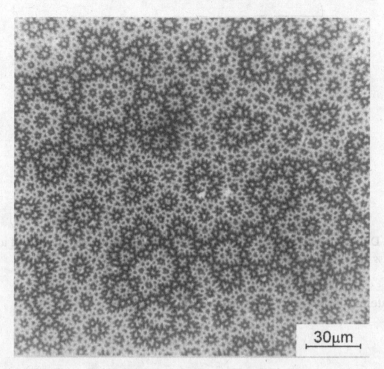

Fig. 9.8. Branching domain pattern of a $Nd_2Fe_{14}B$ single crystals on the (001) plane (perpendicular easy direction) [9.11].

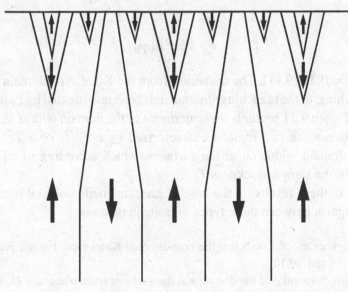

Fig. 9.9. Model of domain branching at the surface of a uniaxial crystal with perpendicular easy direction.

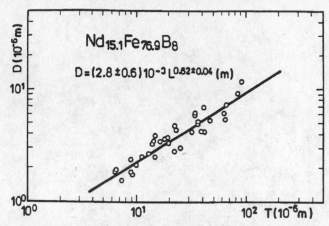

Fig. 9.10. Domain width in the bulk as a function of the thickness of platelets of a sintered $Nd_{15.1}Fe_{7.99}B_8$ magnet according to Szafranska-Miller *et al.* [9.16].

of the platelet follows a $T^{2/3}$-law according to

$$D = c \cdot (\gamma_B T^2)^{1/3}, \tag{9.25}$$

where the constant $c = 0.117$ m/J$^{1/3}$ has been determined by Szafranska-Miller *et al.* [9.16] (see Fig. 9.10) previously for $Nd_{15.1}Fe_{76.9}B_8$. For the surface domain width measured on the plane with a perpendicular easy direction (Fig. 9.8) the relation

$$D_s = c' \cdot 16\pi^2 \mu_0 \gamma_B / J_s^2, \tag{9.26}$$

holds with $c' = 0.31$ [9.11]. The transition from the Kittel-type domain structure, without branching, to the branching structure has been investigated by Pastushenkov *et al.* [9.11]. Figure 9.11 presents measurements of the domain widths showing the transition between the two regions characterized by a $T^{1/2}$- or a $T^{2/3}$-law and the constant domain width D_s at the surfaces, which according to eq. (9.26) is independent of the sample thickness T.

According to these results, in the case of hard magnetic uniaxial materials, we have to distinguish between three types of domain patterns:

1. At small thicknesses of $T < 5$ μm the conventional Kittel-type domain patterns exist with $D \propto T^{1/2}$ (eq. (9.12)).
2. For $T > 5$ μm branching of the domains at the surfaces takes place and $D_s \propto T^{2/3}$, for fields up to macroscopic dimensions eq. (9.25) holds.
3. Within the plane with a perpendicular easy direction the average width, D_s, of the closure domains is independent of the sample thickness and given by eq. (9.26).

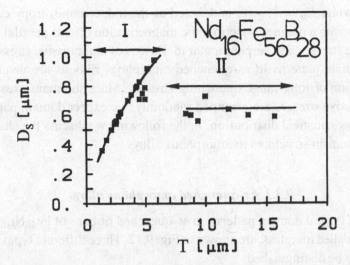

Fig. 9.11. Domain width D_s on the surface of a $Nd_{16}Fe_{56}B_{28}$ sintered magnet as a function of the platelet thickness. The transition from the Kittel structure to the branched structure takes place at $L_{crit} = 8$ μm [9.11].

9.3 Domain patterns in amorphous alloys

Extensive studies of domain patterns in amorphous alloys have been performed by means of the magneto-optical Kerr effect [9.17–9.22] and the Bitter technique [9.23, 9.24]. Reviews of the experimental results have been given previously. Here we concentrate on the most characteristic results of these previous investigations. Due to the noncrystalline amorphous atomic structure, no intrinsic long-range anisotropy exists, i.e., neither the spin–orbit coupling nor the single-ion anisotropy is the origin of long-range anisotropies. Magnetic anisotropies, however, may also be due to long-range internal stresses or to induced anisotropies produced by annealing under applied fields or applied elastic stresses.

In melt-spun materials, in general, long-range quenching stresses exist, the volume average of which vanishes as discussed in Section 8.4.8.4. Introducing partial volume fractions, v_t and v_c, of tensile (σ_t) and compressive stress (σ_c) regions one obtains

$$\sigma_t v_t - |\sigma_c| v_c = 0. \tag{9.27}$$

For a random distribution of domains, $\langle \gamma_i^2 \rangle = 1/2$, the volume average of the total magnetoelastic coupling energy of eq. (2.77) for isotropic magnetostriction is given by

$$\langle \phi_m \rangle = \frac{3}{4} \lambda_s (\sigma_t v_t + |\sigma_c| v_c) = \frac{3}{2} \lambda_s \sigma_c v_c = -\frac{3}{2} \lambda_s \sigma_t v_t. \tag{9.28}$$

In the following, $K_u = \frac{3}{2}\lambda_s\sigma_t v_t$ is denoted as the induced anisotropy constant. In the case of positive magnetostriction, the magnetization orients parallel to the axis of the tensile stresses and perpendicular to the axis of compressive stresses. Therefore the domain patterns in as-quenched amorphous ribbons are determined by the distribution of long-range quenching stresses. Since the main axes of tensile and compressive stresses are arranged randomly, we expect domain patterns with a more or less statistical distribution. In the following we discuss the characteristic features of domain structures in amorphous alloys.

9.3.1 As-quenched amorphous alloys

Examples of typical domain patterns in as-quenched ribbons of $Fe_{40}Ni_{40}P_{14}B_6$, the classical so-called metglass, are shown in Fig. 9.12. Three different types of domain patterns may be distinguished:

1. Wide wavy laminae with a width of 50–100 μm.
2. Patches of narrow, laminar or zig-zag laminae with a domain width of 3–5 μm and a lateral extension of about 150–200 μm. These patches are embraced by wide laminae (Fig. 9.13).
3. Star-like domain patterns where wide laminae are flowing out from one centre (Fig. 9.14).

100μm

Fig. 9.12. Domain structure of an as-quenched $Fe_{40}Ni_{40}P_{14}B_6$ alloy at zero magnetic field showing wide star-like domains and narrow zig-zag laminae.

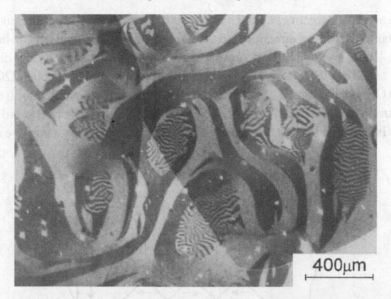

Fig. 9.13. Domain pattern of narrow laminae embraced by wide laminae in $Fe_{40}Ni_{40}P_{14}B_6$.

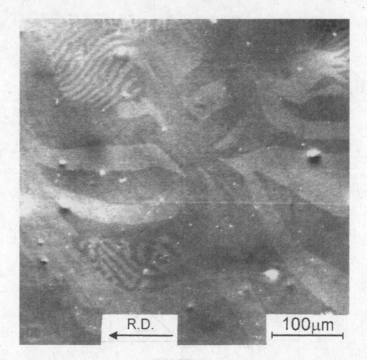

Fig. 9.14. Star-like domain pattern in a region of dilatational stresses (metglas).

After an annealing treatment at 473 K for 20 h the wide laminae are oriented perpendicular to the rolling direction and the number of patches has been considerably
reduced. Under an applied field, at first the wide laminae vanish by dw displacements at a field of 800 A/m (10 Oe), whereas the patches vanish at 1600–3200 A/m
(20–400 Oe) if the magnetic field is applied in-plane and perpendicular to the narrow laminae [9.25]. A model for the domain structure composed of wide and narrow
domains is shown in Fig. 9.15. Since metglass is characterized by a positive magne-

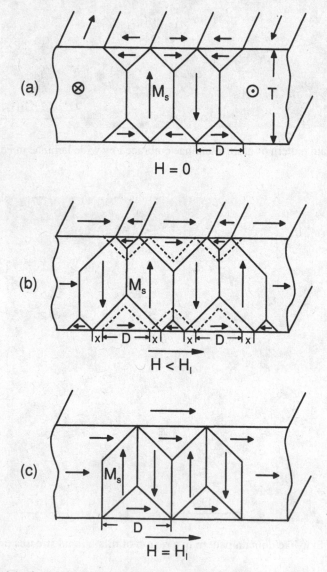

Fig. 9.15. Model of wide and narrow domains with in-plane and perpendicular uniaxial
anisotropy for different states of magnetization [9.25].

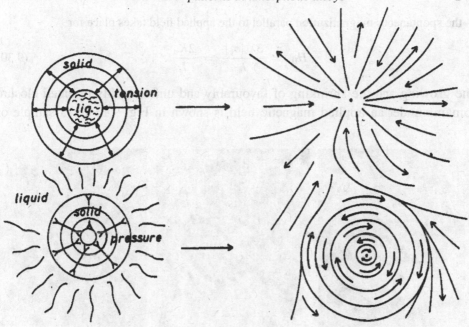

Fig. 9.16. Relation between the solidification process of a rapidly quenched alloy and star-like and narrow domain patterns. Solid nuclei within a liquid melt produce tensile stresses (star-like domains). A liquid island within the solid phase results in compressive stresses (narrow laminae).

tostriction it is assumed that the wide laminae appear in regions of tensile stresses, σ_t, and the narrow domains are observed in regions of compressive stresses, σ_c. A simple model for the formation of tensile and compressive stress sources is shown in Fig. 9.16. A liquid island within a solid matrix leads to radial tensile stresses whereas a solid nucleus within the liquid phase produces a circular pressure.

The domain model shown in Fig. 9.15 gives a simple explanation for the magnetization process within the narrow domain patches. Here we can distinguish three different stages of magnetization:

1. At small fields the magnetization process occurs predominantly by wall displacements of the wide laminae.
2. Within the regions of narrow domains, as a first stage the energetically favourably oriented closure domains grow up to a field of

$$H_I = \frac{3\lambda_s |\sigma_c|}{2J_s} = \frac{K_u}{J_s}. \tag{9.29}$$

3. Above the field H_I further magnetization occurs by rotational processes and a shift of the closure domains to the centre of the ribbon as shown in Fig. 9.15. Full alignment of

the spontaneous magnetization parallel to the applied field takes place for

$$H_{II} = \frac{3\lambda_s |\sigma_c|}{J_s} = \frac{2K_u}{J_s}. \tag{9.30}$$

The growing and the shrinking of favourably and unfavourably oriented closure domains under an applied magnetic field is shown in Fig. 9.17 for a sample of

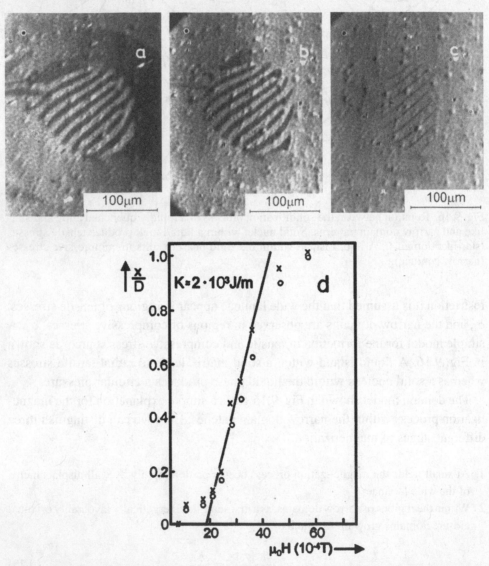

Fig. 9.17. Narrow laminae in $Fe_{80}B_{20}$ under the action of a magnetic field applied perpendicular to the laminae. (a) zero field; (b) Vanished wide laminae, $\mu_0 H = 0.5$ mT; (c) closure domains before vanishing, 4 mT; (d) Relative change of the average domain width of the closure domain as a function of the applied field (see also Fig. 9.15) [9.25].

$Fe_{80}B_{20}$. For the displacement x of the closure domain walls a linear relation holds [9.25]

$$2x/D = J_s(H - H_0)/K_u, \qquad (9.31)$$

where H_0 corresponds to the magnetic field where the wide laminae vanish.

Another test of the domain model is the calculation of the domain width by means of eq. (9.6), which holds for the ideal μ^*-effect and a perpendicular anisotropy. For $H_I = 1750$ A/m and $J_s = 1.6$ T, in the case of $Fe_{40}Ni_{40}P_{14}B_6$ we derive from eq. (9.29) $K_u = 3\lambda_s|\sigma_c|2 = 2800$ J/m^3. For the specific wall energy we then obtain, with $A = 5 \cdot 10^{-12}$ J/m, $\gamma_B = 4\sqrt{AK_u} = 4.75$ m J/m^2. Equation (9.6) now gives for the domain width $D = \sqrt{\frac{2\gamma_B T}{K_u}}$ for $T = 30$ μm a value of 3.2 μm, in full agreement with the experimental results shown in Fig. 9.13. This result also proves that the assumption of a nearly ideal shielding of surface charges by the μ^*-effect is an excellent model. In the case of $Fe_{80}B_{20}$ with $H_I = 4000$ A/m, $J_s = 1.6$ T, K_u is found to be 6400 J/m^3, which leads to $D = 2.6$ μm.

9.3.2 Magnetic annealing of amorphous alloys

In as-quenched materials, the magnetization process in general occurs predominantly by dw displacements, with a certain contribution of rotational processes at larger fields. In order to avoid the hysteresis losses due to dw displacements, special domain patterns may be induced by magnetic annealing under an applied magnetic field below the Curie temperature. In amorphous ribbons, a transverse, laminar domain pattern may be induced by magnetic annealing under a magnetic field applied in-plane and transverse to the ribbon axis. Figure 9.18 represents a characteristic domain pattern after a transverse in-plane magnetic annealing at 300 °C for 50 min, for the amorphous $Fe_{40}Ni_{40}B_{20}$ alloy. The material parameters of this alloy are: $J_s = 1.05$ T, $K_u = 50$ J/m^3, $T_c \simeq 300$ °C, $A = 8.8 \cdot 10^{-12}$ J/m, $\gamma_B = 8.4 \cdot 10^{-5}$ J/m^2. The geometrical data are: $a = 2$ mm, $T = 23$ μm. The experimentally determined domain width according to Fig. 9.16 is $D \sim 40$ μm [9.26]. Applying eq. (9.20) for a domain pattern without closure domain gives a theoretical domain width of 0.06 μm, i.e., a value smaller by a factor of 70 as compared to the experimental result. Taking into account that in soft magnetic materials the μ^*-effect reduces the surface charges, we have to use eq. (9.24) which takes care of the reduction of surface charges. Inserting the experimental parameters into eq. (9.24) gives $D = 27$ μm in fair agreement with the experimental values of 40 μm.

9.3.3 Domain structure and magnetization processes

The type of magnetization processes taking place in amorphous alloys depends on the arrangement of domains, the shape of the specimen and the orientation

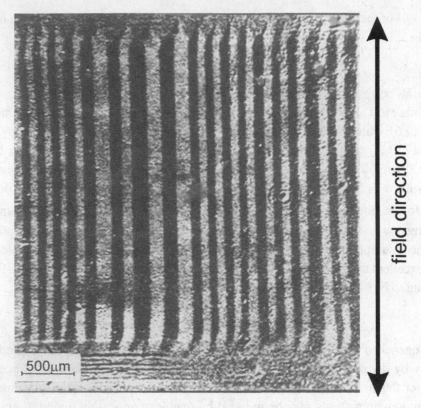

Fig. 9.18. Laminar domain structure of a $Fe_{40}Ni_{40}B_{20}$ alloy after annealing for 50 min at 300 °C in a transverse field. The arrow indicates the rolling direction.

of the applied field. Exclusively rotation processes take place if the internal field $H_i = H_{ext} - N \cdot M_s$ is oriented throughout perpendicular to the resultant magnetization vector $J_R = J_s^I - J_s^{II}$, where $J_s^{I,II}$ denote the polarization adjacent to the domain walls. In this case the magnetostatic force acting on the dw, $J_R \cdot H_i$, vanishes, whereas within the domains a magnetic torque $J_R^{I,II} \times H_i$ produces a rotation of $J_s^{I,II}$. Exclusively dw displacements govern the magnetization process if the magnetic torque vanishes within the domains and if there exists a component of H_i oriented parallel to $J_s^I - J_s^{II}$. Inducing well-defined domain patterns by magnetic annealing, rotational and wall displacement processes may be studied separately. In particular, the high-susceptibility, nonmagnetostrictive alloy $Co_{71}Fe_1Mo_1Mn_4Si_{14}B_9$ corresponds to an interesting case of such an experiment [9.27, 9.28]. Transverse and longitudinal domain patterns have been produced by magnetic annealing at 260 °C in a transverse and a longitudinal magnetic field of 0.1 T (Fig. 9.19). By varying the annealing time, the induced anisotropy, K_u, has been changed from 20 J/m^3 to 110 J/m^3. Figure 9.20 shows the dependence of the longitudinal susceptibilities as a function of K_u for the rotational and wall displacement process. It is of

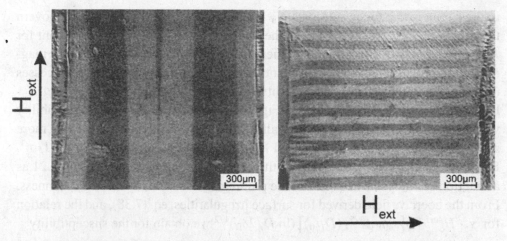

Fig. 9.19. Domain patterns in the nearly nonmagnetostrictive amorphous alloy $Co_{71}Fe_1Mo_1Mn_4Si_{14}B_9$ (Vitrovać 0030) after longitudinal and transverse field annealing ($T_a = 300\,°C/1\,h$, $\mu_0 H_{ext} = 1\,T$).

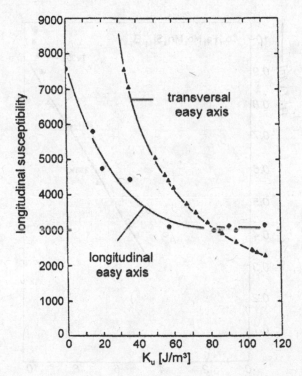

Fig. 9.20. Longitudinal susceptibilities for transverse easy axis (reversible rotations) and longitudinal easy axis (dw displacement) as a function of the induced anisotropy, K_u, for $Co_{71}Fe_1Mo_1Mn_4Si_{14}B_9$.

interest to note that for anisotropies $K_u < 80 \, \text{J/m}^3$, the rotational processes govern the magnetization process, whereas the wall displacements become dominant for $K_u > 80 \, \text{J/m}^3$. This crossover phenomenon is related to the dependence of the magnetization process on the induced anisotropy. In the case of the rotational processes (transversal easy axis) the susceptibility follows the well-known $\chi = J_s^2/2\mu_0 K_u$-law, which is also demonstrated by Fig. 9.20. The decrease of the susceptibility with increasing K_u in the case of the wall displacements takes place somewhat more smoothly, following a $K_u^{-1/2}$-law and leading to the crossover at $K_u = 80 \, \text{J/m}^3$. From the temperature dependence of the coercive field represented in Fig. 9.21 as a function of $K_u^{1/2}$, it is derived that the pinning of dws is due to surface roughness. From the coercive field derived for surface irregularities, eq. (7.38), and the relation for $\chi \cdot H_c^{\text{surf}} = \left[J_s \delta_B \pi^{1/2}/(D\mu_0) \right] (\ln D/2\delta_B)^{1/2}$ we obtain for the susceptibility

$$
\chi = \frac{1}{\pi^{7/2}} \frac{\langle T \rangle^2}{\Delta T a} \frac{1}{(L_2/\lambda_s)^{1/2}} \cdot \frac{J_s^2 \lambda}{\gamma_B} \propto K_u^{-1/2}, \tag{9.32}
$$

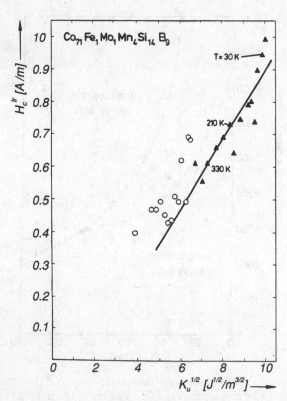

Fig. 9.21. Test of eq. (7.38) for H_c^{surf} for different magnetic annealing treatments and different measuring temperatures. ○ $K_u(293 \, K) = 257 \, \text{J/m}^3$; ▲ $K_u(293 \, K) = 55 \, \text{J/m}^3$ [9.26].

where we have inserted for the domain width, D, eq. (9.24) for $Q \ll 1$. Accordingly the susceptibility governed by the surface pinning of dws follows a $K_u^{-1/2}$-law in agreement with the smoother decrease of χ with increasing K_u in the case of dw displacements.

9.3.4 Stress-induced magnetic anisotropy

Induced anisotropies can be produced either by magnetic field-annealing or by annealing under an applied external stress. Due to the magnetoelastic coupling energy, the magnetization orients parallel to the axis of tensile stress for $\lambda_s > 0$. Furthermore, atomic structures rearrange into configurations with lowest elastic interaction energy. In crystalline materials this process is related to a pair ordering as shown in Fig. 9.22. In amorphous alloys such a rearrangement may take place near so-called free volumes, as shown schematically in Fig. 9.23 [9.29–9.31]. As an example, Fig. 9.24 shows the domain pattern as obtained after an annealing treatment of amorphous and nanocrystalline Finemet, $Fe_{73.5}Cu_1Nb_3Si_{13.5}B_9$, under applied tensile stresses. In the case of the amorphous alloy, the annealing temperature $T_a = 722$ K was below the crystallization temperature. The induced anisotropy

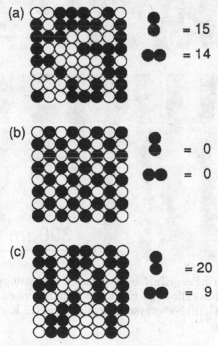

Fig. 9.22. Arrangement of atomic pairs in a binary alloy. (a) disordered, isotropic; (b) ordered; (c) disordered, anisotropic.

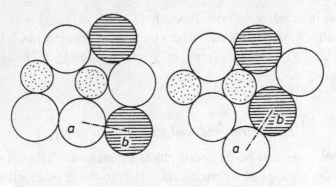

Fig. 9.23. Rearrangement of atomic configurations near free volumes.

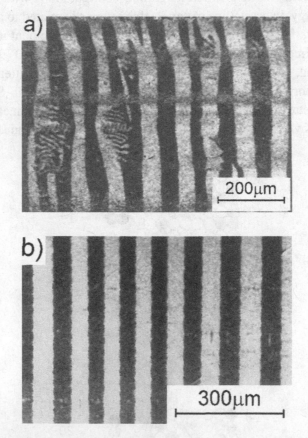

Fig. 9.24. Domain pattern in $Fe_{73.5}Cu_1Nb_3Si_{13.5}B_9$ after annealing treatments under applied tensile streses. (a) Amorphous alloy $T_a = 651$ K, $t = 60$ min, applied stress 508 MPa, induced anisotropy $K_u = 710$ J/m^3. (b) Nanocrystalline alloy $T_a = 813$ K, $t = 60$ min, applied stress 534 MPa, $K_u = 2480$ J/m^3.

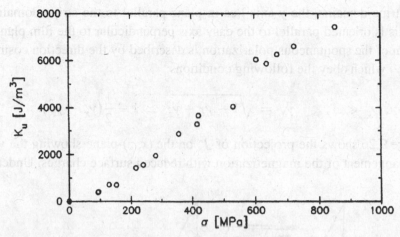

Fig. 9.25. Stress dependence of the induced anisotropy ($T_a = 813$ K, $t = 60$ min) in nanocrystalline $Fe_{73.5}Cu_1Nb_3Si_{13.6}B_9$ [9.32].

results in a transverse laminar domain pattern. In the case of the nanocrystalline Finemet, a similar transverse domain pattern develops at $T_a = 813$ K, independent of whether the annealing under tensile stress starts from the amorphous or from the nanocrystalline state [9.28]. The induced anisotropy increases linearly with the applied stress at a fixed annealing temperature $T_a = 813$ K (Fig. 9.25). The fact that the anisotropy is induced above the Curie temperature of $T_c = 610$ K, clearly proves that the anisotropy is induced by the elastic interactions. The increase of K_u with increasing applied stresses leads to a systematic change of the magnetization curves where the susceptibility is given by $\chi = J_s^2/2\mu_0 K_u$ and the saturation field by $H_s = 2K_u/J_s$ if the field is applied perpendicular to the laminar domains. The coercive fields of these hysteresis loops are rather small, varying between 1 and 6 A/m.

9.4 Stripe domains in thin ferromagnetic films

In thin ferromagnetic films under certain conditions with perpendicular anisotropy so-called stripe domains exist [9.33–9.40]. These stripe domains are periodical oscillations of the magnetization within a laminar conventional domain structure. In films with $Q = 2K_1\mu_0/J_s^2 > 1$ the magnetization is oriented perpendicular to the plane, i.e., J_s is oriented parallel or antiparallel to the easy direction forming a Kittel-type or branched laminar domain pattern. For smaller Q-values the torque exerted by the demagnetizing stray field on J_s rotates the magnetization into the film plane. However, in order to reduce the loss of magnetocrystalline energy, the magnetization oscillates around the in-plane axis. The orientation of the magnetization is described within a cartesian coordinate system with the x-axis lying in-plane perpendicular to the stripe domains, the y-axis lies in-plane parallel to the stripe domains and the z-axis is oriented parallel to the easy axis perpendicular to the film plane. The direction of the spontaneous polarization is described by the direction cosines γ_x, γ_y and γ_z which obey the following conditions:

$$\gamma_x, \gamma_z \ll 1; \qquad \gamma_y = \sqrt{1 - \gamma_x^2 - \gamma_z^2} \simeq 1 - \frac{1}{2}(\gamma_x^2 - \gamma_z^2). \qquad (9.33)$$

Figure 9.26 shows the projection of $\boldsymbol{J}_s$ on the (x, z)-plane showing the vortex-type arrangement of the magnetization with reduced surface charges. Under these

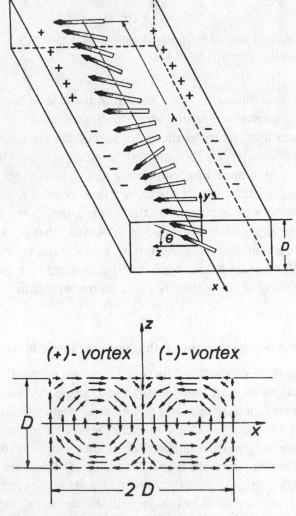

Fig. 9.26. Model of spin configuration of oscillating stripe domains in a thin platelet along the platelet axis and as a projection on the front side.

conditions the micromagnetic equations may be linearized and for a uniaxial anisotropy, $\phi_K = K_1(1 - \gamma_z^2)$, are given by ($H_{ext}$ applied parallel to the y-axis)

$$\Delta\gamma_z + (\kappa_K^2 - \kappa_H^2)\gamma_z - \frac{J_s}{2A}\frac{\partial U_i}{\partial z} = 0,$$

$$\Delta\gamma_x - \kappa_H^2\gamma_x - \frac{J_s}{2A}\frac{\partial U_i}{\partial x} = 0, \qquad (9.34)$$

$$\Delta U_i = \frac{J_s}{\mu_0}\left(\frac{\partial\gamma_x}{\partial x} + \frac{\partial\gamma_z}{\partial z}\right),$$

$$\Delta U_o = 0,$$

with the boundary conditions

$$\left.\begin{array}{c} \partial\gamma_x/\partial z = \partial\gamma_z/\partial z = 0 \\[2mm] U_i = U_o \\[2mm] \dfrac{-\partial U_i}{\partial z} + \dfrac{J_s}{\mu_0}\gamma_z = \dfrac{-\partial U_o}{\partial z} \end{array}\right\} \quad z = \pm D/2. \qquad (9.35)$$

Since the stripe domains represent a periodic distribution of the magnetization, the cosines γ_x and γ_z may be represented by Fourier transforms

$$\gamma_x = \sum_{p=1}^{\infty}\sum_{q_1=1}^{\infty} a_{pq_1}\cos\frac{2\pi p}{\lambda}x\,\sin\frac{(2q_1-1)\pi}{D}z,$$

$$\gamma_z = \sum_{p=1}^{\infty}\sum_{q_2=1}^{\infty} b_{pq_2}\sin\frac{2\pi p}{\lambda}x\,\cos\frac{2\pi q_2}{D}z. \qquad (9.36)$$

This expansion satisfies the micromagnetic boundary conditions (eq. (9.35)) and allows a description of the two-dimensional periodic distribution of the direction of J_s in a film of thickness D. λ denotes the wavelength of the stripe domains. If eq. (9.36) is inserted into the magnetic equations, a homogeneous linear system of equations is obtained for the Fourier coefficients a_{pq_1} and b_{pq_2}. By means of eq. (9.35), the expansion coefficients of U_i and U_o may be expressed by a_{pq_1} and b_{pq_2}. The eigenvalues a_{pq_1} and b_{pq_2} are obtained from the secular determinant of the linearly coupled system of homogeneous equations. A numerical solution of the secular determinant by restricting the summation to finite numbers of p and q_i has been given by Holz and Kronmüller [9.39]. From these calculations three regions of different magnetization configurations were derived, the phase diagram of which is shown in Fig. 9.27.

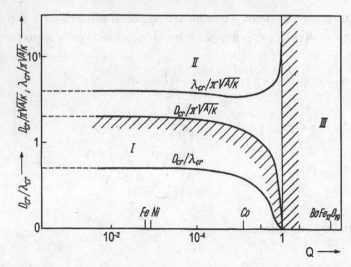

Fig. 9.27. Critical thickness D_{cr} and corresponding wavelength λ_{cr} of stripe domains as a function of the Q-parameter [9.39].

1. In region I $(Q < 1, D < D_{cr})$ the spontaneous magnetization lies in the film plane and domains are separated either by Néel–Bloch or cross-tie walls depending on the film thickness and the material parameters.
2. In region II $(Q < 1, D > D_{cr})$ a periodic stripe domain structure exists. For $D_{cr} \ll D$, the stripe domains structure transforms continuously to the conventional domain structure with domains separated by 180°-Bloch walls.
3. In region III $(Q > 1)$ the magnetization is oriented either parallel or antiparallel to the easy axis which is oriented perpendicular to the film plane.

Whereas in region III from $D = 0$ a homogeneous distribution of domains exists, magnetized either parallel or antiparallel to the easy axis, in region II there exists a continuous transition from a stripe domain structure to a macroscopic domain structure. In particular, in region III the domain structure is built up by nucleation at lattice imperfections or at the edges of the films, whereas in region II the formation of the domain pattern corresponds to a reversible process.

In Fig. 9.27 we have indicated the Q-values of Fe, Ni, Co and of barium ferrite. Stripe domains should exist in the transition metals with $D_{cr}^{Ni} = 278.0$ nm, $D_{cr}^{Fe} = 132$ nm, $D_{cr}^{Co} = 23.3$ nm. At very small Q-values, as in Ni and Fe, D_{cr} approaches the saturation value [9.39]

$$D_{cr} = 2\pi (A/K_1)^{1/2} = 2\delta_B. \qquad (9.37)$$

The independence of D_{cr} on J_s indicates that the formation of stripe domains occurs stray field free. In the region $Q \rightarrow 1$, however, D_{cr} decreases rapidly due to the formation of surface charges. As a further interesting result, for low Q-values the

Table 9.1. *The critical quantities*
$D_{cr}/\pi\sqrt{A/K}$, $\lambda_{cr}/\pi\sqrt{A/K}$ *for stripe domains in uniaxial films as a function of* κ_K *(according to [9.39]).*

Q	D/λ	$D_{cr}/\pi\sqrt{A/K}$	$\lambda_{cr}/\pi\sqrt{A/K}$
0.00795	0.499	1.97	3.95
0.0159	0.497	1.96	3.94
0.0397	0.491	1.91	3.89
0.0795	0.485	1.86	3.83
0.1590	0.461	1.71	3.70
0.5000	0.306	1.14	3.73
0.7950	0.133	0.635	4.77

stripe domain wavelength is found to be

$$\lambda_{cr} = 2D_{cr} = 4\delta_B. \tag{9.38}$$

Numerical results for D_{cr} and λ_{cr} as a function of D are summarized in Table 9.1.

If the stripe domains are exposed to a magnetic field parallel to the y-axis, the formation of stripe domains is suppressed for film thicknesses $D > D_{cr}$. For the case of stray field free stripe domains, i.e., $Q \ll 1$, D_{cr} increases with increasing field according to

$$D_{cr}(H_{ext}) = \frac{2\delta_B}{1 - H_{ext}J_s/2K_1}. \tag{9.39}$$

Also the wavelength $\lambda_{cr}(H_{ext})$ increases according to

$$\lambda_{cr}(H_{ext}) = 2D_{cr}(H_{ext})\sqrt{\frac{2K_1 - H_{ext}J_s}{2K_1 + H_{ext}J_s}} = 4\delta_B\sqrt{\frac{1}{1 - (H_{ext}J_s/2K_1)^2}}. \tag{9.40}$$

The magnetic field where the stripe domains vanish is denoted as the critical field. Table 9.2 summarizes some numerical results for the critical fields and the wavelength dependent on the film thickness for a permalloy film (90% Ni, 10% Fe) with the material parameters $\mu_0 A/J_s^2 = 7 \cdot 10^{-17}$ m^2, $Q = 0.103$. Measurements by Saito *et al.* [9.35] showed a value of $D_{cr}^{exp} = 150$ nm which agrees with the theoretical predictions.

Here it must be noted that actually the Q-value for the above mentioned permalloy film is not well known and Saito used $Q = 0.17$ leading to $D_{cr}^{theor} = 112$ nm.

Stripe domains have also been observed in nanocrystalline $\langle 111 \rangle$-Fe films by Bourret and Dautreppe [9.37]. For this geometry, where the easy direction lies under an angle of $57°$ with respect to the surface normal, the theoretical critical film thickness was $D_{cr}^{theor} = 132$ nm, to be compared with the experimental value

Table 9.2. *The normalized critical field*
$h_{cr} = H_{cr}/M_s$ *and the wavelength* λ_{cr} *of*
stripe domains α_i *and* β_i *for a uniaxial*
permalloy film as a function of the film
thickness (according to [9.39]).

D [10^3 Å]	h_{cr}	λ [10^3 Å]
2	0.33	3.34
3	0.65	3.75
5	0.90	4.55
10	1.08	5.88
20	1.20	8.70

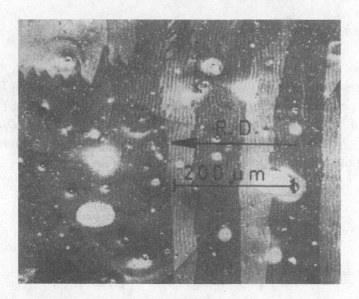

Fig. 9.28. Domain structure of an as-quenched $Co_{70}Fe_5Si_9B_{16}$ ribbon showing wide domains and superimposed weak stripe domains [7.6].

$D_{cr}^{exp} = 130$ nm [9.37]. The theoretical results were obtained with the parameters
for α-Fe: $K_1 = 4.76 \cdot 10^4$ J/m^3, $J_s = 2.15$ T and $A = 2 \cdot 10^{-11}$ J/m. The results
for the critical wavelength were $\lambda_{cr}^{th} = 293$ nm and $\lambda_{cr}^{exp} = 200$ nm. This discrepancy
may be due to the fact that λ depends sensitively on the film thickness [9.39] which
is difficult to determine experimentally.

Stripe domains have also been observed in amorphous alloys. As an example,
Fig. 9.28 shows stripe domains of wavelength $\lambda = 6$ μm within the laminar domain
patterns of an as-quenched amorphous alloy $Co_{70}Fe_5Si_9B_{16}$ [9.22]. In this nearly
nonmagnetostrictive alloy, from the saturation field $H_s = 2K_u/J_s = 400$ A/m and
$J_s = 1.25$ T an anisotropy constant of 250 J/m^3 is obtained. With a reasonable value

of $A = 10^{-11}$ J/m the wavelength is found to be $\lambda = 2.5$ μm. A possible explanation for this discrepancy is the existence of an oblique easy direction which would lead to a reduction of the effective perpendicular anisotropy by $\cos^2\theta$, if θ is the angle of the easy direction with respect to the surface normal. As a consequence, the wavelength is increased by a factor $(\cos\theta)^{-1}$. For $\theta = 70°$ the experimental result of $\lambda = 6$ μm can be quantitatively explained.

9.5 Dislocations and domain patterns

9.5.1 Introduction

In the preceding sections the role of intrinsic material parameters, J_s, K_1, γ_B, concerning the domain structure has been discussed. Whereas in hard magnetic materials the domain patterns are governed by the minimization of the magnetocrystalline energy, in soft magnetic materials the stray field energy is minimized by vortex-type domain patterns. In addition to magnetocrystalline and stray field energy, in magnetostrictive materials the magnetoelastic coupling energy, ϕ_m, becomes important. In the case of homogeneous stresses ϕ_m acts like an additional anisotropy. For a magnetostriction of $\lambda = 50 \cdot 10^{-6}$ and a stress field of $\sigma = 100$ MPa this leads to a magnetoelastic coupling energy of $\phi_m = -(3/2)\lambda\sigma = -7.5 \cdot 10^3$ J/m^3. If the stress field is inhomogeneous, in addition to ϕ_m, the exchange energy, ϕ_{ex}, plays a role. Accordingly the total energy due to inhomogeneous stresses has to be determined on the basis of the micromagnetic equations. In particular, dislocations influence the domain pattern in soft magnetic materials where the magnetoelastic coupling energy becomes comparable with the anisotropy energy. Also Ni and α-Fe are magnetically soft materials with large magnetostriction and moderate magnetocrystalline anisotropy, where dislocations have a significant influence on the domain patterns. These crystals therefore are suitable ferromagnets for the study of characteristic effects of dislocations on the domain patterns.

9.5.2 Domain patterns in plastically deformed Ni-single crystals

In crystalline Ni the easy directions are the $\langle 111 \rangle$-directions. With an anisotropy constant of $K_1 = -4.5 \cdot 10^3$ J/m^3 and large magnetostriction constants $\lambda_{100} = -51 \cdot 10^{-6}$ and $\lambda_{111} = -23 \cdot 10^{-6}$ at RT, Ni may be considered as a soft magnetic material, but very sensitive to internal stresses. For the investigation of the influence of dislocations on the domain pattern, we choose a Ni-plane with in-plane easy directions. Within the $(\bar{1}01)$-planes lie four easy directions, establishing a four-phase domain pattern ($[111]$, $[\bar{1}\,\bar{1}\,\bar{1}]$, $[1\bar{1}1]$, $[\bar{1}1\bar{1}]$) with 180°-, 109°-, and 79°-domain walls as shown in Fig. 9.29 [9.41, 9.42]. In the demagnetized state these four phases are occupied equally. For the study of the effect of dislocations, a so-called middle oriented Ni-single crystal has been plastically deformed in tension.

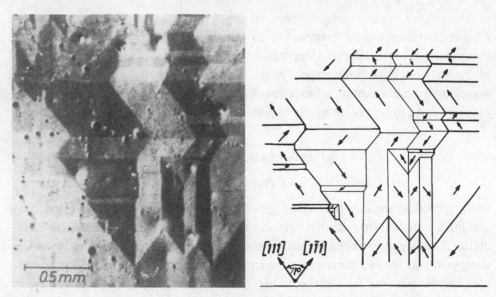

Fig. 9.29. Four-phase domain structure of a Ni single crystal on the (1̄01)-plane containing four easy directions. Image obtained by the magneto-optical Kerr effect [9.41, 9.42].

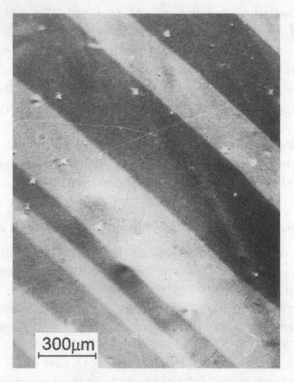

Fig. 9.30. Change of the four-phase domain pattern into a two-phase domain pattern on the (1̄01)-plane by a plastic deformation of 10 MPa [9.41].

For this orientation mainly so-called primary dislocations perform the plastic deformation. The glide plane of these dislocations is the (111)-plane, the Burgers vector, b, the $[\bar{1}01]$-direction and the line direction, l, of the edge dislocations corresponds to the $[1\bar{2}1]$-direction. With respect to the $(\bar{1}01)$-plane of the domain pattern, the geometry of the primary dislocations is the following. The glide plane normal and the line direction lie within the $(\bar{1}01)$-plane, the Burgers vector is oriented perpendicular to the $(\bar{1}01)$-plane. After a plastic deformation up to 10 MPa the four-phase domain pattern is converted into a two-phase domain pattern as shown in Fig. 9.30, with the remaining domains aligned parallel or antiparallel to the $[1\bar{1}1]$-direction which lies nearest to the line direction $[1\bar{2}1]$ of the edge dislocations. The glide plane normal, $\mathbf{n} = [111]$, no longer acts as an easy direction. After a plastic deformation up to 20 MPa the domain pattern again changes drastically. According to Fig. 9.31, a sheet-like arrangement develops, consisting of ladder-shaped stripes of width 200–500 μm, extending over 2500 μm parallel to

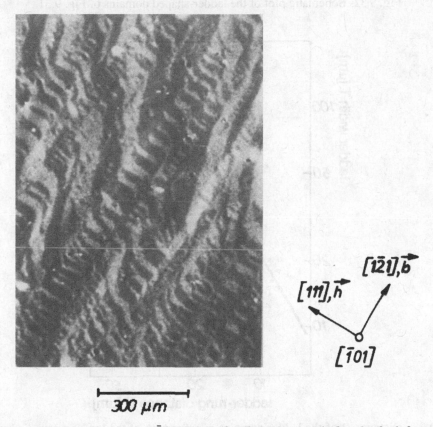

Fig. 9.31. Domain pattern on the $(\bar{1}01)$-plane of a Ni single crystal after plastic deformation up to 20 MPa. The domain ladders are oriented parallel to the $[1\bar{2}1]$ line direction of edge dislocations [9.41].

the line direction [1$\bar{2}$1] of the edge dislocations. The ladder-rungs within the stripes are oriented parallel to the primary glide plane normal [111]. This reappearence of the [111] easy direction will be discussed in Section 9.5.4. The ladder-shaped stripes are separated by the laminar domains which are present after the 10 MPa

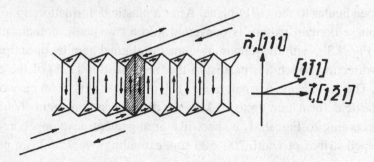

Fig. 9.32. Schematic plot of the ladder-shaped domains of Fig. 9.31.

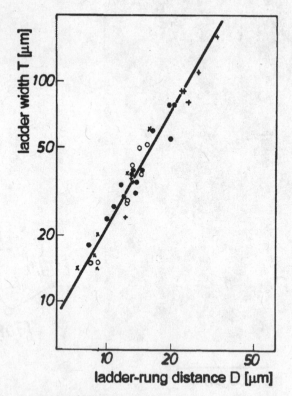

Fig. 9.33. Distance D of the ladder domains as a function of the ladder width T proving the validity of eq. (9.6) with $D \propto T^{1/2}$ [9.41].

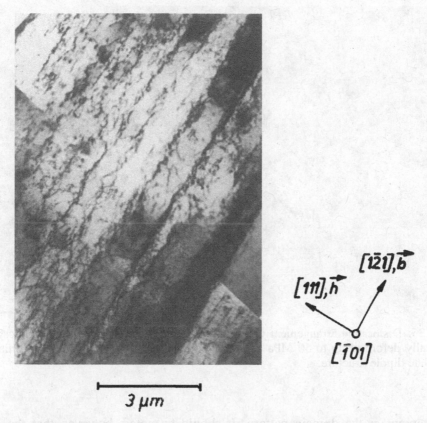

Fig. 9.34. Transmission electron microscopy micrograph of the (Ī01)-plane of a Ni single crystal after plastic deformation up to 20 MPa. The sheet-like dislocation structure is a replica of the ladder domain of Fig. 9.31.

deformation. The domain pattern of Fig. 9.31 is presented schematically in Fig. 9.32. The triangular closure domains allow a stray field free transition of J_s between the ladder-type domains and the laminar domains between the ladders. In Fig. 9.33 the distance, D, between the rungs is plotted as a function of the ladder's width, T. From the plot we derive a $D \propto T^{1/2}$-law as expected for the Landau domain structure of Fig. 9.1a. Using eq. (9.6) in the form $D = \sqrt{\frac{2\gamma_B T}{K_1}}$ we find for $T = 100$ μm a value $D = 23$ μm, in excellent agreement with the experimental results.

The analogy between the domain pattern and the dislocation arrangement is demonstrated by Fig. 9.34, where we show a micrograph of transmission electron microscopy (TEM) as obtained for the crystal plane (Ī01), where b is oriented perpendicular and l in-plane. The dislocation structure shows the same layer-like

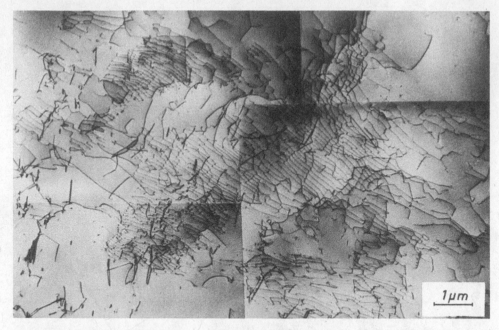

Fig. 9.35. Dislocation arrangement within the primary glide plane of a Ni single crystal plastically deformed up to 50 MPa at 78 K and annealed for 1 h at 660 °C in order to annihilate dipole structure.

arrangement as the domain pattern. It should be noted, however, that the distances between the dislocation layers is a factor of 100 smaller than the ladder thickness T. Some details of the dislocation structure are shown in Fig. 9.35 and Fig. 9.36 by TEM micrographs. Figure 9.35 shows the dislocation structure within the primary glide plane (111) after a deformation up to 50 MPa at 78 K and after an annealing treatment for 1 h at 660 °C. By this annealing treatment, dislocation dipole structures which hide the original primary dislocations have been removed. Figure 9.36 shows the sheeted dislocation structure within the (101)-plane which lies oblique to the primary glide plane, so showing the individual layered dislocation groups within the primary glide plane. The domain pattern remains stable up to annealing temperatures of 740 °C. At annealing temperatures above 740 °C the ladder-shaped domains according to Fig. 9.37 expand parallel to the [111]-direction, losing the ladder-type arrangement and establishing a laminar domain pattern extending parallel to the [111]-direction. Even at an annealing temperature of 980 °C, the original domain pattern of the undeformed single crystal could not be reconstructed because the dislocation structure forms an arrangement of rather stable small angle boundaries which still reflect the original dislocation structure.

Fig. 9.36. TEM micrograph of the (101)-plane of a Ni single crystal plastically deformed up to 50 MPa at 78 K and annealed for 1 h at 740 °C. Here the primary glide plane (111) intersects the (101)-plane.

9.5.3 Domain patterns in plastically deformed Fe-single crystals

In crystalline α-Fe the easy directions are the $\langle 100 \rangle$-directions. With the anisotropy constant $K_1 = 4.8 \times 10^4$ J/m^3 and magnetostrictions $\lambda_{100} = 22.2 \times 10^{-6}$ and $\lambda_{111} = -20.7 \times 10^{-6}$ α-Fe corresponds to a medium soft material, however, it is less sensitive to internal stresses than Ni. The $(\bar{1}01)$-plane and the (210)-plane are especially suitable for the study of the effect of dislocations on the domain pattern [9.43]. Within the $(\bar{1}01)$-plane we have the line direction $[1\bar{2}1]$ of edge dislocations and the Burgers vector [111] of primary dislocations. The (210)-plane contains two easy directions [001] and $[00\bar{1}]$ as well as the line direction $[1\bar{2}1]$. According to Fig. 9.38, within the $(\bar{1}01)$- and the (210)-plane we observe a laminar two-phase domain pattern oriented parallel to the [010]- and the $[0\bar{1}0]$- as well as to the [001]- and the $[00\bar{1}]$-directions in the case of the (210)-plane. After a plastic deformation at 195 K the laminar domain patterns on the $(\bar{1}01)$- as well as the (210)-plane vanish and ribbon-shaped domains parallel to the primary Burgers vector [111] as well as

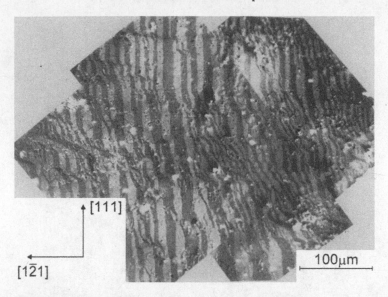

Fig. 9.37. Expansion of the ladder domains into the [111]-direction (glide plane normal) after annealing for 1 h at 780 °C.

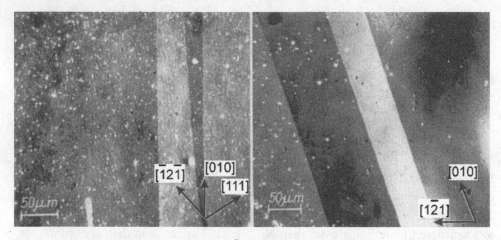

Fig. 9.38. Domain patterns on (left) the (Ī01)- and (right) the (210)-planes of an as-grown iron single crystal [9.43].

parallel to the line direction [1Ī1] of edge dislocations appear (Fig. 9.39). Also in a crystal deformed at 77 K at weak deformation (<6%) ribbon-shaped domain patterns extending parallel to the [111]- and the [1Ī1]-directions exist. At larger deformations (>10%), however, the ribbon-shaped stripes become short and fine and deviate from the [1Ī1]- and the [111]-directions as shown in Fig. 9.40. Figure 9.41 shows the domain pattern after a deformation of 18%. Here the activation of the

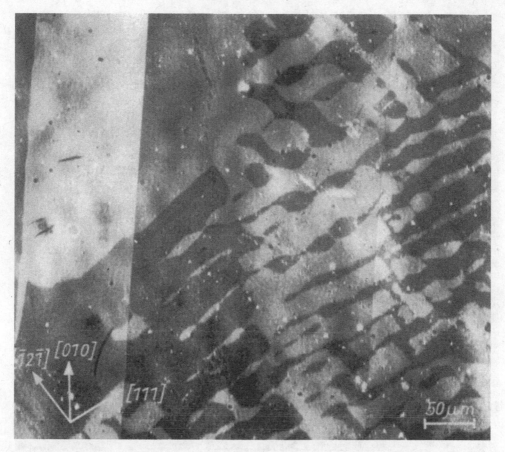

Fig. 9.39. Domain pattern on the ($\bar{1}$01)-plane of an iron single crystal deformed plastically up to 8% strain [9.43].

so-called conjugate slip system transforms the domain pattern into short laminar ribbons extending parallel to the [010]-direction.

9.5.4 Micromagnetic background of the magnetoelastic coupling energy due to dislocations

The effect of internal elastic stresses of dislocations on the direction of the spontaneous magnetization in cubic crystals is described by the magnetoelastic coupling energy

$$\phi'_\sigma = -\frac{3}{2}\lambda_{100}\sum_{i=1}^{3}\sigma_{ii}\gamma_i^2 - \frac{3}{2}\lambda_{111}\sum_{i\neq j}\sigma_{ij}\gamma_i\gamma_j, \qquad (9.41)$$

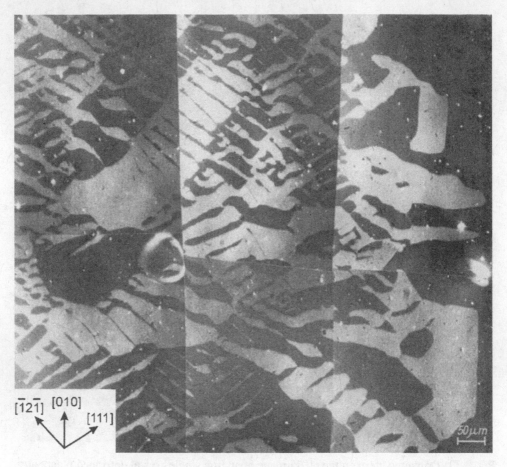

Fig. 9.40. Domain pattern on the $(\bar{1}01)$-plane of an Fe single crystal deformed plastically up to 13% strain [9.43].

which in the case of isotropy is written

$$\phi'_\sigma = -\frac{3}{2}\lambda_s \sum_{i=1}^{3} \sigma_{ii}\gamma_i^2 - \frac{3}{2}\lambda_s \sum_{i \neq j}^{3} \sigma_{ij}\gamma_i\gamma_j. \qquad (9.42)$$

In Section 8.4 we considered the effect of ϕ_σ on the LAFS. In this case the average deviation, ΔJ, from saturation was determined by solving the linearized micromagnetic equations. Due to the anisotropy of the magnetostriction and of the stress tensor, σ, the high-field susceptibility shows a characteristic anisotropy as discussed in Section 8.4.5. Similarly the magnetoelastic coupling energy due to dislocations shows a corresponding anisotropy. The role of elastic stresses may be easily derived from eq. (9.41) by considering homogeneous uniform stress states. As shown schematically in Fig. 9.42, in the case of a positive magnetostriction, $\lambda > 0$,

Fig. 9.41. Domain pattern on the (Ī01)-plane of an Fe single crystal deformed plastically up to 18% strain [9.43].

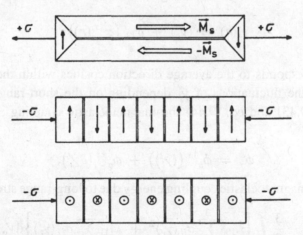

Fig. 9.42. Spin alignment in ferromagnets with positive magnetostriction under tensile ($\sigma > 0$) and compressive ($\sigma < 0$) elastic stresses. Closure domains have been omitted, see however Fig. 9.15.

a tensile stress, $\sigma_{ii} > 0$, aligns the magnetization parallel or antiparallel to the stress axis, thus minimizing ϕ'_σ. Under a compressive stress, $\sigma_{ii} < 0$, the magnetization aligns perpendicular to the axis of compression. In general, internal stresses due to lattice defects are inhomogeneous, showing fluctuations of the principal stress axes as well as of the amplitudes of internal stresses. In the case of large wavelengths of the internal stresses (> 100 nm), the orientation of J_s is fully determined by ϕ'_σ and the magnetocrystalline energy ϕ'_K. However, in the case of smaller wavelengths ($< l_K, l_s, l_H$), inhomogeneous stresses induce inhomogeneities of J_s and therefore the exchange and the stray field energy have to be taken into account. Under these conditions, the total free enthalpy, ϕ_t, must be determined by the concepts of micromagnetism, where the inhomogeneities of the γ_i can be derived from eq. (8.42). An effective approximation to treat the interaction between J_s and internal stresses starts from a separation of $\sigma(r)$ according to

$$\sigma(r) = \sigma^{\text{l.r.}}(r) + \sigma^{\text{s.r.}}(r), \tag{9.43}$$

into a long-range term $\sigma^{\text{l.r.}}(r)$ and a short-range term $\sigma^{\text{s.r.}}(r)$. $\sigma^{\text{l.r.}}(r)$ contains those stress components with wavelengths larger than the domain width, D, and $\sigma^{\text{s.r.}}(r)$ contains all stress contributions with wavelengths smaller than the domain widths. In soft magnetic materials $\sigma^{\text{l.r.}}$ determines the easy direction of J_s as discussed in Section 9.3, and $\sigma^{\text{s.r.}}(r)$ gives rise to fluctuations of the direction of J_s within the domains or in the LAFS. We therefore may write for the direction cosines in the cubic coordinate system

$$\gamma'_i(r) = \langle \gamma'_i \rangle + \delta\gamma'_i \left(\sigma^{\text{s.r.}}(r) \right), \tag{9.44}$$

where $\langle \gamma'_i \rangle$ corresponds to the average direction cosines within the domains and $\delta\gamma'_i$ represents the fluctuations of γ'_i depending on the short-range stresses. By means of eq. (9.43) and eq. (9.44) the magnetoelastic coupling energy may be represented as

$$\phi'_\sigma = \phi'^{\text{l.r.}}_\sigma \left(\langle \gamma'_i \rangle \right) + \phi'^{\text{s.r.}}_\sigma \left(\delta\gamma'_i \right), \tag{9.45}$$

where the total magnetoelastic coupling energy due to long-range stresses is written

$$\phi^{\text{l.r.}}_\sigma = -\frac{3}{2} \int \left\{ \lambda_{100} \sum_{i=1}^{3} \sigma_{ii} \langle \gamma'_i \rangle^2 + \lambda_{111} \sum_{i \neq j} \sigma_{ij} \langle \gamma'_i \rangle \right\} dV, \tag{9.46}$$

and the magnetoelastic coupling energy due to short-range stresses is given by

$$\phi^{\text{s.r.}}_\sigma = -\frac{3}{2} \int \left\{ \lambda_{100} \sum_{i=1}^{3} \sigma^{\text{s.r.}}_{ii} \langle \gamma'_i \rangle \delta\gamma'_i + \lambda_{111} \sum_{i \neq j} \sigma^{\text{s.r.}}_{ij} \langle \gamma'_i \rangle \delta\gamma'_j \right\} dV. \tag{9.47}$$

In deriving eq. (9.47) we have neglected terms quadratic in $\delta\gamma_i'$ and we have taken into account that terms $\sigma_{ij}^{\text{s.r.}}\langle\gamma_i'\rangle\langle\gamma_i'\rangle$ vanish because the volume average of $\langle\sigma_{ij}^{\text{s.r.}}\rangle$ vanishes within each domain.

The average direction cosines $\langle\gamma_i'\rangle$ are obtained from a minimization of the total Gibbs free energy of the homogeneous state, according to

$$\frac{\partial}{\partial\langle\gamma_i'\rangle}\phi_t^{\text{hom}} = \frac{\partial}{\partial\langle\gamma_i'\rangle}\left(\phi_K'(\langle\gamma_i'\rangle) + \phi_S'(\langle\gamma_i'\rangle) + \phi_\sigma'^{\text{l.r.}}\langle\gamma_i'\rangle + \phi_H'(\langle\gamma_i'\rangle)\right) = 0,$$

(9.48)

where $\phi_{\text{ex}}' = 0$ and $\phi_s'(\langle\gamma_i\rangle)$ corresponds to the demagnetization energy $\phi_s' = \frac{1}{2\mu_0}N_{ii}J_s^2\langle\gamma_i'\rangle^2$. The effect of short-range stresses on the magnetization has been treated extensively in Chapter 8. Now we consider the contribution of the short-range stresses to the total Gibbs free energy. In contrast to eq. (9.48), here we have also to take into account the exchange energy and the magnetostatic stray field energy due to magnetic volume charges. For the Gibbs free energy density due to short-range stresses we may write

$$\Delta\phi_t(\sigma^{\text{s.r.}}) = \phi_t(\gamma_i) - \phi_t^{\text{hom}}(\langle\gamma_i\rangle).$$

(9.49)

By subtracting the energy terms of the homogeneous magnetization, eq. (9.49) just takes care of the contribution of $\phi_\sigma'^{\text{s.r.}}(\delta\gamma_i') = \phi_\sigma'(\gamma_i') - \phi_\sigma'^{\text{l.r.}}(\langle\gamma_i'\rangle)$. For the further calculation of $\Delta\phi_t$ we consider the field range in the approach to saturation (LAFS) where the linearized micromagnetic equations hold. From these equations with, $g_i^{\text{el}}(r) = \partial\phi_\sigma^{\text{s.r.}}/\partial\gamma_i|_{\gamma_2=1;\gamma_{1,3}=0}$, we may derive a concise expression for $\Delta\phi_t$. By multiplying the linearized equations with γ_i and partial integration of the exchange term we find within Brown's coordinate system, where the transformation $\delta\gamma_i' \rightarrow \gamma_1, \gamma_3$ and $\langle\gamma_i\rangle \rightarrow \gamma_{1,3} = 0$, $\gamma_2 = 1$ holds,

$$\Delta\phi_t(\sigma^{\text{s.r.}}) = \frac{1}{2}\phi_\sigma(\sigma^{\text{s.r.}}),$$

(9.50)

where $\phi_\sigma^{\text{s.r.}}$ is given by eq. (9.47). The factor $\frac{1}{2}$ appears in eq. (9.50) because in Brown's approximation all energy terms besides $\phi_\sigma^{\text{s.r.}}$ are quadratic terms, whereas $\phi_\sigma^{\text{s.r.}}$ is taken as energy term linear in the $\gamma_i(\delta\gamma_i')$. Within the framework of Brown's coordinate system, we have to transform the direction cosines γ_i' of eq. (9.47) to Brown's coordinate system and then obtain

$$\Delta\phi_t(\sigma^{\text{s.r.}}) = \frac{1}{2}\phi_\sigma^{\text{s.r.}} = \frac{1}{2V}\int\sum_{i=1,3}g_i^{\text{el}}(\sigma_{kl}'')\gamma_i\,dV,$$

(9.51)

with g_i^{el} given by

$$g_i^{\text{el}}(\sigma_{kl}'') = \sum_{k,l}\sum_{m,n}\beta_{km}'\beta_{ln}'C_{mn}(\beta_{2m}\beta_{in} + \beta_{im}\beta_{2n})\cdot\sigma_{kl}'',$$

(9.52)

with

$$C_{mn} = \begin{cases} 3\lambda_{100} & \text{for } m = n \\ 3\lambda_{111} & \text{for } m \neq n. \end{cases} \tag{9.53}$$

The γ_i to be inserted into eq. (9.51) are obtained from the linearized micromagnetic equation (3.37). In the case of magnetostrictive isotropy, $\Delta\phi_t$ may be written as

$$\Delta\phi_t(\sigma^{s.r.}) = \sum_{i=1}^{3} A_{iiii}\beta_i^2(1 - \beta_i^2) - \sum_{i \neq j} A_{iijj}\beta_i^2\beta_j^2 \tag{9.54}$$

$$+ \sum_{i \neq j} A_{ijij}\left(1 - \beta_k^2 - 4\beta_i^2\beta_j^2\right), \qquad k \neq i, j,$$

where the coefficients A_{ijkl} (neglecting the stray field term) are given by

$$A_{ijkl} = \frac{9\lambda_s^2}{8AV} \int_V \frac{\sigma_{ij}''(k)\sigma_{kl}''(k)}{(k^2 + \kappa^2)} d^3k, \tag{9.55}$$

with $\kappa^2 = \kappa_H^2 + \frac{2}{3}K_1/A$. The direction cosines β_i in eq. (9.54) refer to the (b, l, n) coordinate system of an edge dislocation ($x \parallel b$, $y \parallel l$, $z \parallel n$); $\beta_1 = \cos(J_s, b)$, $\beta_2 = \cos(J_s, l)$, $\beta_3 = \cos(J_s, n)$. In eq. (9.55) V corresponds here to the volume of the domain or to the volume occupied by a special dislocation arrangement. For individual dislocations with the stress tensor given by eq. (8.37) we obtain

$$\Delta\phi_t(\sigma^{s.r.}) = \frac{9}{16}\lambda_s^2 \frac{b^2 G^2 N}{4\pi\kappa^2 A} F(\beta_i) \left\{ \ln\frac{1}{2}\kappa R_0 + 0.5772 \right.$$

$$\left. + K_0(\kappa R_0) - I_0(\kappa R_0)K_0(\kappa R_0) \right\}, \tag{9.56}$$

where N denotes the dislocation density per unit area and $F(\beta_i)$ corresponds to an anisotropy function defined by eq. (8.48). Since the β_i refer to the (x'', y'', z'')-coordinate system of the edge dislocations, $F(\beta_i)$ defines an orthorhombic symmetry of $\Delta\phi_t$. Of special interest is the anisotropy within the (111)-plane where ϕ_K of the ideal crystal possesses a six-fold symmetry. In the presence of a straight edge dislocation lying in a {111}-plane and extending parallel to a ⟨121⟩-direction, an additional contribution of four-fold symmetry exists due to $\Delta\phi_t$. The function $F(\beta_i)$ has a minimum for $\beta_2 = 1$, i.e., for a magnetization parallel to the line direction, and maxima for J_s parallel to b or n (see Fig. 9.43). It is noteworthy that these maxima are due to the shear stress component σ_{13}. Accordingly the two-phase domain pattern in Fig. 9.30 is due to the minimum of $\Delta\phi_t \parallel l$. In the case of layered dislocation structures as observed for average plastic deformation (see Fig. 9.31) the components σ_{13}'' and σ_{33}'' are reduced and in the case of vanishing components

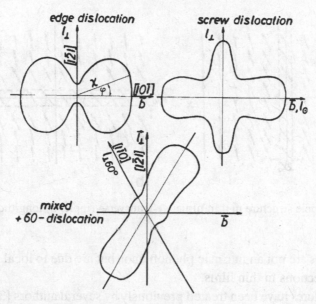

Fig. 9.43. Anisotropy functions $F(\beta_i)$ for edge, screw and mixed dislocations within the corresponding (l, b)-planes.

$F(\beta_i)$ is given by

$$F(\beta_i) = 10\beta_1^2(1 - \beta_1^2) + 16\nu\beta_2^2(1 - \beta_2^2) - 24\beta_1^2\beta_2^2. \qquad (9.57)$$

The anisotropy function now reveals a minimum in all three distinguished directions b, l, n of the dislocations. Within the $(\bar{1}01)$-plane accordingly we have the $[\bar{1}\bar{2}1]$-line direction as an easy direction for a statistical distribution of edge dislocations and two easy directions $[1\bar{2}1]$ and $[\bar{1}11]$ in the case of a layered dislocation structure as shown in Fig. 8.6b. These theoretical results give a straightforward explanation of the modifications of the domain patterns by plastic deformation of Ni-single crystals. The transition to a two-phase domain pattern parallel to the dislocation lines after moderate plastic deformation shown in Fig. 9.30 is a consequence of $\Delta\phi_l$. The formation of the ladder-shaped domains after stronger deformation is a consequence of the minimum of $\Delta\phi_t$ parallel to the n-direction which results from the modification of the stress tensor σ_{13} due to the sheeted dislocation structure shown in Fig. 9.36.

9.5.5 Ripple structures

Ripple structures in thin films correspond to a more or less periodic fine structure of magnetization in polycrystalline films as shown in Fig. 4.9 in the neighbourhood of a cross-tie wall [2.14]. In contrast to stripe domains treated in Section 9.4 the

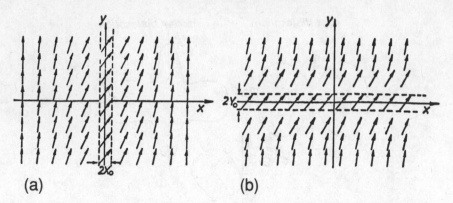

Fig. 9.44. Ripple structure in thin films: a) Transverse ripple, b) longitudinal ripple.

ripple structures are not an intrinsic phenomenon but are due to local perturbations of the easy directions in thin films.

Ripple structures have been treated previously by several authors [3.6, 3.8, 9.44–9.46] and critical assessments have been given by Brown [9.47] and Kronmüller [3.6]. The latter authors have shown that the neglect of a rigorous treatment of magnetostatic interactions introduces errors because the long-range character of the magnetostatic interaction is an essential feature of ripple structures. In particular, the replacement of long-range forces by short-range forces leads to exponential decays of spin perturbations whereas inverse power laws are correct.

For a model calculation a thin film of thickness D with an easy direction parallel to the y-axis is considered. $\mathbf{J}_s$, is constrained to be independent of z and to lie in the (x, y)-plane of the film. Two orientations of locally perturbed stripes of the easy directions are investigated. As shown in Fig. 9.44 the perturbed stripe may be oriented parallel to the easy direction, defining a so-called transverse ripple, or a perturbed stripe oriented perpendicular to the easy direction, defining a longitudinal ripple. It turns out that a treatment of these two configurations may suitably be performed in terms of the scalar field $\psi_1 = \operatorname{div} \mathbf{M}_s$ and the vector field $\boldsymbol{\psi}_2 = \operatorname{rot} \mathbf{M}_s$ as introduced in Section 3.3.

In the case of the transverse ripple structure ψ_1 and ψ_2 are given by

$$\psi_1 = M_s \frac{d\gamma_1}{dx}; \qquad \psi_2 = i_3 M_s \frac{d\gamma_2}{dx}. \tag{9.58}$$

If the deviations from the easy direction (y-axis) are small, i.e., $\gamma_1 \ll 1, \gamma_2 \sim 1$, and the torque within the perturbed stripe is described by a δ-like function

$$g_1 = i_1 A_0 \exp(-x^2/x_0^2) \tag{9.59}$$

the micromagnetic equations can be linearized and are given by

$$\Delta\psi_1 - \left(\kappa_s^2 + \kappa_K^2\right)\psi_1 = \frac{M_s}{2A}\,\mathrm{div}\,\boldsymbol{g}_1, \qquad \psi_2 = 0. \tag{9.60}$$

For the longitudinal ripple the scalar and the vector field are given by

$$\psi_1 = M_s\frac{\mathrm{d}\gamma_2}{\mathrm{d}y}, \qquad \boldsymbol{\psi}_2 = -i_3 M_s\frac{\mathrm{d}\gamma}{\mathrm{d}y}, \tag{9.61}$$

and $\boldsymbol{g}_1$ is obtained from eq. (9.59) by substituting x by y and x_0 by y_0. The micromagnetic equation is given by

$$\Delta\boldsymbol{\psi}_2 - \kappa_K^2\boldsymbol{\psi}_2 = \frac{M_s}{2A}\,\mathrm{rot}\,\boldsymbol{g}_1, \qquad \psi_1 = 0. \tag{9.62}$$

In the case of very thick films $D > \kappa_{K,S}^{-1}$ the solutions for $\psi_1, \boldsymbol{\psi}_2$ and γ_1 are well known, leading to exponential functions with decay distances $\kappa_{K,S}^{-1}$. The different decay distances appearing in eq. (9.60) and eq. (9.62) are related to the fact that the longitudinal ripple corresponds to a vortex-type structure with $\psi_1 = 0$, whereas the transverse ripple is related to magnetic charges and $\boldsymbol{\psi}_2 = 0$. For the two field variables ψ_1 and ψ_2 in the case of thin films with $\kappa_i D < 1$ and $\kappa_i x_0 \ll 1$ the following results are obtained:

$$\psi_1 = \frac{M_s a_1 D(\kappa_K^2 + \kappa_s^2)^{1/2}}{4A\sqrt{\pi}} K_1\left[(\kappa_K^2 + \kappa_s^2)^{1/2} \cdot x\right] \tag{9.63}$$

for the transerse ripple, and

$$\boldsymbol{\psi}_2 = i_3\frac{M_s a_1 D\kappa_K}{4A \cdot \sqrt{\pi}} K_1(\kappa_K y) \tag{9.64}$$

for the longitudinal ripple, with the parameter $a_1 = x_0^2 A_0$ for the transverse ripple and $y_0^2 A_0$ for the longitudinal ripple. The calculations of the potentials U and A according to eq. (3.27) and the calculation of M_s according to eq. (3.26) gives for the transverse ripple:

$$U(x) = (M_s a_1 D^2/32A)\left[I_0\left\{(\kappa_K^2 + \kappa_s^2)^{1/2}x\right\} - L_0\left\{(\kappa_K^2 + \kappa_s^2)^{1/2}x\right\}\right], \tag{9.65}$$

$$\gamma_1(x) = (1/M_s)\,\mathrm{d}\,U/\mathrm{d}x = \left[a_1(\kappa_K^2 + \kappa_s^2)^{1/2}D^2/32A\right]\left[I_1\left\{(\kappa_K^2 + \kappa_s^2)^{1/2}x\right\}\right.$$
$$\left. - L_1\left\{(\kappa_K^2 + \kappa_s^2)^{1/2}x\right\} - 2/\pi\right]. \tag{9.66}$$

The result for the longitudinal ripples is obtained by replacing $\kappa_K^2 + \kappa_s^2$ by κ_s^2 and x by y in eqs. (9.65) and (9.66). The functions I_n and L_n denote the modified Bessel and Struve functions of order n. It is important that the function $I_0 - L_0$ decreases as $1/x$ at large arguments, whereas the function $I_1 - L_1 - 2/\pi$ decreases as $1/x^2$. Accordingly exponential decay distances with κ_K^{-1} and κ_s^{-1} values smaller than 0.05 µm lead to incorrect results, which is supported by the fact that according to Fig. 4.9 the quasiperiodicity of the ripples is of the order of 1µm.

References

[9.1] Kittel, C., 1946, *Phy. Rev.* **70**, 96; 1949, *Rev. Mod. Phys.* **21**, 541.

[9.2] Néel, L., 1944, *J. de Physique* **5**, 265.

[9.3] Málek, Z., and Kamberský, V., 1958, *Czech. Z. Phys.* **8**, 416.

[9.4] Kronmüller, H., Moser, N., and Reininger, T., 1990, *Proc. SMM 9, Anales de Fisica* **B86**, 1.

[9.5] Williams, H.J., Bozorth, R.M., and Shockley, W., 1949, *Phys. Rev.* **75**, 155.

[9.6] Szymczack, R., 1973, *Acta Phys. Polon.* **A43**, 571.

[9.7] Kaczér, J., 1964, *Sov. Phys. JETP* **19**, 1204.

[9.8] Kaczér, J., and Gemperle, R., 1960, *Czech. J. Phys.* 1961, **B10**, 505; *ibid.* **B11**, 510.

[9.9] Wyslocki, B., 1963, *Phys. Stat. Sol.* **3**, 1333.

[9.10] Wyslocki, B., 1968, *Acta Phys. Polon.* **34**, 327.

[9.11] Pastushenkov, J., Forkl, A., and Kronmüller, H., 1997, *J. Magn. Magn. Mater.* **174**, 278; *ibid.* 1991, **101**, 363.

[9.12] Durst, K.-D., and Kronmüller, H., 1986, *J. Magn. Magn. Mater.* **59**, 86.

[9.13] Hubert, A., 1967, *Phys. Stat. Sol.* **24**, 699.

[9.14] Bodenberger, R., and Hubert, A., 1977, *Phys. Stat. Sol. (A)* **44**, K7.

[9.15] Plusa, D., Wyslocki, J.J., Wyslocki, B., and Pfranger, R., 1986, *J. Appl. Phys.* **A40**, 167.

[9.16] Szafranska-Miller, B., Wyslocki, J.J., Plusa, D., Wyslocki, B., and Pfranger, R., 1987, *Proc. 5th Int. Symp. on Magnetic Anisotropy and Coercivity in Rare-Earth Transition Metal Alloys* (Eds. C. Herget, H. Kronmüller and R. Poerschke, DPG-GmbH, Bad Honnef) p. 289.

[9.17] Kronmüller, H., Schäfer, R., and Schröder, G., 1977, *J. Magn. Magn. Mater.* **6**, 61.

[9.18] Schröder, G., Schäfer, R., and Kronmüller, H., 1978, *Phys. Stat. Sol. (A)* **50**, 475.

[9.19] Hubert, A., 1977, *J. Magn. Magn. Mater.* **6**, 38.

[9.20] Gröger, B., and Kronmüller, H., 1978, *J. Magn. Magn. Mater.* **9**, 203; 1981, *Appl. Phys.* **24**, 287.

[9.21] Kronmüller, H., and Fernengel W., 1981, *Phys. Stat. Sol. (A)* **64**, 593.

[9.22] Kronmüller, H., Fähnle, M., Domann, M., Grimm, H., Grimm, R., and Gröger, B., 1974, *J. Magn. Magn. Mater.* **13**, 53.

[9.23] Fujimori, H., 1983, 'Magnetic Anisotropy'. In *Amorphous Metallic Alloys* (Ed. F.E. Luborsky, Butterworth, London–Boston) p. 300.

[9.24] Obi, Y., Fujimori, H., and Saito, H., 1976, *Jpn. J. Appl. Phys.* **15**, 611.

[9.25] Fernengel, W., and Kronmüller, H., 1983, *J. Magn. Magn. Mater.* **37**, 167.

[9.26] Reininger, T., and Kronmüller, H., 1991, *Phys. Stat. Sol. (A)* **128**, 491; *ibid.* 1992, **129**, 247.

[9.27] Reininger, T., Moser, N., Hofmann, A., and Kronmüller, H., 1988, *Phys. Stat. Sol. (A)* **110**, 243.

[9.28] Reininger, T., Vazquez, M., and Kronmüller, H., 1992, *Phys. Stat. Sol. (A)* **132**, 477.

[9.29] Egami, T., 1983, 'Atomic Short-range Ordering in Amorphous Metallic Alloys'. In *Amorphous Metallic Alloys* (Ed. F.E. Luborsky, Butterworth, London–Boston) p. 100.

[9.30] Berry, B.S., and Pritchet, W.C. 1975, *Phys. Rev. Lett.* **34**, 1022.

[9.31] Luborsky, F.E., and Walter, J.L., 1977, *Mater. Sci. Eng.* **28**, 77.

[9.32] Hofmann, B., and Kronmüller, H., 1995, *J. Magn. Magn. Mater.* **152**, 91.

[9.33] Kaczér, J., Zelený, M., and Šuda, P., 1963, *Czech. J. Phys.* **B13**, 579.

[9.34] Spain, R.J., 1963, *Appl. Phys. Lett.* **3**, 208.

[9.35] Saito, N., Fujiwara, H., and Sugita, Y., 1964, *J. Phys. Soc. Japan* **19**, 1116.

[9.36] De Blois, R.W., 1965, *J. Appl. Phys.* **36**, 1647.

[9.37] Bourret, A., and Dautreppe D., 1966, *Phys. Stat. Sol.* **13**, 559; *ibid.* 1967 **23**, 207.

[9.38] Murayama, Y., 1966, *J. Phys. Soc. Japan* **21**, 2253.

[9.39] Holz, A., and Kronmüller, H., 1969, *Phys. Stat. Sol.* **31**, 787.

[9.40] Müller, M.W., 1961, *Phys. Rev.* **1961**, 1485.

[9.41] Willke, H., 1972, *Philos. Mag.* **25**, 397.

[9.42] Kronmüller, H., 1981, *Atomic Energy Rev.* (Int. Atomic Energy Agency, Vienna 1981) p. 255.

[9.43] Umakoshi, Y., and Kronmüller, H., 1981, *Phys. Stat. Sol. (A)* **68**, 159.

[9.44] Rother, H., 1964, *Z. Physik* **179**, 229.

[9.45] Krey, U., 1967, *Physik kondens. Materie* **6**, 218.

[9.46] Harte, K.J., 1968, *J. Appl. Phys.* **39**, 1503.

[9.47] Brown, W.F., Jr., 1970, *IEEE Trans. Magn.* **6**, 121.

10

Magnetic after-effects in amorphous alloys

10.1 Introduction

Since the discovery that amorphous ferromagnetic alloys are high-permeability magnetic materials with lower energetic losses than the crystalline permalloys and FeSi steels [10.1–10.3] their thermal structural stability has been a topic of numerous papers [10.4–10.21]. Melt-spun amorphous alloys in general show relaxation phenomena for all characteristic parameters of the hysteresis loop. The stability of magnetic properties, however, is a prerequisite for the technical applications of high-permeability amorphous alloys. Whereas in crystalline magnetic materials magnetic after-effects have previously been studied and analysed extensively [10.11, 10.12], such investigations have also been extended to amorphous alloys. In contrast to crystalline alloys, however, the type of defect structures that are the origins of magnetic after-effects in amorphous alloys is not so clear. Well-known defect structures in crystalline magnets are carbon or nitrogen interstitial atoms in α-Fe [10.22], intrinsic interstitial atoms [10.11], binary complexes of impurity and interstitial atoms, clusters of vacancies, etc. Since in amorphous alloys atomic defects are not as well-defined as in crystalline materials, investigations of magnetic relaxation by defects may give valuable information with respect to the type of mobile atomic defects in these materials. Magnetic after-effects basically are due to thermally activated processes which may be related either to irreversible Barkhausen jumps or to reversible atomic rearrangements of local defect configuration. In the latter case the relaxation phenomena can be effectively influenced by annealing treatments, which result in a stabilization of the amorphous structures.

A number of concepts that have been developed for the analysis of magnetic after-effects of crystalline magnets may also be applied for amorphous alloys. For example, the stabilization energy of domain walls (dws) as introduced by Néel [10.23] and others [10.9, 10.12] proves to be a suitable way of describing magnetic after-effects in amorphous ferromagnets. Furthermore, the model of double-well systems (DWS), developed originally for the interpretation of the specific heat of amorphous alloys, has been found to be suitable to describe the relaxation phenomena in amorphous alloys [10.24]. Accordingly, it is obvious to apply the double-well model to amorphous alloys. This model is based on the assumption that in amorphous alloys a large number of isostructural configurations exist and transitions between these configurations are possible in the neighbourhood of so-called free volumes [10.14, 10.25–10.27]. In general, there exists a wide spectrum of such DWS, which are characterized by a wide spectrum of atomic rearrangement giving rise also to such systems for the thermal activation spectrum. This model should be able to explain magnetic after-effects as well as the formation of uniaxial induced anisotropies.

10.2 Double-well model of magnetic after-effects in amorphous alloys

A basic requirement for the existence of magnetic after-effects is the presence of mobile defect structures with an anisotropic interaction energy with the spontaneous magnetization. In the case of binary crystalline alloys such defects are atom pairs in the neighbourhood of vacancies. Due to the interaction energy with the spontaneous magnetization, these mobile atom pairs try to orient their anisotropy axis into the most favourable orientation. Whereas in crystalline materials a reorientation of atom pairs requires the presence of vacancies, in amorphous alloys so-called free volumes are present. These play a similar role. Amorphous alloys in general have a lower packing density than fcc crystals because of the steric misfits between atoms of different ionic radii. Therefore, the potential minima occupied by the atoms are not as deep as in crystalline materials. Therefore, even at low temperatures small rearrangements of atoms are possible by jumps of atoms into neighbouring free volumes. Here it is useful to distinguish between rearrangements between nearly energetically equivalent configurations and those rearrangements which lead to more stable configurations. In both cases the two configurations are separated by a potential barrier which has to be overcome either by a thermally activated process or by a tunnelling process. As an example, Fig. 10.1 shows two-dimensional hard sphere models of double-well configurations which may give rise to the following relaxation processes:

1. Isomorphic transitions corresponding to a reorientation of the atomic pair axes within an atomic cluster giving rise to reversible relaxations where the free volume remains unchanged.
2. Polymorphic transitions where the reorientation of the pair axis takes place by an exchange of participating atoms. Irreversible processes are connected with long-range diffusion of free volumes leading to an irreversible decrease of the number of mobile

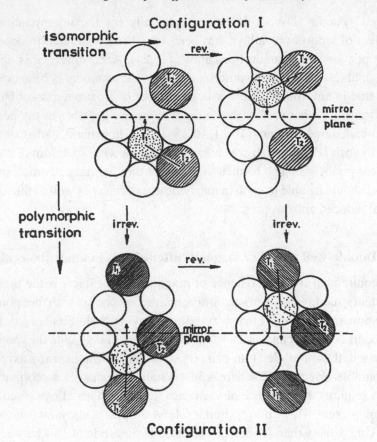

Fig. 10.1. Planar hard sphere model of a DWS showing isomorphic reorientations of atom pair axes (reversible process) as well as polymorphic transitions between two configurations with exchange of atoms T_2.

atom pairs. Due to anisotropic magnetic interactions the potential energy of an atom pair depends on the orientation of the spontaneous magnetization.

According to Fig. 10.2 the double-well system is characterized by an activation enthalpy, Q, of a hypothetical nonferromagnetic alloy which, however, is modified by structural and magnetic interaction energies, $\varepsilon_{i,j}$, of the atom pair in orientations $j = 1, 2$. The activation enthalpies for the transitions $1 \rightarrow 2$ and $2 \rightarrow 1$ are given by

$$Q_{i,1} = Q_i - \varepsilon_{i,1}; \qquad Q_{i,2} = Q_i - \varepsilon_{i,2}. \tag{10.1}$$

Since a large number of DWS exist we may apply Bose statistics for calculating the thermodynamic equilibrium configurations. Introducing the so-called splitting energy

$$2\Delta_i = \varepsilon_{i,2} - \varepsilon_{i,1} = 2\Delta_{i,s}^{s} + 2\Delta_{i,m}^{m}, \tag{10.2}$$

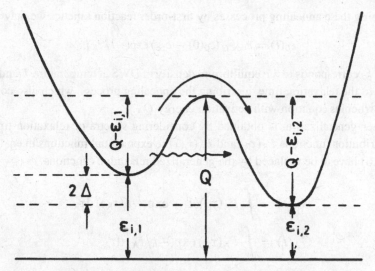

Fig. 10.2. Model of an asymmetric DWS showing the modification of activation enthalpy by the local interaction energies $\varepsilon_{i,j}$.

the equilibrium concentrations in configurations 1 and 2 are given by

$$c_{i,j\infty} = \frac{c_0 \cdot \exp[\pm\Delta_i/kT]}{2\,\mathrm{ch}[\Delta_i/(kT)]}, \qquad j = 1, 2, \tag{10.3}$$

where the $(+)$ sign holds for the lower energy level, and the $(-)$ sign for the upper energy level, and c_0 denotes the concentration of DWS with splitting energy Δ_i. In general Δ_i is composed of a structural term Δ_i^s and a purely magnetic term Δ_i^m, which takes care of the magnetic interactions. Δ_i^s plays a role if the two configurations $j = 1, 2$ are not fully symmetric. Assuming an equidistribution of the double-well systems on the configurations $j = 1, 2$ at $t = 0$, the approach to thermal equilibrium may be calculated on the basis of rate equations of first-order reaction kinetics. For an ensemble of double-well functions of concentration c_0, splitting energy Δ_i, and relaxation time τ_R, the concentration c_j occupying the jth double-well is given by

$$c_{i,j}(\Delta, t) = \frac{c_0}{2}\{1 \pm \tanh(\Delta_i/(kT)) \cdot (1 - \exp(-t/\tau_R))\}, \tag{10.4}$$

where the relaxation time follows an Arrhenius equation

$$\tau_R = \tau_{0R} \exp(Q_R/(kT)). \tag{10.5}$$

Equation (10.4) may be generalized with respect to two aspects:

1. The number of DWS with splitting energy Δ may be time-dependent due to irreversible annealing processes such as agglomeration of free volumes or a change of configuration.

Describing these annealing processes by first-order reaction kinetics we may write

$$c_0(t) = c_\infty + (c_0(0) - c_\infty)\exp(-t/\tau_A), \tag{10.6}$$

where c_∞ corresponds to the equilibrium density of DWS at temperature T and τ_A corresponds to the relaxation time describing the annealing process, which also corresponds to an Arrhenius equation with activation energy Q_A.

2. A further generalization is obtained by considering spectra of relaxation times. With the distribution functions $P_R(\tau_R)$ and $P_A(\tau_A)$ the exponential functions in eq. (10.4) and eq. (10.6) have to be replaced by the generalized relaxation functions

$$G_R(t) = \int P_R(\tau_R)(1 - \exp(-t/\tau_R))\,d\tau_R,$$

$$G_A(t) = \int P_A(\tau_A)\exp(-t/\tau_A)\,d\tau_A. \tag{10.7}$$

Inserting eq. (10.6) and eq. (10.7) into eq. (10.4) gives

$$c_{i,j}(t) = \frac{1}{2}c_\infty + \frac{1}{2}\Delta c\,G_A(t) \pm \frac{1}{2}\tanh\frac{\Delta_i}{kT}\,(c_\infty G_R(t) + \Delta c\,G_R(t)G_A(t)). \tag{10.8}$$

Here $\Delta c = c_0 - c_\infty$ denotes the excess density of mobile DWS which are able to anneal at a given temperature T. Equation (10.8) transforms into eq. (10.4) for $\tau_A \to \infty$ i.e., $G_A(t) \equiv 1$ and $P_R(\tau_R) = \delta(\tau_R)$.

10.3 Stabilization energy of domain walls

The total stabilization energy of a dw due to the interaction of a dw with DWS is defined as

$$E_s(t) = \sum_i \sum_{j=1,2} (c_j(t) - c_j(0)) \cdot \varepsilon_{i,j}(t). \tag{10.9}$$

In general, the splitting energy Δ is much smaller than kT. Under these conditions we have for a unique relaxation process

$$c_j(t) - c_j(0) = \pm\frac{\Delta}{2kT}(1 - \exp(-t/\tau_R)). \tag{10.10}$$

The sum in eq. (10.9) extends over all DWS i and orientations j of the defects. Figure 10.3 presents schematically the formation of the stabilization energy by the reorientation of preferred anisotropy axes of atom pairs. Starting from an isotropic distribution of anisotropy axes, with increasing time the axes orient into energetically preferred directions thus reducing the total interaction energy. For the determination of $E_s(t)$ we assume the dw to be displaced at time t, by a

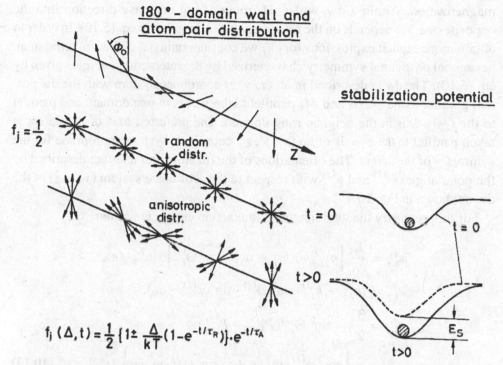

Fig. 10.3. Schematic model of magnetic after-effects due to reorientation of anisotropic defects within a dw. The original random distribution of pair axes changes into a fan-like distribution with the fan orientation parallel to the local spin orientation. The right side shows the formation of the stabilization energy.

distance U from its original position. The total interaction energy is then determined by the original concentrations $c_j(t)$ and the new interaction energies $\varepsilon_{i,j}(z_j - U)$, where z_j denotes the position of the ith DWS. Replacing the sums in eq. (10.9) by integrals over the different variables gives for the stabilization energy per unit area of the dw,

$$E_s(U, t) = -\frac{c_0}{kT}\left\langle \frac{1}{ch^2(\Delta^s/(kT))} \right\rangle (1 - \exp(-t/\tau_R))$$
$$\int \langle \Delta^m(z) \cdot \Delta^m(z - U) \rangle dz. \tag{10.11}$$

Here the angle brackets indicate an averaging process over the statistical parameters of the interaction energy $\varepsilon_{i,j}$ which is composed of three contributions,

$$\varepsilon_{i,j} = \varepsilon_{i,j}^{ex} + \varepsilon_{i,j}^{K} + \varepsilon_{i,j}^{el}, \tag{10.12}$$

which result from the exchange, magnetocrystalline and magnetoelastic interaction energies of the DWS. These energy terms depend on the orientation of magnetization. Within a dw, where M_s rotates from one easy direction into the opposite one, $\varepsilon_{i,j}$ depends on the rotation angle φ_0 as given in eq. (5.10). In order to obtain more explicit expressions for $\varepsilon_{i,j}$ we consider uniaxial defects of tetragonal, hexagonal or trigonal symmetry characterized by the interaction energies given by eq. (5.10). The dw is described in an (x, y, z) coordinate system with the dw normal parallel to the z-axis and M_s parallel to the y-axis in one domain and parallel to the $(-)y$-axis in the neighbouring domain. The preferred axis of the defect is taken parallel to the z'-axis of an (x', y', z')-coordinate system appropriate for the symmetry of the defect. The orientations of the ith atom pair axes are described by the polar angles $\Theta_i^{(j)}$ and $\varphi_i^{(j)}$ with respect to the coordinate system (x, y, z) of the dw as shown in Fig. 10.4.

For this geometry the total magnetic interaction energy is written

$$\varepsilon_{i,j}^{ex} = \frac{K_1}{A}\left\{a_{11,i}^{(j)}\sin\Theta_i^j + a_{33,i}^{(j)}\cos^2\Theta_i^{(j)}\right\}\sin^2\varphi_0(z),$$

$$\varepsilon_{i,j}^K = (k_{3,i}^{(j)} - k_{1,i}^{(j)})\sin^2\Theta_i^{(j)}\cdot\sin^2(\varphi_i^{(j)} - \varphi_0),$$

$$\varepsilon_{i,j}^m = \frac{3}{2}\lambda_s\cdot\sin^2\Theta_i^{(j)}(P_{33,i}^{(j)} - P_{11,i}^{(j)})$$

$$\times \left\{\cos^2\varphi_i^{(j)}\sin^2\varphi_0 + \frac{1}{2}\sin^2\varphi_i^{(j)}\sin^2\varphi_0\right\}. \tag{10.13}$$

Introducing eq. (10.13) into Δ of eq. (10.2) and averaging over the random

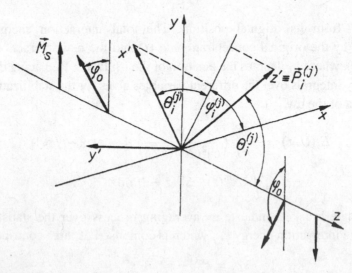

Fig. 10.4. Definition of coordinate system of the 180°-dw (x, y, z) and of the atom pairs of uniaxial symmetry (x', y', z'). The atom pair axis is chosen parallel to the z'-axis and is characterized by the polar angle $\Theta_i^{(j)}$ and the azimuthal angle $\varphi_i^{(j)}$ [10.16].

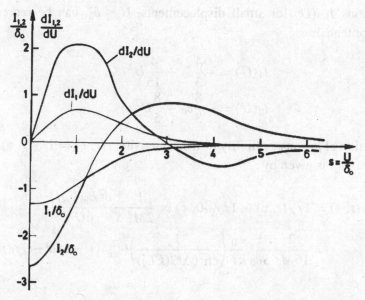

Fig. 10.5. The integrals $I_1(U)$ and $I_2(U)$ as a function of the dw displacement [10.9].

distributions of $\varphi_i^{(j)}$ and $\Theta_i^{(j)}$ gives

$$E_s(U, t) = \frac{1}{15} \frac{c_0}{kT} \left\langle \frac{1}{\mathrm{ch}^2(\Delta^s/(kT))} \right\rangle \{ \langle \varepsilon_1^2 \rangle I_1(U) + \langle \varepsilon_2^2 \rangle I_2(U) \}$$
$$\times \{ 1 - \exp(-t/\tau_R) \}, \qquad (10.14)$$

where the integrals $I_{1,2}$ shown in Fig. 10.5 are defined for a 180°-dw with $\sin \varphi_0(z) = 1/\mathrm{ch}(z/\delta_0)$:

$$I_1(U) = -\int_{-\infty}^{\infty} \sin^2(\varphi_0(z)) \cdot \sin^2(\varphi_0(z - U)) \, dz$$
$$= 4\delta_0 \cdot \frac{1 - (U/\delta_0)\mathrm{ch}(U/\delta_0)}{\mathrm{sh}^2(U/\delta_0)}, \qquad (10.15)$$

$$I_2(U) = -\int_{-\infty}^{\infty} \sin(2\varphi_0(z)) \cdot \sin(2\varphi_0(z - U)) \, dz$$
$$= 4I_1(U/\delta_0) - 8\delta_0 \frac{U/\delta_0}{\mathrm{sh}(U/\delta_0)}. \qquad (10.16)$$

The averaged interaction constants $\langle \varepsilon_{1,2}^2 \rangle$ are defined as

$$\langle \varepsilon_1^2 \rangle = \frac{2}{3} \langle (a_{33,i} - c_{11,i})^2 \rangle + \langle (k_{3,i} - k_{1,i})^2 \rangle + \frac{9}{4} \lambda_s^2 \langle (P_{33,i} - P_{11,i})^2 \rangle$$
$$\langle \varepsilon_2^2 \rangle = \langle (k_{3,i} - k_{1,i})^2 \rangle + \frac{9}{4} \lambda_s^2 \langle (P_{33,i} - P_{11,i})^2 \rangle. \qquad (10.17)$$

The functions $I_{1,2}(U)$ for small displacements, $U \leq \delta_0$, can be represented as harmonic potentials

$$I_1(U) = -\frac{4}{3}\delta_0 + \frac{4}{15}U^2/\delta_0,$$

$$I_2(U) = -\frac{8}{3}\delta_0 + \frac{8}{5}U^2/\delta_0. \tag{10.18}$$

The relaxation of the susceptibility or the reluctivity, $r(t, \tau) = 1/\chi(t, \tau)$, between $t = 0$ and time t is given by

$$\Delta r(t, \tau) = 1/\chi(t, \tau) - 1/\chi_0(0, \tau) = \frac{1}{2M_s^2}\frac{1}{S}\frac{\mathrm{d}^2 E_s(U, t)}{\mathrm{d}U^2}$$

$$= \frac{2}{15M_s^2}\frac{1}{S\delta_0}\frac{c_0}{kT}\left\langle\frac{1}{\mathrm{ch}^2(\Delta^s/(kT))}\right\rangle\langle\varepsilon_{\text{eff}}^2\rangle(1 - \exp(-t/\tau_R)), \tag{10.19}$$

where S denotes the dw area per unit volume and χ_0 corresponds to the initial susceptibility of the unrelaxed material. The effective interaction constant is defined as

$$\langle\varepsilon_{\text{eff}}^2\rangle = \frac{2}{15}\langle\varepsilon_1^2\rangle + \frac{1}{5}\langle\varepsilon_2^2\rangle. \tag{10.20}$$

If we introduce into eq. (10.19) the generalized relaxation functions $G_R(t)$ and $G_A(t)$, eq. (10.19) is written

$$\Delta r(t) = \frac{2}{15M_s^2}\frac{1}{S\delta_0}\left\langle\frac{1}{\mathrm{ch}^2(\Delta^s/(kT))}\right\rangle\frac{\langle\varepsilon_{\text{eff}}^2\rangle}{kT}\{c_\infty G_R(t) + \Delta c G_A(t)G_R(t)\}. \tag{10.21}$$

If the spectra of relaxation times τ_R and τ_A correspond to box-type distributions, the generalized relaxation functions $G_R(t)$ and $G_A(t)$ are given by

$$G_R(t) = 1 - \frac{1}{\ln(\tau_{2R}/\tau_{1R})}\left\{E_i(-\frac{t}{\tau_{1R}}) - E_i\left(-\frac{t}{\tau_{2R}}\right)\right\}$$

$$G_A(t) = \frac{1}{\ln(\tau_{2A}/\tau_{1A})}\left\{E_i\left(-\frac{t}{\tau_{1A}}\right) - E_i\left(-\frac{t}{\tau_{2A}}\right)\right\}. \tag{10.22}$$

The relaxation times $\tau_{i,R,A}$ correspond to the lower and upper bounds of the spectra of relaxation times. At small times, $t \leq \tau_{R,A}$, the relaxation function is linear in t and at large times, $t \geq \tau_{R,A}$, exponential time laws hold. At intermediate times, $\tau_{1R,A} \leq t \leq \tau_{2R,A}$, a logarithmic time law is valid. Furthermore, it is of interest to note that the first term in the brackets of eq. (10.21) describes the reversible part of the relaxation, whereas the second term corresponds to the irreversible contribution. At all times the reversible after-effects are important, whereas at larger times the irreversible after-effects dominate.

10.4 Formation of induced anisotropy

In crystalline binary alloys the formation of an induced anisotropy is related to the directional ordering of atom pairs [10.28]. This ordering process is governed by the diffusional properties of vacancies, i.e., by the energy of self-diffusion. Similarly, in amorphous alloys atom pairs may reorient their axes by means of free volumes as visualized in Fig. 10.1. Under the influence of magnetic interactions, atom pairs occupy such orientations where their interaction energies $\varepsilon_{i,j}$ are lowest. This thermally activated reorientation process in general takes place under an applied magnetic field, which guarantees a homogeneous magnetization.

In contrast to the situation of the magnetic after-effect of dws, therefore, exchange and magnetoelastic energies do not contribute to the reorientation process because $\Delta \gamma_i \equiv 0$ and no applied elastic stress σ^{el} exists. The total magnetic interaction energy after an annealing with M_s parallel to the z-direction as shown in Fig. 10.6 is given by [10.29]

$$\phi_0 = \sum_{i,j} c_j(\Delta, t) \varepsilon_{i,j}^K(\Theta_{i,0}^{(j)}, \varphi_{i,0}^{(j)}), \tag{10.23}$$

where $(c_j(\Delta, t)$ is given by eq. (10.10) and $\Theta_{i,0}^{(j)}$ and $\varphi_{i,0}^{(j)}$ denote the polar angles of the atom pair axis. The induced anisotropy is defined as the change of the total interaction energy if the magnetization rotates from the z-axis into an arbitrary direction defined by the angles Θ and φ as shown in Fig. 10.6. The energy

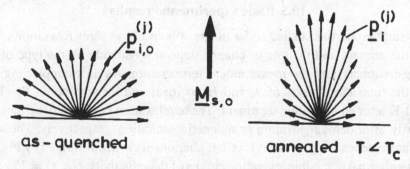

Fig. 10.6. Isotropic distribution of the orientations of pair axes in as-quenched materials and the anisotropic textured distribution after magnetic annealing. The lengths of the arrows indicate the probability of occupation of certain directions [10.15].

change is defined as

$$\phi_u = \sum_{i,j} c_j(\Delta_i, t)\left(\varepsilon_{i,j}^K(\Theta, \varphi) - \varepsilon_{i,j}^K\left(\Theta_{i,0}^{(j)}, \varphi_{i,0}^{(j)}\right)\right) \tag{10.24}$$

where

$$\Delta_i = \frac{1}{2}\left\{\varepsilon_{i,2}^K(\Theta_{i,0}^{(2)}, \varphi_{i,0}^{(2)}) - \varepsilon_{i,1}^K(\Theta_{i,0}^{(1)}, \varphi_{i,0}^{(1)})\right\}$$

corresponds to the distribution function of the pair axes established during the annealing process with M_s parallel to the z-axis at a temperature T_a. Replacing in eq. (10.24) the sum by integrations and performing the averaging over the statistical parameters $\Theta_i^{(j)}$ and $\varphi_i^{(j)}$ we finally obtain

$$\phi_u = \frac{1}{15}c_0(t)\frac{1}{ch^2(\Delta^s/(kT))}\frac{\langle|\varepsilon_{0,i}^K|^2\rangle}{kT_a}(1 - \exp(-t/\tau_R))\sin^2\Theta, \tag{10.25}$$

with an induced anisotropy constant

$$K_u(t) = \frac{1}{15}c_0(t)\left\langle\frac{1}{ch^2(\Delta^s/(kT))}\right\rangle\frac{\langle(\varepsilon_{0,i}^K)^2\rangle}{kT_a}(1 - \exp(-t/\tau_R)), \tag{10.26}$$

and

$$\varepsilon_{0,i}^K = k_{3,i} - k_{1,i}. \tag{10.27}$$

Similar to the magnetic-after effect, $c_0(t)$ may be reduced by the generalized annealing function $G_A(t)$ and the relaxation function by $G_R(t)$ thus giving

$$K_u(t) = \frac{1}{15}\left\langle\frac{1}{ch^2(\Delta^s/kT)}\right\rangle\frac{\langle(\varepsilon_{0,i}^K)^2\rangle}{kT_a}\{c_\infty G_R(t) - \Delta c G_R(t)G_A(t)\}. \tag{10.28}$$

10.5 Basic experimental results

All physical properties studied so far in amorphous alloys show relaxation effects, where the amount and the rate of change depend sensitively on the type of thermal pre-treatment and the measurement temperature. As an example, Fig. 10.7 shows the relaxation curves of χ_0 and H_c as measured for amorphous $Fe_{80}B_{20}$ at $T = 351$ K after an annealing treatment. These relaxation curves may be measured arbitrarily after demagnetization or magnetic saturation, respectively. The strong temperature dependence of the relaxation phenomena is demonstrated by Fig. 10.8 where isothermal reversible relaxation curves of the reluctivity, $r(t, \tau) = 1/\chi(t, \tau)$, of $Fe_{80}B_{20}$ are presented for measurement temperatures between 260 K and 420 K [10.17, 10.30].

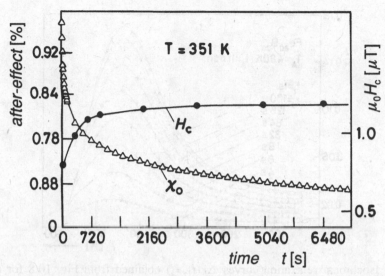

Fig. 10.7. Isothermal relaxation curves of χ_0 and H_c of $Fe_{80}B_{20}$ at 351 K as measured after an annealing treatment of 220 min at 480 K.

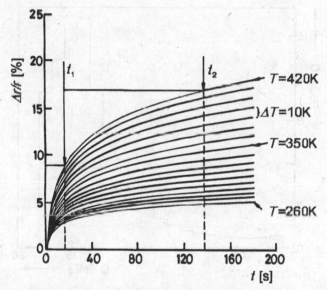

Fig. 10.8. Isothermal relaxation curves of the reluctivity, $\Delta r(t) = 1/\chi$, of amorphous $Fe_{80}B_{20}$ as measured after an annealing treatment of 220 min at 480 K.

From these isothermal relaxation curves, so-called isochronal relaxation curves are constructed by plotting the quantity in Fig. 10.9,

$$\Delta r(t_1, t_2, T) = \frac{1}{\chi(t_2, T)} - \frac{1}{\chi(t_1, T)}, \qquad (10.29)$$

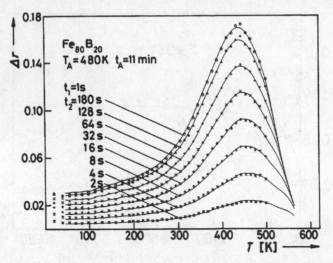

Fig. 10.9. Isochronal relaxation curves $\Delta r(t_1, t_2)$ obtained from Fig. 10.8 for different measurement times t_2 and $t_1 = 1$ s. The solid lines correspond to a computer fit using the spectra of activation energies of Fig. 10.10 and $\tau_{0R} = \tau_0 = 1.2 \cdot 10^{-15}$ s.

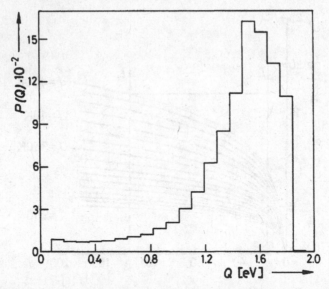

Fig. 10.10. Spectrum of activation energies of the isochronal relaxation curves of Fig. 10.9 using the generalized relaxation function $G_R(t)$ of eq. (10.22) using 20 box-type distributions of Q_R and $\tau_{0R} = \tau_0 = 1.2 \cdot 10^{-15}$ s.

as measured between the measuring times t_1 and t_2 as a function of temperature. In Fig. 10.9 the solid lines correspond to a computer fit of the isochronal relaxation curves by means of a spectrum of activation energies shown in Fig. 10.10 with a pre-exponential factor $\tau_{0R} = 1.2 \cdot 10^{-15}$ s. The low-energy tail of this spectrum indicates that there exist a large number of DWS with only a small energy barrier

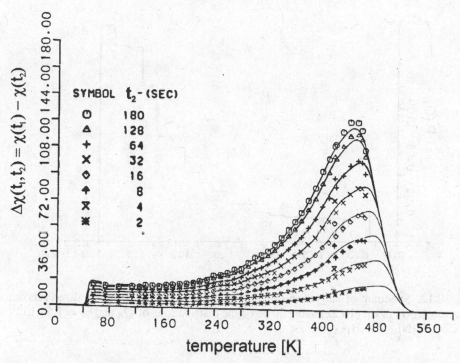

Fig. 10.11. Isochronal relaxation curves $\Delta\chi(t_1, t_2) = \chi(t_1) - \chi(t_2)$ of $Co_{58}Ni_{10}Fe_5Si_{11}B_{16}$ ($t = 1$ s).

giving rise to low-energy relaxation effects. Figure 10.11 shows the relaxation spectra for the nearly nonmagnetostrictive amorphous alloy $Co_{58}Fe_5Ni_{10}Si_{11}B_{16}$ and Fig. 10.12 presents the spectrum of activation energies [10.31]. As in the case of $Fe_{80}B_{20}$ there exists an extended tail for low temperatures due to a wide spectrum of activation energies below 1 eV.

Figures 10.13 and 10.14 show the formation of an induced anisotropy K_u in $Co_{58}Fe_5Ni_{10}Si_{11}B_{16}$ for an as-quenched and a pre-annealed ribbon [10.32, 10.33]. The magnetocrystalline anisotropy was induced by annealing in a transversal applied field oriented parallel to the ribbon plane and perpendicular to the ribbon axis. The difference between these two types of isothermal annealing curves may be interpreted as the irreversible contribution due to the nonequilibrium mobile free volumes in the as-quenched specimen. These excess free volumes have been annealed in the pre-annealed specimen at 668 K. According to the time dependence of K_u as derived in Section 10.4, $G_R(t)$ and $G_A(t)$ may be determined from the annealing curves shown in Figs. 10.13 and 10.14. Subtracting the annealing curves of the as-quenched and the pre-annealed specimen gives the function $G_R(t) \cdot G_A(t)$. Thus dividing the difference of the two curves by $G_R(t)$ gives the function $G_A(t)$ shown in Fig. 10.15.

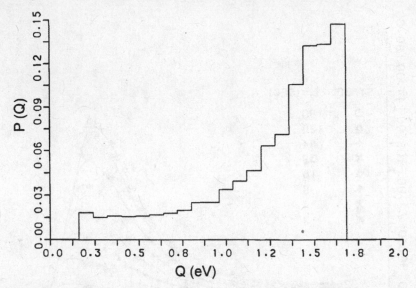

Fig. 10.12. Spectrum of activation energy derived by means of the generalized relaxation function $G_R(t)$ of eq. (10.22) using 20 box-type distributions of Q_R and $\tau_0 = 1.5 \cdot 10^{-15}$ s, for $Co_{59}Fe_5Ni_{10}Si_{10.4}B_{15.6}$.

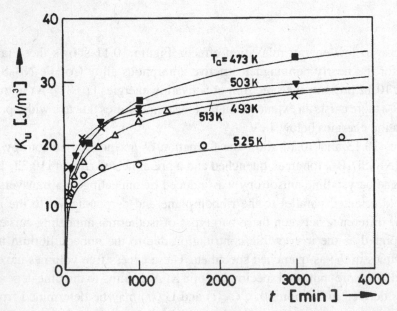

Fig. 10.13. Isothermal annealing curves of the induced anisotropy in a transverse field of $\mu_0 H = 0.8\,T$ for an as-quenched $Co_{58}Ni_{10}Fe_5Si_{11}B_{16}$ alloy. The full curves correspond to computer fits using eq. (10.28) and the spectra of activation energies presented in Fig. 10.16 [10.33].

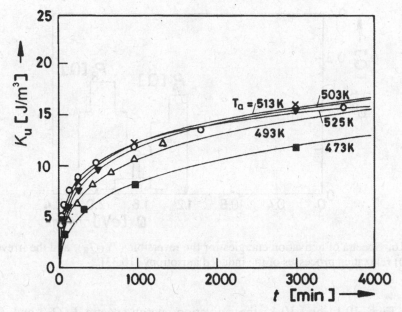

Fig. 10.14. Isothermal annealing curves of the induced anisotropy in a transverse magnetic field of $\mu_0 H = 0.8$ T for a $Co_{58}Ni_{10}Fe_5Si_{11}B_{16}$ alloy after annealing for 1.5 h at 668 K. The full curves correspond to computer fits using eq. (10.28) with $G_A(t) \equiv 1$ and the spectra of activation energies presented in Fig. 10.16 [10.33].

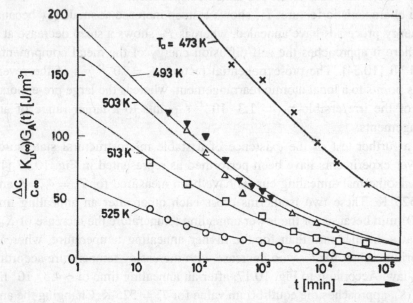

Fig. 10.15. The irreversible relaxation process of K_u as derived from the difference in the annealing curves of the as-quenched (Fig. 10.13) and the pre-annealed (Fig. 10.14) specimens. The full curves correspond to a data fit using the spectrum, $P_A(Q_A)$, of activation energies shown in Fig. 10.16 [10.33].

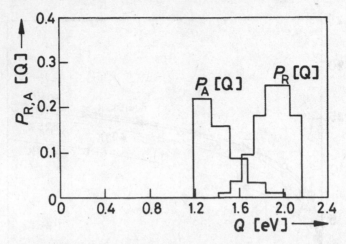

Fig. 10.16. Spectra of activation energies for the reversible ($P_R(Q_R)$) and the irreversible ($P_A(Q_A)$) relaxation processes of the induced anisotropy [10.33].

From Figs. 10.14 and 10.15 the activation energy spectra $P(Q_R)$ and $P(Q_A)$ have been determined by considering box-type distributions of activation energies with the generalized relaxation functions of eq. (10.22). Figure 10.16 shows that the activation energies of the irreversible process are considerably smaller than that of the reversible process of the pre-annealed specimen.

As a characteristic feature, P_A shows a steep increase at small Q_A because all low-energy processes have annealed, whereas P_R shows a steep decrease at large Q_R, where it approaches the self-diffusion energy of the metal components Co, Fe and Ni [10.34]. The pre-exponential factor $\tau_{0R} = 10^{-15\pm2}$ s of the reversible process points to a local atomic rearrangement, whereas the large pre-exponential factor of the irreversible $\tau_{0A} = 1.3 \cdot 10^{-12}$ s points to a larger range of atomic rearrangements.

For a further test of the existence of a stable microstructural state so-called crossover experiments have been performed as represented in Fig. 10.17 [10.33, 10.35]. Isothermal annealing curves have been measured for $T_a = 473$ K and for $T_a = 525$ K. These two isotherms cross each other after an annealing time of $\sim 2 \cdot 10^4$ min because for the lower annealing temperature the increase of K_u proceeds at a smaller rate than for the higher annealing temperature, whereas the equilibrium values for $t \to \infty$ decrease with increasing temperature according to a $1/T_a$-law. According to Fig. 10.17, after an annealing time of $\sim 4.5 \cdot 10^4$ min at 473 K, K_u approaches the equilibrium value for $T_a = 525$ K. Changing the annealing temperature to 525 K results in a steep decrease of K_u within a time interval of $\sim 10^3$ min. For larger annealing times, K_u again increases, approaching now the smaller equilibrium value for $T_a = 525$ K. The inverse annealing behaviour of K_u is

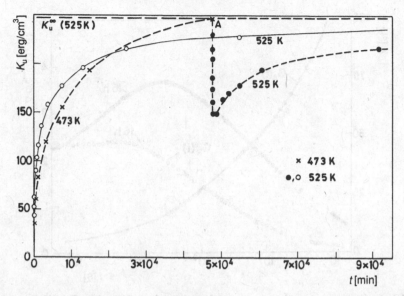

Fig. 10.17. Crossover behaviour of the induced anisotropy studied by a change of the annealing temperature from 473 K to 525 K at $t = 4.5 \cdot 10^4$ min. The equilibrium value of $K_u(\infty)$ for 525 K is given by the upper broken line.

due to the existence of a spectrum of activation energies. During the annealing time up to $t = 4.5 \cdot 10^4$ min all DWS with small relaxation times τ_R (or small activation energy) have approached their thermal equilibrium distribution, whereas those with the larger relaxation times (or larger activation energies) have not. Increasing the annealing temperature, those DWS which are in a thermal equilibrium have to re-arrange, i.e., for an increased annealing temperature the orientational order of pair axes is reduced, leading to a steep decrease of K_u. DWS with large activation energy after the temperature change proceed to reorient and to approach their thermal equilibrium, but with a larger rate. This process corresponds to the increase of K_u for $t \geq 4.5 \cdot 10^4$ min. It should be noted that the inverse annealing behaviour of K_u is quantitatively described by eq. (10.28) and the spectrum $P_R(Q_R)$ of activation energies for the reversible relaxation process is presented in Fig. 10.16. According to these results, the crossover experiment is a direct proof for the existence of a quasiequilibrium state and a spectrum of activation energies in amorphous alloys. In order to demonstrate the equivalence of the relaxation behaviour of $K_u(t)$ and of the reluctivity $\Delta r(t)$, the latter was measured for an as-quenched and a pre-annealed (1.5 h at 668 K) specimen of the alloy $Co_{58}Fe_5Ni_{10}Si_{11}B_{16}$. After further annealing of both specimens for 16 h at 513 K in a magnetic field applied parallel to the ribbon axis, isothermal relaxation curves were measured for both specimens at 513 K after a demagnetization treatment up to measurement times of $1.1 \cdot 10^5$ s (Fig. 10.18).

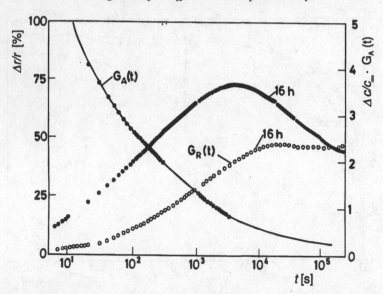

Fig. 10.18. Isothermal relaxation curves of the reluctivity of $Co_{58}Ni_{10}Fe_5Si_{11}B_{16}$ measured at 513 K after different pretreatments ($\bullet$ as-quenched; $\circ$ $T_a = 668$ K, 1.5 h). Prior to the relaxation measurements both specimens were annealed for 16 h at 513 K in a longitudinal magnetic field. The irreversible contribution $(\Delta c/c_\infty)\,G_A(t)$ was determined according to the procedure outlined in this section.

The larger relaxation amplitude measured for the as-quenched specimen is due to the excess concentration of mobile DWS (free volumes), which, however, anneal out after a measurement time of $t \approx 10^5$ s, where both isotherms approach the same equilibrium value determined by $c_\infty G_R(t)$. The difference between both relaxation curves therefore may be identified with the term $\Delta c G_A(t) \cdot G_R(t)$. The quantity $(\Delta c/c_\infty)\,G_A(t)$ therefore is determined by the difference of both relaxation curves divided by the amplitude of the pre-annealed relaxation curve. According to Fig. 10.18 $G_A(t)$ corresponds to a monotonously decreasing annealing curve as expected for an annealing process of first-order reaction kinetics.

10.6 Concluding remarks

The preceding sections have clearly shown that the model of DWS effectively describes relaxation phenomena in amorphous alloys. Reversible and irreversible relaxation processes are attributed to the reorientation of mobile atom pairs in the neighbourhood of free volumes. The centres of the spectra of activation energies lie around 1.6–2 eV and the pre-exponential factors of $\tau_0 \approx 10^{-15}$ s point to local atomic processes. Within the investigated time intervals only DWS with splitting parameters $\Delta^s \geq kT$ contribute to the reversible relaxation process. Irreversible relaxation processes may result from different origins:

1. Long-range migration of free volumes that sinks at surfaces, or agglomerates of free volumes.
2. Change of the configuration of DWS by a local rearrangement leading to a change of the topological and chemical short-range order and of the structural splitting parameter Δ^s. From the measurements of long-time isotherms of H_c, χ_0, K_u and from crossover experiments it may be concluded that in pre-annealed specimens, below the crystallization temperature, defect structures are present in a thermodynamic quasiequilibrium.

References

[10.1] Tsuei, C.C., and Duwez, P., 1966, *J. Appl. Phys.* **37**, 435.

[10.2] Duwez, P., and Liu, S.C.H., 1967, *J. Appl. Phys.* **38**, 4096.

[10.3] Luborsky, F.E., and Walter, J.L., 1977, *IEEE Trans. Magn.* **13**, 1635.

[10.4] Egami, T., *J. Mater. Sci.* 1978, **13**, 2587.

[10.5] Spielermann, H.H., Graham Jr. C.D., and Flandres, P.J., 1977, *IEEE Trans. Magn.* **13** 1541.

[10.6] Fujimori, H., Morita, H., Obi, Y., and Ohta, S., 1977, *Amorphous Magnetism II* (Eds. R.A. Levy and R. Hasegawa, Plenum Press, New York–London) p. 393.

[10.7] Taub, A.I., and Spaepen, F., 1981, *J. Mater. Sci.* **16**, 3087.

[10.8] Gibbs, M.R.J., Evetts, J.E., and Leake, J.A., 1983, *J. Mater. Sci.* **18**, 278.

[10.9] Kronmüller, H., and Moser, N., 1983. In *Amorphous Metallic Alloys* (Ed. F. Luborsky, Butterworth, London) pp. 341–359.

[10.10] Tsao, S.S., and Spaepen, F., 1982. In *Rapidly Quenched Metals IV* (eds. T. Masumuto and K. Suzuki, Japan Institute of Metals, Sendai) p. 463.

[10.11] Blythe, H., Kronmüller, H., Seeger, A., and Walz, F., 2000, *Phys. Stat. Sol. (A)* **181**, 233–345.

[10.12] Kronmüller, H., 1968, *Nachwirkung in Ferromagnetika* (Springer-Verlag, Berlin–Heidelberg–New York).

[10.13] Moser, N., and Kronmüller, H., 1980, *J. Magn. Magn. Mater.* **19**, 275; 1982, *Phys. Lett. A* **93**, 101.

[10.14] Kronmüller, H., 1983, *Philos. Mag. B* **48**, 127.

[10.15] Kronmüller, H., 1984, *J. Magn. Magn. Mater.* **41**, 366.

[10.16] Kronmüller, H., 1985, *Phys. Stat. Sol. (b)* **127**, 531.

[10.17] Kronmüller, H., Moser, N., and Rettenmeier, F., 1984, *IEEE Trans. Magn.* **20**, 1388.

[10.18] Kisdi-Koszò, F., Vojtanik, P., and Potocky, L., 1980, *J. Magn. Magn. Mater.* **19**, 159.

[10.19] Allia, P., Mazzetti, P., and Vinai, F., 1980, *J. Magn. Magn. Mater.* **19**, 281.

[10.20] Allia, P., and Vinai, F., 1982, *Phys. Rev. B* **26**, 6141.

[10.21] Boas, A., Potocky, L., Novák, L., and Kisdi-Koszò, 1981, *IEEE Trans. Magn.* **17**, 2712.

[10.22] Richter, G., 1937, *Ann. Phys.* **29**, 605; 1938, ibid. **32**, 683.

[10.23] Néel, L., 1950, *J. Phys. Rad. Paris* **11**, 49; 1951, *ibid.* **12**, 339; 1952, *ibid.* **13**, 249.

[10.24] Halperin, B.I., and Varma, C.M., 1972, *Philos. Mag.* **25**, 1.

[10.25] Turnbull, D., and Cohen, M.H., 1961, *J. Chem. Phys.* **34**, 120; 1970, *ibid.* **52**, 3088.

[10.26] Egami, T., Maeda, K., and Vitek, V., 1980, *Philos. Mag. A* **41**, 883.

[10.27] Egami, T., Maeda, K., Srolovitz, D., and Vitek, V., 1980, *J. Phys. (Paris)* **41**, C8-272.

[10.28] Graham, C.D., 1959. In *Magnetic Properties of Metals and Alloys* (Amer. Soc. for Metals, Cleveland) p. 288.

[10.29] Kronmüller, H., 1983, *Phys. Stat. Sol. (B)* **118**, 661.

[10.30] Rettenmeier, F., Kisdi-Koszò, E., and Kronmüller, H., 1986, *Phys. Stat. Sol. (A)* **93**, 597.

[10.31] Rettenmeier, F., and Kronmüller, H., 1986, *Phys. Stat. Sol. (A)* **93**, 221.

[10.32] Guo, H.-Q., Fernengel, W., Hofmann, A., and Kronmüller, H., 1984, *IEEE Trans. Magn.* **20**, 1394.

[10.33] Kronmüller, H., Guo, H.-Q., Fernengel, W., Hofmann, A., and Moser, N., 1985, *Cryst. Latt. Def. Amorph. Mat.* **11**, 135.

[10.34] Kronmüller, H., 1998. In *Elements of Rapid Solidification* (Ed. M.A. Otooni, Springer-Verlag, Berlin–Heidelberg), p. 93.

[10.35] Reininger, T., and Kronmüller, H., 1991, *Phys. Stat. Sol. (A)* **125**, 327.

11

Magnetostriction in amorphous and polycrystalline ferromagnets

It was outlined in Chapter 3 that the shape magnetostriction of a magnetic material is determined by the competition between spontaneous magnetostrictive strains ε_{ij}^Q and elastic strains ε_{ij}^{el} as soon as ε_{ij}^Q represents a spatially inhomogeneous tensor field. Such a situation arises, for instance, in a perfect crystalline ferromagnet if there are inhomogeneous magnetic states (domain walls, spin waves etc.). In a polycrystalline or amorphous material, inhomogeneous magnetostrictive strains exist even for a homogeneously magnetized state due to the spatial fluctuations of the orientation of the grains of the polycrystal or of the atomic scale structural units of the amorphous matrix. In the present chapter we extend the theory outlined in Chapter 3 to such materials. After an introduction to the problem (Section 11.1) the polycrystalline model of amorphous ferromagnets is introduced in Section 11.2, which enables us to treat polycrystalline and amorphous ferrromagnets on an equal footing. In Section 11.3 the basic computational ideas are introduced, and the mathematical formalism is described in Section 11.4. The results for the macroscopic magnetostriction tensor of magnetically saturated systems are discussed in Section 11.5, and the field dependence of the magnetostriction in the field regime of approach to ferromagnetic saturation is calculated in Section 11.6.

11.1 Outline of the problems

Amorphous ferromagnetic alloys based on 3d-transition metals exhibit a very small macroscopic magnetic anisotropy, which is one of the reasons for their sometimes excellent soft magnetic properties [11.1]. These technologically important properties are limited by the magnetoelastic interactions of domain walls with internal stresses [11.2, 11.3] introduced during the fabrication procedure or originating from the bending of ribbons in technical devices, because, in spite of the low magnetic anisotropy, the saturation magnetostriction constants, λ_s, are often comparable to those of crystalline counterparts ($|\lambda_s| \approx$ some 10^{-5} [11.1]). Exceptions are some near-zero magnetostrictive Co-rich alloys with $|\lambda_s| \approx$ some 10^{-6}.

The investigation of the magnetostrictive properties of amorphous alloys is also interesting from a fundamental point of view. It will be shown in the next section that in these materials magnetostriction primarily arises from the strain dependence of the local atomic-scale magnetocrystalline anisotropy rather than from the strain dependence of the low macroscopic (i.e., volume-averaged) magnetocrystalline anisotropy. It is therefore a fingerprint of the local symmetry of the structural units of the amorphous matrix, and it reacts rather sensitively to small structural modifications induced by changes of the composition, by an annealing treatment or by application of external stresses.

To extract information about the local symmetry from λ_s one has to establish the relation between the local magnetostrictive strains associated with the atomic-scale structural units and the macroscopic magnetostriction tensor which exhibits only one independent component, the magnetostriction constant λ_s, if the macroscopic magnetic anisotropy is zero (see Chapter 2). A first attempt to establish this relation was made by O'Handley and Grant [11.4] who developed the polycrystalline model of amorphous ferromagnets. They assumed that due to the strong structural short-range order amorphous alloys are composed of very small structural units ('grains'), which exhibit similar symmetry and chemistry but which are arranged in such a way that there is no long-range order. For simplicity, all the grains are supposed to be identical, and they are described on a phenomenological level by local material tensors, e.g., the local tensor of elastic constants, $c_{ijkl}(r)$, or of elastic compliances, $s_{ijkl}(r)$, with $s = c^{-1}$, and the local magnetostriction tensor $\lambda_{ijkl}(r)$. These local tensors have the supposed symmetry in the local coordinate systems of the grains adapted to the local symmetry, but they exhibit spatial fluctuations in a fixed external coordinate system due to the different orientations of the grains (Section 11.2). On this phenomenological level amorphous ferromagnets may be treated on the same footing as polycrystalline ferromagnets. In a monocrystalline material all the structural units have the same orientation and are strained coherently, and thus the macroscopic magnetostrictive strains are identical to the local ones. In contrast, in an amorphous or polycrystalline material the grains are strained incoherently because of the different grain orientations, and because the grains are coupled by elastic interactions the local deformations are different from the ones the grains would exhibit if they were isolated. It is therefore rather complicated to extract information about the local symmetries from the measurement of the macroscopic magnetostriction constant λ_s. O'Handley and Grant [11.4] calculated λ_s for the polycrystalline model by neglecting the elastic couplings, i.e., by taking the volume average, $\langle \lambda_{ijkl}(r) \rangle$, which may be calculated from the local magnetostriction tensor of the grains and from the statistics of the grain orientation. For uniaxial symmetry of the grains this volume average is in general nonzero for random grain orientation, whereas the volume average of the related local magnetic anisotropy vanishes. On the other hand it may happen that $\lambda_s = 0$ although $\lambda_{ijkl}(r) \neq 0$, i.e., the material does not show a macroscopic magnetostrictive strain although it is magnetostrictive on a local scale, so that materials with $\lambda_s = 0$ may still exhibit magnetoelastic interactions with internal stresses on a local scale. It is one objective of Chapter 11 to go beyond the noninteracting grain model of O'Handley and Grant by taking into account the elastic interactions between the grains when calculating λ_s.

11.2 Polycrystalline model of amorphous ferromagnets

The shape magnetostriction arises from the strain dependence of the magnetocrystalline anisotropy energy density $\phi_k(r)$, which encompasses all contributions to the magnetic energy depending on the orientation of the magnetization in the material, e.g., anisotropic exchange couplings and spin–orbit couplings (which dominate in most cases). For disordered or amorphous ferromagnets the lowest-order term of $\phi_k(r)$ is given by (Chapter 2)

$$\phi_k(r) = \sum_{i \neq j} k_{ij}(r)\, \gamma_i(r)\, \gamma_j(r), \tag{11.1}$$

where $k_{ij}(r)$ is the second-order anisotropy tensor and the quantities $\gamma_i(r)$ denote the direction cosines characterizing the orientation of the magnetization with respect to an external coordinate system. Apart from Section 11.6 we consider in the following only the fully saturated state, i.e., $\gamma_i(r) = $ const., which may be obtained by applying a strong external magnetic field.

In an amorphous ferromagnet $k_{ij}(r)$ represents a spatially fluctuating tensor field with Fourier components of all wavelengths. For an ideal isotropic amorphous structure the volume average $\langle k_{ij}(r) \rangle$ is zero. Values of $\langle k_{ij}(r) \rangle \approx (10^2 - 10^4)$ erg/cm^3 may be induced by stress, or field-annealing of the samples. Fluctuations with wavelength of some μm and local values of some 10^4 erg/cm^3 arise from long-wavelength internal stresses of several 10^7 N/m^2 and have an influence (Chapter 9) on the domain structure [11.2]. Fluctuations with wavelengths of about 100 Å and local values of about 10^5 erg/cm^3 originate from medium-range microstructural stresses of some 10^8 N/m^2 and contribute to the deviations from the ferromagnetic saturation (Chapter 8). Finally, atomic-scale fluctuations are intrinsically related to the structural incompatibilities of the amorphous material which cause atomic-level stresses of about 10^{10} N/m^2 [11.5, 11.6]. They are especially important for amorphous rare earth magnets, but it has been shown by semiempirical Hartree–Fock calculations [11.7] that even in Fe-based amorphous ferromagnets there are atomic-scale anisotropies of the order of several 10^7 erg/cm^3, with a totally random distribution of the local easy axis directions. Magnetostriction in amorphous ferromagnets arises mainly from the strain dependence of the atomic-scale magnetocrystalline anisotropy energy, i.e., from the local magnetoelastic tensor,

$$B_{klij}(r) = \partial k_{ij}(r)/\partial \varepsilon_{kl}. \tag{11.2}$$

To describe the atomic-scale fluctuations on a phenomenological level the polycrystalline model of amorphous ferromagnets is adopted, i.e., the amorphous material is conceived as being composed of extremely small 'grains' consisting basically of an atom and its nearest neighbour atoms. For simplicity it is assumed that all the grains are identical within their own local coordinate systems adapted to the local symmetry, and that the corresponding local material tensors, generally denoted as T_{ij}^{local} or T_{ijkl}^{local}, exhibit uniaxial, for simplicity, hexagonal symmetry. (For the form of the material tensors in hexagonal systems see Appendix.)

Because the orientations of the grains vary from site to site, the representation of a local material tensor in a fixed external coordinate system exhibits spatial fluctuations and may be calculated via

$$T_{ijkl}(r) = a_{i\mu}(r) a_{j\nu}(r) a_{k\kappa}(r) a_{l\lambda}(r) T_{\mu\nu\kappa\lambda}^{local}, \tag{11.3}$$

where the quantity $a_{i\mu}(r)$ is the transformation matrix between the coordinates of the local and the external coordinate system given by

$$a_{i\mu}(r) = \begin{pmatrix} -\sin\varphi(r) & -\cos\theta(r)\cos\varphi(r) & \sin\theta(r)\cos\varphi(r) \\ \cos\varphi(r) & -\cos\theta(r)\sin\varphi(r) & \sin\theta(r)\sin\varphi(r) \\ 0 & \sin\theta(r) & \cos\theta(r) \end{pmatrix}. \tag{11.4}$$

Here $\theta(r)$ and $\varphi(r)$ denote the polar and azimuthal angles describing the orientation of the six-fold hexagonal axis in the external coordinate system. Within this model the statistical information is given by the set of n-point correlation functions for which we assume the statistically decoupled form [11.8]

$$\langle a_{i_1\mu_1}(r_1) a_{i_2\mu_2}(r_2) \dots a_{i_n\mu_n}(r_n) \rangle$$
$$= \langle a_{i_1\mu_1}(r_1) a_{i_2\mu_2}(r_1) \dots a_{i_n\mu_n}(r_1) \rangle g^{(n)}(r_1, r_2, \dots r_n). \tag{11.5}$$

This ansatz implies that correlations between the orientations of the internal coordinate systems and the shapes of the grains are neglected [11.8]. This may be a good approximation for polycrystals with large grains, but for the atomic-scale grains of amorphous ferromagnets the shapes of the grains should be correlated to the orientations of the internal coordinate systems. The prefactors in front of $g^{(n)}$ are calculated from the distribution function density, $P(\cos\theta, \varphi)$, for the angles θ

and φ, for which we use

$$P(\cos\theta, \varphi) = \frac{1}{4\pi}\left[1 + \delta(3\cos^2\theta - 1)\right]. \tag{11.6}$$

Here δ is an anisotropy parameter which is zero for ideal isotropic systems but which attains nonzero values for most real amorphous ferromagnets, especially after field or stress annealing. It turns out that in isotropic systems only the property $g^{(n)}(r_1, r_1, \ldots, r_1) = 1$ is required for the final calculations based on a perturbation approach (Section 11.4). For macroscopically anisotropic systems with uniaxial symmetry we confine ourselves to the first-order perturbation approach where we need just $g^{(2)}$, which we can write most generally as

$$g^{(2)} = g^{(2)}\left[\left(\frac{x^2 + y^2}{\zeta_\perp^2} + \frac{z^2}{\zeta_\parallel^2}\right)^{1/2}\right] = g^{(2)}\left[\frac{r}{\zeta_\perp}(1 - \alpha\cos^2\theta)^{1/2}\right], \tag{11.7}$$

with two different correlation lengths $\zeta_\perp$ and $\zeta_\parallel$ perpendicular and parallel to the macroscopic anisotropy axis and the anisotropy parameter

$$\alpha = 1 - \frac{\zeta_\perp^2}{\zeta_\parallel^2}. \tag{11.8}$$

For the final calculation the analytical form of eq. (11.7) is not required, just the value of α.

11.3 Basic computational ideas

To illustrate the basic physical idea we perform a gedanken-experiment according to Fig. 11.1 (similar to the gedanken-experiment of Chapter 2), which sketches the following states.

State a: We consider a polycrystalline model according to Section 11.2, i.e., either a real polycrystal or an amorphous material described by this model. Furthermore, we assume a more or less random grain orientation with concomitant random orientations of the six-fold hexagonal axes. We switch off the exchange interactions, yielding a system without magnetization and without magnetostrictive deformations, which defines the reference state for the following calculations. It should be noted that for amorphous materials this reference state is not stress-free, but there are the atomic-level structural stresses [11.5, 11.6] discussed in Section 11.2.

State b: We cut the system into the single grains. Thereby, we do not allow structural relaxation processes which would remove the atomic-level structural stresses.

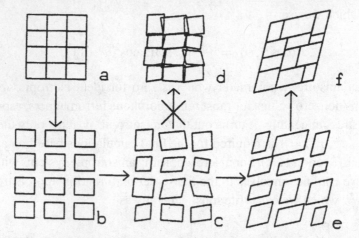

Fig. 11.1. Schematic illustration of the magnetostriction process in the polycrystalline model (see text).

State c: Now the exchange interactions and a strong external magnetic field are switched on, but we do not allow for rigid rotations of the grains for the moment. Then all the grains will exhibit spontaneous magnetostrictive strains (Chapter 2),

$$\varepsilon_{ij}^Q = \lambda_{ijkl}\gamma_k\gamma_l = -s_{ijmn}B_{mnkl}\gamma_k\gamma_l, \tag{11.9}$$

which differ from grain to grain because of the random grain orientation. As a result of these strains, the magnetocrystalline anisotropy tensor k_{ij} of the grain is modified by an additive term $B_{klij}\varepsilon_{kl}^Q$. Then the representation of k_{ij} in the original internal coordinate system of the unstrained grain with polar axis parallel to the original easy axis direction will no longer have the simple hexagonal form (11.69). However, it is still possible to find a new internal system for which the modified tensor is diagonal (with modified anisotropy strength k_{33}) and where the new polar axis defines the new easy axis direction [11.9]. This strain-induced modification of the easy axis direction is visualized in Fig. 11.2. The strain-induced modification of the local anisotropy strength and of the local easy axis direction is denoted as the 'conventional mechanism', because it also determines magnetostriction in monocrystalline ferromagnets. The so-called 'reorientation mechanism' which occurs in addition for polycrystalline and amorphous ferromagnets will be discussed below.

States d–f: After the introduction of the spontaneous magnetostrictive strains ε_{ij}^Q, the grains no longer fit together into a compact material (d). Additional elastic strains ε_{ij}^{el} of the grains must be performed (e) in order to restore compatibility of the material (f), and the total magnetostrictive strain is $\varepsilon_{ij}^m = \varepsilon_{ij}^Q + \varepsilon_{ij}^{el}$. It is this

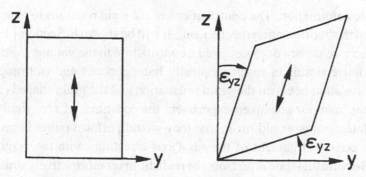

Fig. 11.2. Schematic illustration of the strain-induced reorientation of the easy-axis direction.

competition between the gain of magnetocrystalline anisotropy energy due to ε_{ij}^Q and the cost of elastic energy due to ε_{ij}^{el} which determines the effective macroscopic magnetostriction tensor λ_{ijkl}^{eff}. If no elastic strains were required, the volume average of the total magnetostrictive strain could be calculated as a simple average over the local spontaneous magnetostrictive strains, yielding $\lambda_{ijkl}^{eff} = \langle \lambda_{ijkl}(\boldsymbol{r}) \rangle$ as in the noninteracting grain model of O'Handley and Grant [11.4]. Because of the elastic coupling effects the grains can no longer deform as if they were isolated. As an effect, λ_{ijkl}^{eff} is different from $\langle \lambda_{ijkl}^{eff}(\boldsymbol{r}) \rangle$. Furthermore, the elastic strains are accompanied by elastic stresses

$$\sigma_{ij}^{el}(\boldsymbol{r}) = c_{ijkl}(\boldsymbol{r})\varepsilon_{kl}^{el}(\boldsymbol{r}). \tag{11.10}$$

Here $c_{ijkl}(\boldsymbol{r})$ denotes the tensor of elastic constants which may be obtained from the hexagonal tensor c_{ijkl}^{local} in the symmetry adapted local coordinate systems of the grains via eq. (11.3). It is obvious that the existence of a more or less random strain field, $\varepsilon_{ij}^m = \frac{1}{2}(\partial_i s_j + \partial_j s_i)$ with s denoting the displacement vector, implies a random field of rigid rotations, $w_{ij}^m = \frac{1}{2}(\partial_i s_j - \partial_j s_i)$. This means that although so far the primary spontaneous deformation is a pure strain deformation (we did not allow for rigid rotations of the isolated grains) we obtain a rotation field w_{ij}^m concomitant to the strain field ε_{ij}^m.

We now finally take into account that a rigid rotation of a grain also changes the magnetic anisotropy energy, and allow that the units may perform rigid rotations not only when being compelled to restore compatibility but also 'intentionally' to obtain a better alignment of the local easy axis directions with the magnetization in order to lower the local anisotropy energy. This means that the energy functional of the system does not depend just on the strains ε_{ij}^m but it also contains a term depending on w_{ij}^m (see Section 11.4). If the grain orientation is correlated with the shape of the grain, then a better alignment of the easy axis directions should induce a macroscopic deformation. The contributions of the rigid rotations to magnetostriction is called the 'reorientation mechanism'. It will be shown in Section 11.5 that the reorientation mechanism does not yield a contribution to the volume average of the magnetostrictive strains in macroscopically homogeneous and isotropic systems even if a correlation between the grain orientation and the grain shape is admitted.

Removing from our gedanken-experiment the constraint of zero rigid rotations for the isolated grains would mean that they would perform rather large rotations to obtain a perfect alignment of the easy axis directions with the magnetization direction. Because this state is far from the realistic magnetostrictively strained state, it does not make sense to introduce the concept of spontaneous magnetostrictive rotations w_{ij}^Q. As a result, the effect of rigid rotations cannot be described by the incompatibility method which is closely related to our gedanken-experiment, but we have to consider from the very beginning the total rotations w_{ij}^m by a method based on the minimization of an energy functional (Section 11.4).

11.4 Mathematical formalism

In this chapter we outline two methods [11.10] for the calculation of the effective magnetostriction tensor $\lambda_{ijkl}^{\text{eff}}$ defined by

$$\langle \varepsilon_{ij}^m \rangle = \lambda_{ijkl}^{\text{eff}} \gamma_k \gamma_l, \tag{11.11}$$

the balance-of-force method and the incompatibility method. These two methods yield results in the form of two different infinite perturbation series. If only a few terms of these series are calculated, respectively, two different approximate solutions are obtained, and from a comparison of the two methods for the same order of the perturbation theory a numerical uncertainty may be estimated.

11.4.1 Balance-of-force method

In this method we do not decompose the total deformations into spontaneous and elastic deformations as in the gedanken-experiment of Section 11.3, but we consider from the very beginning the total strains ε_{ij}^m and the total rotations w_{ij}^m defined with respect to the nonmagnetic reference state (a) of Section 11.3. The total displacement field s, which is related to ε_{ij}^m and w_{ij}^m via

$$\varepsilon_{ij}^m = \frac{1}{2}(\partial_i s_j + \partial_j s_i), \tag{11.12}$$

$$w_{ij}^m = \frac{1}{2}(\partial_i s_j - \partial_j s_i), \tag{11.13}$$

is obtained by minimizing the total energy functional

$$\phi_t = \int d^3r \{ \phi_{el} [\varepsilon^m(r)] + \phi_k [\varepsilon^m(r), w^m(r)] \}. \tag{11.14}$$

Here ϕ_{el} is the elastic energy density, which in a local approximation may be written as

$$\phi_{el} = \frac{1}{2} c_{ijkl}(r) \varepsilon_{ij}^m(r) \varepsilon_{kl}^m(r). \tag{11.15}$$

It is well known [11.11] that elasticity theory can be extended to the atomic scale (in the present case to the atomic-scale 'grains' of the polycrystalline model of amorphous ferromagnets), but in this case in principle a nonlocal theory should be applied. The first nonlocal corrections to eq. (11.15) have been taken into account in the balance-of-force method in [11.12]. It has been shown that nonlocal corrections become relevant for systems with $|\lambda_{ijkl}(r)| \gg \lambda_{ijkl}^{eff}$ (type 2 materials, see Section 11.5) but can be neglected otherwise. In the following the local approximation is applied.

The quantity ϕ_k is the magnetocrystalline anisotropy energy density. Its strain-dependent part $\phi_k^{(\varepsilon)}$ is given by

$$\phi_k^{(\varepsilon)} = B_{ijkl} \varepsilon_{ij}^m \gamma_k \gamma_l, \tag{11.16}$$

and it contains the two contributions to the conventional mechanism discussed in Section 11.3. The rotation-dependent part, responsible for the reorientation mechanism, may be written as

$$\phi_k^{(w)} = 2\gamma_i w_{ik}^m k_{kj} \gamma_j. \tag{11.17}$$

This contribution is derived in the following way. If a considered grain rotates rigidly, the tensor k_{ij} remains unchanged in a co-rotating coordinate system. To describe the situation in a global system, we must perform a coordinate transformation between the rotated and the global system. Inserting the transformed k-tensor into $\phi_k = k_{ij} \gamma_i \gamma_j$ and taking into account only the rotation-dependent part yields in linear approximation eq. (11.17).

The displacement field $s(r)$ is obtained by minimizing ϕ_t, subject to the constraint that there are no macroscopic rotations of the sample, i.e.,

$$\int d^3r \, w_i(r) = 0. \tag{11.18}$$

Here $w_i(r)$ is the rotation vector, which for small rotations is related to the tensor $w_{ij}^m(r)$ via

$$w_i(r) = \frac{1}{2} \varepsilon_{ijk} w_{jk}^m(r), \tag{11.19}$$

where ε_{ijk} are the components of the permutation tensor. This task is accomplished by minimizing

$$\tilde{\phi}_t = \phi_t + \alpha_i \int d^3r \, \frac{1}{2} \varepsilon_{ijk} w^m_{jk}(\mathbf{r}), \tag{11.20}$$

with respect to s, where α_i are Lagrangian parameters, yielding for the balance-of-force equation for the bulk (surface effects are neglected in the following) the relation

$$\partial_j c_{ijkl}(\mathbf{r}) \partial_k s_l(\mathbf{r}) = -f_i^{\text{magn}}(\mathbf{r}) = -\partial_j \tau_{ij}^{\text{magn}}(\mathbf{r}), \tag{11.21}$$

with

$$\tau_{ij}^{\text{magn}} = B_{ijkl} \gamma_k \gamma_l + \varepsilon_{ijk} \left[(\boldsymbol{\gamma} \times k\boldsymbol{\gamma})_k - \frac{1}{2}\alpha_k \right]. \tag{11.22}$$

The Lagrangian parameters α_k are most conveniently obtained from the demand that the volume average of the torque $\mathbf{D}$ exerted by the forces $f_i^{\text{magn}}(\mathbf{r})$ vanishes, i.e.,

$$\mathbf{D} = \int d^3r \left(\mathbf{r} \times f_i^{\text{magn}}(\mathbf{r}) \right) = 0, \tag{11.23}$$

which is equivalent to the restriction in eq. (11.18), yielding

$$\alpha_k = 2\langle (\boldsymbol{\gamma} \times k\boldsymbol{\gamma})_k \rangle, \tag{11.24}$$

and hence

$$\tau_{ij}^{\text{magn}} = B_{ijkl} \gamma_k \gamma_l + \varepsilon_{ijk} [\boldsymbol{\gamma} \times (\mathbf{k} - \langle \mathbf{k} \rangle) \boldsymbol{\gamma}]. \tag{11.25}$$

In eq. (11.25) the first term describes the conventional magnetostriction mechanism, whereas the second part accounts for the reorientation mechanism. The divergence of τ_{ij}^{magn} enters the balance-of-force equation (11.21) in the form of 'external' forces $f_i^{\text{magn}}(\mathbf{r})$, whereas the stress tensor $\sigma_{ij}^{\text{el}} = c_{ijkl}\varepsilon_{kl}$ describes the elastic reaction of the system on these 'external' forces.

Equation (11.21) is solved by the Green's function method, yielding

$$s_i(\mathbf{r}) = \int d^3r \, G_{ij}(\mathbf{r}, \mathbf{r}') f_j^{\text{magn}}(\mathbf{r}'), \tag{11.26}$$

with the Green's function $G_{ij}(\mathbf{r}, \mathbf{r}')$ defined via

$$\partial_j \, c_{ijkl}(\mathbf{r}) \, \partial_k G_{ln}(\mathbf{r}, \mathbf{r}') = -\delta_{in} \, \delta(\mathbf{r}, \mathbf{r}'). \tag{11.27}$$

Inserting $f_j^{\text{magn}} = \partial_i \tau_{ji}^{\text{magn}}$ into eq. (11.26) yields after partial integration for the distortion tensor

$$\beta_{ij}^{\text{m}}(\boldsymbol{r}) = \partial_i s_j(\boldsymbol{r}) = \int \mathrm{d}^3 r' \, \Gamma_{ijkl}(\boldsymbol{r}, \boldsymbol{r}') \, \tau_{kl}^{\text{magn}}(\boldsymbol{r}'), \tag{11.28}$$

with the strain Green's function [11.13]

$$\Gamma_{ijkl}(\boldsymbol{r}, \boldsymbol{r}') = -\frac{\partial^2 G_{jl}(\boldsymbol{r}, \boldsymbol{r}')}{\partial x_i \partial x_k'}. \tag{11.29}$$

If $c_{ijkl}(\boldsymbol{r})$ exhibits arbitrary spatial fluctuations then it is impossible to calculate G_{ij} and Γ_{ijkl} exactly. For an iterative solution of the problem we decompose $c_{ijkl}(\boldsymbol{r})$ according to

$$c_{ijkl}(\boldsymbol{r}) = \langle c_{ijkl}(\boldsymbol{r}) \rangle + \delta c_{ijkl}(\boldsymbol{r}), \tag{11.30}$$

define the homogeneous Green's function G_{ij}^0 [11.8, 11.13, 11.14] by the equation

$$\partial_j \langle c_{ijkl} \rangle \partial_k G_{ln}^0(\boldsymbol{r}, \boldsymbol{r}') = -\delta_{in} \delta(\boldsymbol{r}, \boldsymbol{r}'), \tag{11.31}$$

and represent $G_{ij}(\boldsymbol{r}, \boldsymbol{r}')$ by a Neumann series which is formally written as

$$G = G^0 - G^0 \delta L G^0 + G^0 \delta L G^0 \delta L G^0 - G^0 \delta L G^0 \delta L G^0 \delta L G^0 + \cdots, \tag{11.32}$$

where δL stands for $\partial_j \delta c_{ijkl}(\boldsymbol{r}) \partial_k$. We then obtain up to second order

$$\begin{aligned}
\beta_{ij}^{\text{m}}(\boldsymbol{r}) = &\int \mathrm{d}^3 r' \, \Gamma_{ijkl}^0(\boldsymbol{r} - \boldsymbol{r}') \tau_{kl}^{\text{magn}}(\boldsymbol{r}') \\
&+ \iint \mathrm{d}^3 r' \, \mathrm{d}^3 r'' \, \Gamma_{ijkl}^0(\boldsymbol{r} - \boldsymbol{r}') \Gamma_{mnpq}^0(\boldsymbol{r}' - \boldsymbol{r}'') \delta c_{klmn}(\boldsymbol{r}') \tau_{pq}^{\text{magn}}(\boldsymbol{r}'') \\
&+ \iiint \mathrm{d}^3 r' \, \mathrm{d}^3 r'' \, \mathrm{d}^3 r''' \, \Gamma_{ijkl}^0(\boldsymbol{r} - \boldsymbol{r}') \Gamma_{mnpq}^0(\boldsymbol{r}' - \boldsymbol{r}'') \Gamma_{rstu}^0(\boldsymbol{r}'' - \boldsymbol{r}''') \\
&\times \delta c_{klmn}(\boldsymbol{r}') \delta c_{pqrs}(\boldsymbol{r}'') \tau_{tu}^{\text{magn}}(\boldsymbol{r}''').
\end{aligned} \tag{11.33}$$

Here $\Gamma_{ijkl}^0(\boldsymbol{r} - \boldsymbol{r}')$ is obtained from $G_{jl}^0(\boldsymbol{r} - \boldsymbol{r}')$ according to eq. (11.29) and exhibits for a macroscopically isotropic material the general structure [11.8, 11.13, 11.14]

$$\Gamma_{ijkl}^0(\boldsymbol{r}) = E_{ijkl} \delta(\boldsymbol{r}) + F_{ijkl}(\theta, \varphi)/r^3, \tag{11.34}$$

where the tensor components E_{ijkl} and F_{ijkl} are related to the two independent components of $\langle c_{ijkl} \rangle$, the bulk modulus K_0 and the shear modulus G_0. The average value $\langle c_{ijkl} \rangle$ can be calculated from the hexagonal tensor c_{ijkl}^{local} of the polycrystalline model using eq. (11.3) and the distribution function density of eq. (11.6). Finally, from $\langle \varepsilon_{ij}^{\text{magn}} \rangle = \frac{1}{2} \langle \beta_{ij}^{\text{magn}} + \beta_{ji}^{\text{magn}} \rangle$ the effective magnetostriction tensor $\lambda_{ijkl}^{\text{eff}}$ may be obtained via eq. (11.11).

11.4.2 Incompatibility method

As discussed in Section 11.3, the reorientation mechanism of magnetostriction cannot be conveniently described by the incompatibility method. It will be shown in Section 11.5 that in most cases this mechanism yields at most a negligibly small contribution. Therefore, it is reasonable to neglect for the moment the reorientation mechanism and to deal with the conventional mechanism by a second type of approach, the incompatibility method. This method is closely related to the gedanken-experiment performed in Section 11.3 of cutting the system into the individual grains that exhibit in an external magnetic field spontaneous but incompatible magnetostrictive strains, and restoring compatibility by additional elastic strains. Please note the different reference states for the two types of calculations. In the balance-of-force method the reference state is the nonmagnetic state, and the strains ε^m obtained by switching on the magnetization are defined with respect to this nonmagnetic state. In the incompatibility method the starting point is the incompatible state with the spontaneous magnetostrictive strains

$$\varepsilon_{ij}^Q(r) = \lambda_{ijkl}(r)\gamma_k\gamma_l. \tag{11.35}$$

The isolated magnetostrictively deformed grains are thereby free of stress, because a subdivision of the grains into smaller pieces would not induce any relaxation. In contrast, the elastic strains which are required to restore compatibility give rise to elastic stresses

$$\sigma_{ij}^{el}(r) = c_{ijkl}(r)\varepsilon_{kl}^{el}(r), \tag{11.36}$$

and they must be determined in such a way that

$$\text{Ink}(\varepsilon^m) = \text{Ink}(\varepsilon^Q + \varepsilon^{el}) = 0, \tag{11.37}$$

yielding

$$\text{Ink}(\varepsilon^{el}) = -\text{Ink}(\varepsilon^Q) = \eta. \tag{11.38}$$

Because a tensor field is determined by its incompatibilities and its sources, we have to supplement eq. (11.38) with the fundamental equation of elastostatics (for zero external stress)

$$\text{Div}\,\sigma_{ij}^{el}(r) = \partial_i c_{ijkl}(r)\varepsilon_{kl}^{el}(r) = 0. \tag{11.39}$$

Equation (11.39) is solved by the general ansatz

$$\sigma^{el}(r) = \text{Ink}(\chi(r)), \tag{11.40}$$

with

$$\varepsilon_{ij}^{el} = s_{ijkl}\sigma_{kl}^{el}, \tag{11.41}$$

where

$$s = (c)^{-1} \tag{11.42}$$

is the tensor of elastic compliances. Equation (11.38) then may be written as

$$\text{Ink}[s(r) \cdot (\chi(r))] = \eta(r). \tag{11.43}$$

Equation (11.43) is the counterpart of eq. (11.21) from the balance-of-force method. Again, it may be solved by the Green's function method, yielding

$$\chi_{ij}(r) = \int d^3r' \, D_{ijkl}(r, r') \, \eta_{kl}(r'). \tag{11.44}$$

Inserting eq. (11.44) and $\eta = -\text{Ink}\,\varepsilon^Q$ into eq. (11.40) and performing partial integrations yields

$$\sigma_{ij}^{el}(r) = \int d^3r' \, \Delta_{ijkl}(r, r') \, \varepsilon_{kl}^Q(r'), \tag{11.45}$$

which is the counterpart to eq. (11.28), with the stress Green's function [11.13, 11.14]

$$\Delta_{ijkl}(r, r') = -\varepsilon_{imn}\varepsilon_{jpq}\varepsilon_{krs}\varepsilon_{ltu} \frac{\partial^4 D_{nqsu}(r, r')}{\partial x_m \partial x_p \partial x_r' \partial x_t'}. \tag{11.46}$$

If $s_{ijkl}(r)$ exhibits arbitrary spatial fluctuations then it is impossible to calculate D_{ijkl} and Δ_{ijkl} exactly. Therefore, again the Green's function is represented by a Neumann series analogous to eq. (11.32), where D^0 corresponds to the Green's function of a homogeneous material with tensor $\langle s \rangle$ of elastic compliances, with $s = \langle s \rangle + \delta s$. This yields up to second order

$$\sigma_{ij}^{el}(r) = \int d^3r' \, \Delta_{ijkl}^0(r - r')\varepsilon_{kl}^Q(r')$$

$$+ \iint d^3r' \, d^3r'' \, \Delta_{ijkl}^0(r - r')\Delta_{mnpq}^0(r' - r'') \, \delta s_{klmn}(r') \, \varepsilon_{pq}^Q(r'')$$

$$+ \iiint d^3r' \, d^3r'' \, d^3r''' \, \Delta_{ijkl}^0(r - r')\Delta_{mnpq}^0(r' - r'')\Delta_{rstu}^0(r'' - r''')$$

$$\times \delta s_{klmn}(r') \, \delta s_{pqrs}(r'') \, \varepsilon_{tu}^Q(r'''). \tag{11.47}$$

Here Δ^0_{ijkl} is obtained from D^0_{nqsu} according to eq. (11.46) and exhibits for a macroscopically isotropic material the general structure [11.8, 11.13]

$$\Delta^0_{ijkl}(\boldsymbol{r}) = U_{ijkl}\delta(\boldsymbol{r}) + \frac{V_{ijkl}(\theta, \varphi)}{r^3}, \qquad (11.48)$$

where the tensor components U_{ijkl} and V_{ijkl} are related to the two independent components κ_0 and β_0 of $\langle s \rangle$.

With $\varepsilon^{el}_{ij} = s_{ijkl}\sigma^{el}_{kl}$ we then can calculate $\langle \varepsilon^m_{ij} \rangle = \langle \varepsilon^Q_{ij} + \varepsilon^{el}_{ij} \rangle$ and obtain λ^{eff}_{ijkl} from eq. (11.11).

11.4.3 Zeroth- and first-order terms

The explicit forms of the zeroth- and first-order terms in the perturbation series for λ^{eff}_{ijkl} are given by:

1. Balance-of-force method

$$
\begin{aligned}
\lambda^{eff}_{ijkl} = & \int \mathrm{d}^3 r' \, \Gamma^0_{ijpq}(\boldsymbol{r} - \boldsymbol{r}')\langle B_{pqkl}(\boldsymbol{r}')\rangle \\
& + \iint \mathrm{d}^3 r' \, \mathrm{d}^3 r'' \, \Gamma^0_{ijpq}(\boldsymbol{r} - \boldsymbol{r}')\Gamma^0_{rstu}(\boldsymbol{r}' - \boldsymbol{r}'') \\
& \times \big\{ \langle \delta c_{pqrs}(\boldsymbol{r}')B_{tukl}(\boldsymbol{r}'')\rangle + \varepsilon_{tuv}\varepsilon_{vkm} \\
& \times \langle \delta c_{pqrs}(\boldsymbol{r}')[k_{ml}(\boldsymbol{r}'') - \langle k_{ml}(\boldsymbol{r}'')\rangle]\rangle \big\} + \cdots .
\end{aligned}
\qquad (11.49)
$$

2. Incompatibility method

$$\lambda^{eff}_{ijkl} = \langle \lambda_{ijkl}(\boldsymbol{r})\rangle + \int \mathrm{d}^3 r' \, \Delta^0_{pqrs}(\boldsymbol{r} - \boldsymbol{r}')\langle \delta s_{ijpq}(\boldsymbol{r})\lambda_{rskl}(\boldsymbol{r}')\rangle + \cdots . \qquad (11.50)$$

In the following we will consider materials which are macroscopically isotropic or nearly isotropic. Therefore the tensors $\boldsymbol{\Gamma}^0$ and $\boldsymbol{\Delta}^0$ for isotropic materials are always used and anisotropy effects are accounted for by anisotropic distribution function densities (eq. (11.6)) or anisotropic two-point correlation functions (eq. (11.7)). Taking into account the general structure of the tensor $\boldsymbol{\Gamma}^0$ (eq. (11.34)) and of the tensor $\boldsymbol{\Delta}^0$ (eq. (11.48)), which both consist of a δ-function part and an angular-dependent part, it becomes obvious that we have to evaluate two different types of integrals:

1. Calculation of moments, e.g., $\langle B_{pqkl}(0)\rangle$, $\langle \delta c_{pqrs}(0)B_{tukl}(0)\rangle$, $\langle \delta s_{ijpq}(0)\lambda_{rskl}(0)\rangle$ etc. In the polycrystalline model of Section 11.2 this reduces to the calculation of the prefactors in eq. (11.5) by use of the distribution function density (eq. (11.6)).

2. Calculation of integrals over Γ^0 or Δ^0 and of integrals over products of Γ^0 or Δ^0 and correlation functions $g^{(n)}(r_1, r_2, \ldots, r_n)$, with [11.8]

$$\int d^3r' \, \Gamma^0_{ijkl}(r') = -\langle c_{ijkl}\rangle^{-1} \tag{11.51}$$

and

$$\int d^3r' \, \Delta^0_{ijkl}(r') = 0. \tag{11.52}$$

For materials with isotropic correlation functions (which depend only on the relative angles between the vectors r_1, r_2, ..., r_n) it can be shown [11.10] that integrals containing angular parts of Γ^0 or Δ^0 are zero. The only nonvanishing contributions are those containing exclusively the δ-function parts of Γ^0 or Δ^0, yielding simply $g^{(n)}(0, 0, \ldots, 0) \equiv 1$, i.e., for macroscopically isotropic materials the explicit form of the correlation functions is irrelevant for the calculation of $\lambda^{\text{eff}}_{ijkl}$. This means that within the polycrystalline model of amorphous ferromagnets the same results for $\lambda^{\text{eff}}_{ijkl}$ are obtained for amorphous and for polycrystalline materials, and the size of the grains does not enter. For anisotropic correlation functions we confine ourselves to the determination of the first-order results, where only the correlation function $g^{(2)}$ given by eq. (11.7) enters. Now integrals containing the angular-dependent parts of the Green's functions yield nonvanishing contributions. By evaluating $g^{(2)}$ of eq. (11.7) into a power series of the anisotropy parameter δ and neglecting higher than linear terms it can be shown that again the explicit form of $g^{(2)}$ does not enter, i.e., we again obtain the same results for amorphous and polycrystalline ferromagnets. The explicit forms of the correlation functions are required only when calculating, by use of eq. (11.33) or eq. (11.47), the correlations of type $\langle \varepsilon^m_{ij}(r)\varepsilon^m_{kl}(r')\rangle$, e.g., which yield information about the spatial fluctuations of the magnetostrictive strains. Such calculations have been performed in [11.15] (and some results will be summarized in Section 11.5).

The final explicit results for $\lambda^{\text{eff}}_{ijkl}$ from the balance-of-force method and the incompatibility method up to the first order of the perturbation theory are summarized in the appendix of [11.10] for a macroscopically isotropic polycrystalline model with hexagonal grains. In an isotropic system the effective magnetostriction tensor is totally determined by the magnetostriction constant $\lambda_S = \langle \varepsilon^m_\parallel\rangle = -\langle \varepsilon^m_\perp\rangle/2 = \frac{2}{3}(\langle\varepsilon^m_\parallel\rangle + \langle\varepsilon^m_\perp\rangle)$, where $\varepsilon^m_\parallel$ and $\varepsilon^m_\perp$ denote the magnetostrictive strains parallel and perpendicular to the magnetization direction, and it attains the form

$$\lambda^{\text{eff}}_{ijkl} = \begin{pmatrix} \lambda_S & -\frac{1}{2}\lambda_S & -\frac{1}{2}\lambda_S & 0 & 0 & 0 \\ -\frac{1}{2}\lambda_S & \lambda_S & -\frac{1}{2}\lambda_S & 0 & 0 & 0 \\ -\frac{1}{2}\lambda_S & -\frac{1}{2}\lambda_S & \lambda_S & 0 & 0 & 0 \\ 0 & 0 & 0 & \frac{3}{2}\lambda_S & 0 & 0 \\ 0 & 0 & 0 & 0 & \frac{3}{2}\lambda_S & 0 \\ 0 & 0 & 0 & 0 & 0 & \frac{3}{2}\lambda_S \end{pmatrix}. \tag{11.53}$$

Table 11.1. *Results for λ_s according to the balance-of-force method (bfm) and the incompatibility method (im).*

	Co		Gd	
	bfm	im	bfm	im
0 order	$-7.04 \cdot 10^{-5}$	$-6.47 \cdot 10^{-5}$	$3.63 \cdot 10^{-6}$	$1.07 \cdot 10^{-5}$
1 order	$-6.54 \cdot 10^{-5}$	$-6.79 \cdot 10^{-5}$	$7.52 \cdot 10^{-6}$	$7.52 \cdot 10^{-6}$
2 order	$-6.71 \cdot 10^{-5}$	$-6.85 \cdot 10^{-5}$	$7.25 \cdot 10^{-6}$	$7.36 \cdot 10^{-6}$

The zeroth-order term of the incompatibility method then yields for hexagonal grains

$$\lambda_s^0 = \frac{2}{5}\lambda_{11} + \frac{2}{15}\lambda_{33} - \frac{2}{15}\lambda_{12} + \frac{4}{15}\lambda_{44}, \tag{11.54}$$

where the λ_{ij} are defined by eq. (11.71).

11.5 Results for the saturation magnetostriction of amorphous and polycrystalline ferromagnets

To obtain quantitative results for $\lambda_{ijkl}^{\text{eff}}$ the material has to be characterized by the values of the local material tensors c_{ijkl}^{local}, s_{ijkl}^{local}, B_{ijkl}^{local}, $\lambda_{ijkl}^{\text{local}}$ and k_{ml}^{local} and by the anisotropy parameters δ of the distribution function density (eq. (11.6)) and α of the two-point correlation function $g^{(2)}$ (eq. (11.7)). For a polycrystalline material the values of the local material tensors may be known from measurements on the monocrystalline counterparts. In contrast, for amorphous materials (for which the polycrystalline model is a rough approximation anyway) we must start from some realistic values and then modify these values in a reasonable range. As realistic values we insert those [11.16, 11.17] for crystalline Co at $T = 300$ K and crystalline Gd at $T = 4$ K. Note that this does not mean that we consider Co-based or Gd-based amorphous materials. An Fe-based amorphous ferromagnet, for example, may exhibit local anisotropy tensors similar to those of crystalline Co [11.7], and the local values for the near-zero magnetostrictive Co-based alloys may be totally different from those of crystalline Co. The results for λ_s in various orders of the perturbation theory for the balance-of-force method and the incompatibility method are shown in Table 11.1 for macroscopically isotropic materials with the local material parameters of Co and Gd.

The following questions will be attacked in the present chapter:

1. How strong are the deviations from the noninteracting grain model of O'Handley and Grant [11.4], i.e., how important are the elastic coupling effects between the grains on $\lambda_{ijkl}^{\text{eff}}$?

2. How large are the contributions of the conventional and the reorientation mechanisms?
3. How strong are the spatial fluctuations of the local magnetostrictive strains?

Concerning the first question it becomes obvious from eq. (11.50) that the zeroth-order term, $\langle \lambda_{ijkl} \rangle$, of the incompatibility method corresponds to the noninteracting grain model, whereas the higher-order terms describe the effect of the elastic couplings. Taking into account eq. (11.51), the zeroth-order term of the balance-of-force method yields $\lambda_{ijkl}^{\text{eff},0} = -\langle c_{ijpq} \rangle^{-1} \langle B_{pqkl} \rangle$, which in general is different from $\langle \lambda_{ijkl} \rangle = -\langle c_{ijpq}^{-1} B_{pqkl} \rangle$. In both methods the elastic coupling effects are related to spatial fluctuations δc_{ijkl}, showing that for a hypothetical material with elastically isotropic grains there would be no effect of the elastic couplings, i.e., the noninteracting grain model would yield the exact results. For elastically anisotropic grains the following two types of material may exist:

1. Type 1 materials for which the components λ_{11}, λ_{12}, λ_{33}, and λ_{44} of the local magnetostriction tensor of the grains are comparable in magnitude to λ_s.
2. Type 2 materials for which these components are considerably larger than λ_s. This may arise from a partial cancellation of the various contributions to the zeroth-order result (eq. (11.54)) or from a partial cancellation of the zeroth-order result and the higher-order terms describing the elastic coupling effects. These materials then may be strongly magnetostrictive on a local scale but weakly magnetostrictive on a macroscopic scale. The question arises whether the near-zero magnetostrictive Co-based alloys belong to type 1 (then these materials would be near-zero magnetostrictive also on a local scale) or to type 2 (then the materials would be magnetostrictive on a local scale).

For the local material parameters of Co a type 1 material is found (Table 11.1). The dominant contribution is thereby given by the zeroth-order term, i.e., in this case the effect of the elastic couplings (characterized by $(\lambda_s \,(2\,\text{order}) - \lambda_s\,(0\,\text{order}))/\lambda_s\,(2\,\text{order})$ in Table 11.1) is rather small. In contrast, for the local material parameters of Gd (which exhibits a larger elastic anisotropy) the local λ-values (Table 11.1) are similar to those of Co, but there is a partial cancellation of various terms in eq. (11.54), so that a type 2 material arises, and the elastic coupling effects are much stronger (Table 11.1). (It should be recalled (Section 11.4.1) that in this case the elastic nonlocality should be taken into account in order to get quantitative results). Playing around with the detailed values of the λ parameters while fixing their order of magnitude we find situations for which $\lambda_s^0 = 0$ while $\lambda_s \neq 0$, i.e., in this case the whole macroscopic magnetostriction is related to the elastic coupling effects and the noninteracting grain model fails totally. On the other hand, we can find materials with $\lambda_s = 0$ but nonzero values of λ_{11}, λ_{12}, λ_{33} and λ_{44}, i.e., type 2 materials which are zero magnetostrictive on a macroscopic scale but strongly magnetostrictive on a local scale. This result may be relevant when trying to find new zero-magnetostrictive alloys. If $\lambda_s = 0$ arises from the fact that the local magnetostriction tensor is zero (type 1) then it would correspond to a very special electronic situation for the grains which possibly could only be achieved around some special composition of Co-based materials. Because there are in principle infinitely many sets of local material tensors leading to a type 2 zero-magnetostrictive system, it may be that zero-magnetostrictive materials of totally different compositions, e.g., Fe-based alloys, can be produced. Considering the temperature dependence of λ_s of near-zero magnetostrictive Co-based amorphous alloys, arguments have been given [11.18] for the conjecture that these materials are of type 2.

Concerning the magnetostriction mechanism, it can be shown [11.10] that for statistically homogeneous and isotropic systems the reorientation mechanism does not contribute at all to λ_s. For the first-order term this becomes obvious from eq. (11.49). Because ε_{tuv} is antisymmetric in t and u and δc_{pqrs} is symmetric in r and s, only that part of Γ^0_{rstu} which is symmetric in r, s and antisymmetric in t, u contributes, and this part is given by $F_{(rs)[tu]}$. For a homogeneous and isotropic material the remaining integral is zero because of $\int F_{(rs)[tu]}(\theta, \varphi)\,d\Omega = 0$. (Please note that this result holds for arbitrary homogeneous and isotropic correlation functions $\langle \delta c_{pqrs}(r')[k_{ml}(r'') - \langle k_{ml}(r'')\rangle]\rangle$ and is not bound to the special ansatz in eq. (11.5) which would eliminate the correlation between grain orientation and grain shape.) A finite but in any case extremely small contribution of the reorientation mechanism arises for anisotropic correlation functions. We thus conclude that magnetostriction in polycrystalline and amorphous ferromagnets is dominated by the conventional mechanism for which magnetostriction originates from the strains rather than from the rigid rotations of the grains. Please note (Section 11.3) that on the other hand this conventional mechanism includes the modification of the local anisotropy strengths k_{33} and the strain-induced modifications of the local easy axis directions. It has been shown [11.9] that in some materials (e.g., for the Co parameters) this latter aspect of the conventional mechanism is more important than the modification of k_{33}, whereas in other materials (e.g., for the Gd parameters) it is the other way round. Altogether, we conclude that the modification of the easy axis directions is important for polycrystalline and amorphous ferromagnets but that the rigid rotations of the grains are not relevant for λ_s.

Concerning the spatial fluctuations of the magnetostrictive deformations [11.15], they are moderate for the Co parameters but rather strong for the Gd parameters and especially strong for near-zero magnetostrictive systems of type 2. For instance, for the Gd parameters we obtain $\sqrt{(\langle(\varepsilon^m_\parallel)^2\rangle - \langle\varepsilon^m_\parallel\rangle^2)/\langle\varepsilon^m_\parallel\rangle^2} = 5$. However, it can be shown [11.15] that for materials with atomic-scale fluctuations of the material parameters there are fluctuations of the magnetostrictive strains also only on an atomic scale, as long as effects of elastic nonlocality can be neglected. This has important consequences for the micromagnetic theories of domain wall pinning in amorphous ferromagnets (Chapter 5). On the one hand there may be a strong amplitude of the magnetostrictive fluctuations, on the other hand these fluctuations are of short range. Therefore, they must be taken into account only when discussing the anyway rather weak influence of atomic-scale internal structural stresses on the domain wall pinning. When calculating the strong effect of internal stresses of extension 50–100 Å, however, the magnetoelastic properties of the material may be well represented by the macroscopic magnetostriction constant λ_s.

Although the contributions of the reorientation mechanism average out for $\langle \varepsilon_{ij}^m \rangle$, there is an effect of this mechanism on $\langle (\varepsilon_{ij}^m)^2 \rangle$, which, however, is appreciable only for near-zero magnetostrictive materials of type 2 with strong local anisotropy strength k_{33} [11.15]. The data for $\langle (w_{ij}^m)^2 \rangle$ are comparable to those for $\langle (\varepsilon_{ij}^m)^2 \rangle$. This results from the fact (Section 11.3) that the existence of a spatially more or less random strain field ε_{ij}^m implies a random field of rigid rotations w_{ij}^m.

Finally, it should be noted that any process which changes the structure of the material will change the values of the local material tensors and the statistical properties, i.e., the anisotropy parameters δ and α of eq. (11.6) and eq. (11.7), and hence will change magnetostriction. For instance, application of external tensile stress changes λ_s of Co-based near-zero magnetostrictive glasses according to [11.19–11.22]

$$\lambda_s(\sigma) = \lambda_s(0) + k\sigma, \tag{11.55}$$

with $k \approx -$ some 10^{-10} MPa^{-1}. This phenomenon has been attributed to an elastic process (directional ordering of atom pairs or local easy axes [11.23]), to the coexistence of two amorphous phases [11.24], and it can be explained within the present polycrystalline model by a strain dependence of B_{ijkl}^{local} [11.25]. Furthermore, a field or stress annealing [11.26, 11.27] changes λ_s of near-zero magnetostrictive Co-based alloys by $|\Delta\lambda_s| \approx 10^{-7}$, i.e., by an amount which is comparable to $|\lambda_s|$ before annealing. The effect of field annealing is commonly interpreted [11.28–11.31] in terms of thermally activated changes of the atomic short-range order. Within the present polycrystalline model the field annealing can be accounted for [11.32] by the development of an anisotropy parameter δ in eq. (11.6). A critical review of experiments and theories on the effect of tensile stress and field annealing is given in [11.33].

11.6 Field dependence of magnetostriction

Experimentally, a magnetic field dependence of the magnetostriction constant $\lambda(H)$ with $\lim\limits_{H \to \infty} \lambda(H) = \lambda_s$ has been found [11.34] for strongly magnetostrictive

Fe-based alloys ($\lambda_s \approx$ some 10^{-5}) in the field regime of approach to ferromagnetic saturation and in near-zero magnetostrictive Co-based alloys [11.22, 11.35]. Whereas in the first case $|\lambda_s|$ increases with increasing H, a decrease of $|\lambda_s|$ was reported for the second case. In the latter case the field dependence is probably [11.22, 11.35] a thermodynamic effect, arising from the field dependence of the thermodynamic spin-correlation functions which determine the single-ion and two-ion contributions [11.36] to magnetostriction in near-zero magnetostrictive alloys. In contrast, the field effect in strongly magnetostrictive systems arises from the spatial scatter in the magnetization direction even at technical saturation due to the magnetoelastic coupling of the magnetization to medium-range internal stresses (see Chapter 8 on the law of approach to ferromagnetic saturation). This will be demonstrated in the present chapter, where the balance-of-force method is generalized to the case of nonperfect magnetic saturation ($\gamma = \gamma(r)$).

The difference in energy between the reference state (hypothetical amorphous state with internal stresses but without magnetization and magnetostrictive strains) and the same amorphous structure with magnetization and magnetostrictive strains ε_{ij}^m is given by (reorientation mechanism is neglected in the following)

$$
\phi = \int d^3 r \left\{ \frac{1}{2} c_{ijkl}(r)\, \varepsilon_{ij}^m \varepsilon_{kl}^m \right.
$$
$$
+ \left[k_{ij,0}(r) + B_{mnij}(r)\, s_{klmn}(r)\, \sigma_{kl}^{\text{def}} + B_{klij}(r)\, \varepsilon_{kl}^m \right] \gamma_i \gamma_j
$$
$$
\left. + A_{mn} \partial_m \gamma_i \partial_n \gamma_i - |M| H_{\text{ext}} \cdot \gamma_3 \right\}. \tag{11.56}
$$

Here the first term describes the elastic energy in local approximation. The second term is the magnetic anisotropy energy. It includes the atomic-scale fluctuations $k_{ij,0}(r)$ inherently related to the atomic structure (Section 11.2), the magnetoelastic coupling energy due to the internal stresses σ^{def} (see eq. (2.75) with eq. (11.9)) and the modification of the anisotropy energy due to the magnetostrictive strain ε^m. The third term is the expression for the exchange energy in an amorphous material (Chapter 2) and A_{mn} denotes the tensor of exchange constants. In the following we neglect the spatial fluctuations of A_{mn} and assume $A_{mn} = A\delta_{mn}$. The last term represents the magnetostatic coupling term (eq. (2.36)) to a constant external field H_{ext} in the z-direction, and we assume $|M| = \text{const.} = $ saturation magnetization.

Minimizing eq. (11.56) with respect to $s(r)$ with $\varepsilon_{ij}^m = \frac{1}{2}(\partial_i s_j^m + \partial_j s_i^m)$ and adopting Brown's approximation $\gamma_{1,2} \ll \gamma_3$ yields [11.37] the following system of coupled differential equations:

$$
\partial_i c_{ijkl}(r) \varepsilon_{kl}^m(r) = -\partial_i B_{ijkl}(r) \gamma_k(r) \gamma_l(r), \tag{11.57}
$$

$$2A\partial_m\partial_m\gamma_i(r) - 2\tilde{k}_{i3}(r) - |M|H_{\text{ext}}\gamma_i(r) = 0; \qquad i = 1, 2, \tag{11.58}$$

$$\tilde{k}_{ij} = k_{ij,0} + B_{mnij}s_{klmn}\sigma_{kl}^{\text{def}} + B_{klij}\varepsilon_{kl}^{\text{m}}. \tag{11.59}$$

In general, the modification of the anisotropy tensor by the magnetostrictive strain $\varepsilon_{kl}^{\text{m}}$ will have only a very small effect on $\gamma_i(r)$. Therefore, we neglect this effect in the following, which decouples eq. (11.58) from eq. (11.57). We then assume for the moment that eq. (11.58) has been solved, yielding $\gamma(r)$. For this $\gamma(r)$ eq. (11.58) has to be solved iteratively by the Green's function approach of Section 11.4.1 with the fluctuations $\delta c_{ijkl}(r)$ as perturbation parameters. Considering only the zeroth-order result we derive

$$\langle\varepsilon_{ij}^{\text{m}}(r)\rangle = -\langle c_{ijkl}\rangle^{-1}\langle B_{klmn}(r)\gamma_m(r)\gamma_n(r)\rangle. \tag{11.60}$$

It should be recalled that the use of this result is equivalent to the neglect of the elastic coupling effects between the grains, which is a good approximation for strongly magnetostrictive systems but which is not valid for materials of type 2. Starting from this equation and inserting reasonable ansatzes for the correlation between $B_{klmn}(r)$ and $k_{ij,0}(r)$, the influence of the atomic-scale anisotropy fluctuations $k_{ij}^0(r)$ on $\langle\varepsilon_{ij}^{\text{m}}\rangle$ and hence on λ_s via eqs. (11.11) and (11.53) has been calculated in [11.37], and it has been shown that the resulting deviations of $\lambda(H)$ from λ_s are rather weak for the field regime of approach to ferromagnetic saturation (similarly, the effect of spatial fluctuations of $k_{ij,0}(r)$ on the law of approach to ferromagnetic saturation for the magnetization is very weak, see Chapter 8). To consider the effect of medium-range inhomogeneities of $\gamma(r)$ (due to internal stresses σ^{def} or magnetic stray fields at the edges of the sample) we can neglect the correlations between $\gamma(r)$ and the atomic-scale fluctuations of $B_{klmn}(r)$, yielding from eq. (11.60)

$$\Delta\langle\varepsilon_{ij}^{\text{m}}(r)\rangle = \lambda_{ijmn}^{\text{eff}}\Delta\langle\gamma_m(r)\gamma_n(r)\rangle. \tag{11.61}$$

With

$$\lambda = \langle\varepsilon_{33}^{\text{m}}(r)\rangle = \lambda_s + \lambda_{33mn}^{\text{eff}}\Delta\langle\gamma_m(r)\gamma_n(r)\rangle \tag{11.62}$$

and Brown's approximation $\gamma_1, \gamma_2 \ll \gamma_3$, i.e.,

$$\gamma_3 = 1 - \frac{1}{2}(\gamma_1^2 + \gamma_2^2), \tag{11.63}$$

and taking into account the structure of the tensor $\lambda_{ijmn}^{\text{eff}}$ in isotropic systems, we obtain the final result [11.37, 11.38]

$$\lambda(H_{\text{ext}}) = \lambda_s\left(1 - 3\frac{\Delta M(H_{\text{ext}})}{M_s}\right), \tag{11.64}$$

with $\Delta M / M_s = \Delta \gamma_3$. We thus obtain the important result that the relative deviations $(\lambda_s - \lambda(H_{ext}))/\lambda_s$ of the field-dependent magnetostriction constant $\lambda(H_{ext})$ in the regime of approach to ferromagnetic saturation are three times larger than the deviations $(M_s - M(H_{ext}))/M_s$ of the magnetization. Therefore, the relations obtained in Chapter 8 for $\Delta M(H_{ext})$ can be directly applied to $\lambda(H_{ext})$, as long as medium- and long-range magnetic inhomogeneities are considered.

It should be recalled that eq. (11.64) was derived for strongly magnetostrictive materials where the elastic couplings between the strains have only a weak influence on λ_s. In these materials $|\lambda(H_{ext})|$ always increases with increasing H_{ext}. A decrease of $|\lambda(H_{ext})|$ may only arise if the elastic couplings are so strong that the higher-order terms of the perturbation series are comparable to or larger than the zeroth-order term. In this case the delicate cancellation of various terms at low fields may be disturbed in larger fields, and this may lead both to an increase and to a decrease of $\lambda(H_{ext})$. Therefore, if a material exhibits a decrease of $\lambda(H_{ext})$ at very low temperatures (where we can exclude the thermodynamic contributions discussed above), it must be concluded that it is a material of type 2.

References

[11.1] O'Handley, R.C., 1987, *J. Appl. Phys.* **62**, R 15.

[11.2] Kronmüller, H., Fähnle, M., Domann, M., Grimm, H., Grimm, R., and Gröger, B., 1979, *J. Magn. Magn. Mater.* **13**, 53.

[11.3] Kronmüller, H., 1981, *J. Appl. Phys.* **52**, 1859.

[11.4] O'Handley, R.C., and Grant, N.J., 1985, *Proc. Conf. on Rapidly Quenched Metals*, (Eds. S. Steeb and H. Warlimont, Elsevier Science, Amsterdam) p. 1125.

[11.5] Egami, T., and Srolovitz, D., 1982, *J. Phys. F: Met. Phys.* **12**, 2141.

[11.6] Pawellek, R., and Fähnle, M., 1989, *J. Phys. Condens. Matter* **1**, 7257.

[11.7] Elsässer, C., Fähnle, M., Brandt, E.-H., and Böhm, M.C., 1988, *J. Phys. F: Met. Phys.* **18**, 2463.

[11.8] Kröner, E., 1988. In *Modelling Small Deformations of Polycrystals* (Eds. J. Gittus and J. Zarke, Elsevier, Barking) p. 229.

[11.9] Fähnle, M., and Furthmüller, J., 1988, *J. Magn. Magn. Mater.* **72**, 6.

[11.10] Furthmüller, J., Fähnle, M., and Herzer, G., 1987, *J. Magn. Magn. Mater.* **69**, 79; *ibid.*, 89.

[11.11] Kröner, E., and Datta, B.K., 1966, *Z. Phys.* **196**, 203.

[11.12] Beuerle, T., and Fähnle, M., 1990, *J. Magn. Magn. Mater.* **88**, 7.

[11.13] Kröner, E., and Koch, H., 1976, *SM Archives* **1**, 183.

[11.14] Zeller, R., and Dederichs, P.H., 1973, *Phys. Stat. Sol. (B)* **55**, 831.

[11.15] Pawellek, R., Furthmüller, J., and Fähnle, M., 1988, *J. Magn. Magn. Mater.* **75**, 225.

[11.16] Carr, W.J., 1969. In *Magnetism and Metallurgy 1* (Eds. A.E. Berkowitz and E. Kneller, Academic Press, New York, London) p. 79.

[11.17] Simmons, G., and Wang, H., 1971. In *Single Crystal Elastic Constants and Calculated Aggregate Properties, a Handbook* (MIP Press, Cambridge, MA, London).

[11.18] Fähnle, M., Furthmüller, J., Pawellek, R., and Beuerle, T., 1991, *Appl. Phys. Lett.* **59**, 2049.

[11.19] Herzer, G., 1985. In *Proc. of Soft Magnetic Materials* **7**, edited by the European Physical Society (Cardiff, Wolfson Centre for Magnetics Technology) p. 355.

[11.20] Barandiarán, J.M., Hernando, A., Madurga, V., Nielsen, O.V., Vázquez, M., and Vázquez-López, M., 1987, *Phys. Rev. B* **35** 5066.

[11.21] Lachowicz, H.K., Siemko, A., Moser, N., Forkl, A., and Kronmüller, H., 1988, *Phys. Stat. Sol. (A)* **105**, 597.

[11.22] Kraus, L., and Duhaj, P., 1990, *J. Magn. Magn. Mater.* **83**, 337.

[11.23] Szymczak, H., and Lachowicz, H.K., 1988, *IEEE Trans. Magn.* **24**, 1747.

[11.24] Hernando, A., 1989. In *Physics of Magnetic Materials* (Eds. W. Gorzkowski, II.K. Lachowicz, and II. Szymczak, World Scientific, Singapore) p. 248.

[11.25] Fähnle, M., and Furthmüller, J., 1989, *Phys. Stat. Sol. (A)* **116**, 819.

[11.26] Warlimont, H., and Hilzinger, H.R., 1982. In *Proc. Conf. on Rapidly Quenched Metals* (Eds. T. Masumoto and K. Suzuki, Japan Inst. of Metals, Sendai) p. 1167.

[11.27] Hernando, A., Madurga, V., Núñez de Villavicencio, C., and Vázquez, M., 1984, *Appl. Phys. Lett.* **45**, 802.

[11.28] González, J., and Kulakowski, K., 1990, *J. Magn. Magn. Mater.* **86**, 207.

[11.29] Kronmüller, H., 1983, *Phys. Stat. Sol. (B)* **118**, 661.

[11.30] Kronmüller, H., Guo, H.-Q., Fernengel, W., Hofmann, A., and Moser, N., 1985, *Cryst. Latt. Def. Amorph. Mat.* **11**, 135.

[11.31] Gibbs, M.R.J., 1990, *J. Magn. Magn. Mater.* **83**, 329.

[11.32] Fähnle, M., and Furthmüller, J., 1990, *Phys. Stat. Sol. (A)* **117**, K71.

[11.33] Fähnle, M., Furthmüller, J., Pawellek, R., Beuerle, T., and Elsässer, C., 1991. In *Physics of Magnetic Materials* (Eds. W. Gorzkowski, M. Gutowski, H.K. Lachowicz and H. Szymczak, World Scientific, Singapore) p. 204.

[11.34] Grössinger, R., Herzer, G., Kerschnagel, J., Sassik, H., and Spindler, H., 1989. In *Proc. Conf. on Soft Magnetic Materials* **9**, Madrid, Spain.

[11.35] Hernando, A., Vázquez, M., Barandiarán, J.M., and van Hattum, W.J., 1988, *J. Physique* **49**, C8, 1333.

[11.36] du Trémolet de Lacheisserie, E., 1993, *Magnetostriction: Theory and Applications of Magnetoelasticity* (CRC Press, Boca Raton, USA)

[11.37] Pawellek, R., and Fähnle, M., 1989, *Phys. Stat. Sol. (A)* **111**, 617.

[11.38] Fähnle, M., Pawellek, R., and Kronmüller, H., 1989, *Phys. Stat. Sol. (A)* **112**, 189.

Appendix

The definitions of the tensors k_{ij} and k_{ijkl} from eq. (2.29), B_{ijkl} from eq. (11.2) and λ_{ijkl} from eq. (11.9) are not unique. For instance, we can multiply the magnetocrystalline anisotropy energy density ϕ_k by

$$1 = (1 - a) + a\delta_{ij}\gamma_i\gamma_j, \tag{11.65}$$

with an arbitrary real number a, using $\gamma_1^2 + \gamma_2^2 + \gamma_3^2 = 1$. This yields

$$\phi_k = k_{ij}\gamma_i\gamma_j = \tilde{k}_{ij}\gamma_i\gamma_j + \tilde{k}_{ijkl}\gamma_i\gamma_j\gamma_k\gamma_l, \tag{11.66}$$

with

$$\tilde{k}_{ij} = (1 - a)k_{ij}, \qquad \tilde{k}_{ijkl} = ak_{ij}\delta_{kl}, \tag{11.67}$$

without changing ϕ_k. Furthermore, we can add to B_{ijkl} any tensor of the form $B_{ij}\delta_{kl}$ without changing the magnetoelastic coupling energy density $\phi_{\mathrm{me}} = B_{ijkl}\varepsilon_{ij}\gamma_k\gamma_l$ with respect to its dependence on the magnetization direction, and we can add to λ_{ijkl} any tensor of the form $\lambda_{ij}\delta_{kl}$ without changing the spontaneous magnetostriction $\varepsilon_{ij}^{\mathrm{Q}} = \lambda_{ijkl}\gamma_k\gamma_l$ with respect to its dependence on the magnetization direction. The tensors used in Chapter 11 are traceless with respect to the third and fourth index, i.e.,

$$k_{ijkl}\delta_{kl} = B_{ijkl}\delta_{kl} = \lambda_{ijkl}\delta_{kl} = 0. \tag{11.68}$$

For hexagonal systems we then have

$$k^{\mathrm{local}} = k_{33} \begin{pmatrix} -\frac{1}{2} & 0 & 0 \\ 0 & -\frac{1}{2} & 0 \\ 0 & 0 & 1 \end{pmatrix}, \tag{11.69}$$

with

$$k_{33} = -\frac{2}{21}(7K_1 + 8K_2 + 8K_3), \tag{11.70}$$

where K_1, K_2 and K_3 are the hexagonal anisotropy constants defined by eq. (2.32). The tensor λ has the form

$$\lambda = \begin{pmatrix} \lambda_{11} & \lambda_{12} & -\lambda_{11} - \lambda_{12} & 0 & 0 & 0 \\ \lambda_{12} & \lambda_{11} & -\lambda_{11} - \lambda_{12} & 0 & 0 & 0 \\ -\frac{\lambda_{33}}{2} & -\frac{\lambda_{33}}{2} & \lambda_{33} & 0 & 0 & 0 \\ 0 & 0 & 0 & \lambda_{44} & 0 & 0 \\ 0 & 0 & 0 & 0 & \lambda_{44} & 0 \\ 0 & 0 & 0 & 0 & 0 & \lambda_{11} - \lambda_{12} \end{pmatrix}, \tag{11.71}$$

with

$$\lambda_{11} = \frac{2}{3}\lambda_{\mathrm{A}} - \frac{1}{3}\lambda_{\mathrm{B}}, \tag{11.72}$$

$$\lambda_{12} = \frac{2}{3}\lambda_{\mathrm{B}} - \frac{1}{3}\lambda_{\mathrm{A}}, \tag{11.73}$$

$$\lambda_{33} = -\frac{2}{3}\lambda_{\mathrm{C}}, \tag{11.74}$$

$$\lambda_{44} = 2\lambda_{\mathrm{D}} - \frac{1}{2}\lambda_{\mathrm{A}} - \frac{1}{2}\lambda_{\mathrm{C}}, \tag{11.75}$$

where λ_{A}, λ_{B}, λ_{C}, and λ_{D} are Mason's constants (W.P. Mason, *Phys. Rev.* **96**, 302 (1954)). The tensor B has the same form as λ except for B_{66} which is given by $\frac{1}{2}(B_{11} - B_{12})$ and is calculated from λ and c via $B_{ijkl} = -c_{ijpq}\lambda_{pqkl}$ and from the prescriptions for the transformation of the tensors from the full index notation to the Voigt notation. Finally, the tensor of elastic constants has the form

$$
c = \begin{pmatrix}
c_{11} & c_{12} & c_{13} & 0 & 0 & 0 \\
c_{12} & c_{11} & c_{13} & 0 & 0 & 0 \\
c_{13} & c_{13} & c_{33} & 0 & 0 & 0 \\
0 & 0 & 0 & c_{44} & 0 & 0 \\
0 & 0 & 0 & 0 & c_{44} & 0 \\
0 & 0 & 0 & 0 & 0 & \left[\frac{1}{2}(c_{11} - c_{12})\right]
\end{pmatrix}. \tag{11.76}
$$

12

Micromagnetic theory of phase transitions in spatially disordered spin systems

In the preceding chapters the magnetic and magnetoelastic properties of inhomogeneous magnets have been described by a micromagnetic approach, thereby neglecting the thermal excitations at finite temperatures. The present chapter is devoted to the micromagnetic theory of disordered magnets close to a magnetic phase transition. It should be stated explicitly at the beginning of this chapter that the micromagnetic theory will not be able to determine the critical behaviour, i.e., it cannot calculate asymptotic critical exponents, the correction terms to the asymptotic critical behaviour or possible crossover phenomena between various fixed points of the renormalization group theory. The objective of the micromagnetic theory is to describe the (noncritical) crossover regime from the critical behaviour of the spin system (which results from the couplings of critical spin fluctuations) to the mean field regime (where the correlations between spin fluctuations can be totally neglected). It will be demonstrated that in this regime the magnetic behaviour is determined by the interplay between the noncritical correlations in the spin system and the structural inhomogeneities.

In Section 12.1 a classification of the spin systems with spatial disorder which will be considered is given. The main part of Chapter 12 concentrates on phase transitions in random exchange ferromagnets, especially in amorphous ferromagnets. Section 12.2 will give a short overview on phase transitions in random exchange ferromagnets, whereas in Sections 12.3–12.5 the micromagnetic theory for phenomena outside the critical regime is developed. In Section 12.3 the molecular field theory and the Landau–Ginzburg theory of disordered ferromagnets are described in which the correlations between the spin fluctuations are totally neglected. It will be shown that the experimentally observed phenomena cannot be explained by these approaches. In contrast, the correlated molecular field theory (Section 12.4) which takes into account in a phenomenological manner the interplay between the noncritical spin fluctuations and the structural inhomogeneities yields results in good agreement with the experimental data. In Section 12.5 the correlated molecular field theory is extended to dynamic phenomena, and in Section 12.6 it is applied to describe the behaviour of spin glasses and random anisotropy magnets outside the critical regime.

12.1 Classification of disordered spin systems

In a spatially disordered spin system the local values of the parameters characterizing the material on an atomic scale (e.g., eq. (12.3)) or on a continuum scale (e.g., eq. (12.24)) deviate from their volume averages, and these spatial fluctuations are sometime called 'impurities'. It is customary to distinguish between 'annealed' and 'quenched' impurities [12.1]. For annealed impurities there is a thermal equilibrium between the spins and the impurities, whereas for quenched impurities the impurity variables are fixed forever and the spins distribute themselves to fit the condition set up by the quenched impurities. In the following only systems with quenched impurities are considered. Examples are amorphous ferromagnets which are quenched from the liquid state at very high temperatures and afterwards are thoroughly annealed at temperatures below the crystallization temperature in order to obtain a metastable configuration with no further irreversible structural changes.

For a further classification of the disordered spin systems considered in Chapter 12 we consider a classical atomic-scale Hamiltonian with spins S_i localized at sites i,

$$S_i = (S_{i1}, S_{i2}, \ldots, S_{in}),$$ (12.1)

and associated magnetic moments

$$m_i = g\mu_B S_i.$$ (12.2)

An Ising, XY, or Heisenberg system corresponds to $n = 1$, 2 or 3. When dealing with localized magnetic moments formed by itinerant electrons, the magnetic moments $|m_i|$ may attain noninteger multiples of μ_B which then correspond formally to noninteger values $|S_i|$ in eq. (12.2). In amorphous alloys the magnetic moments may differ from site to site because the various sites of an amorphous structure are crystallographically different and because the sites may be occupied by different atoms. For instance, in amorphous FeNiB systems the Fe atoms exhibit large magnetic moments of about $2.2\mu_B$ whereas the Ni atoms carry small magnetic moments of about $0.4\mu_B$ [12.2]. Such a situation can be described formally by writing the local spin vector as $S_i \cdot c_i$, where $|S_i| = S_i^{max} = |m_i|^{max}/(g\mu_B)$ is the 'spin quantum number' corresponding to the maximum value of $|m_i|$ occuring in the system, and the occupation numbers c_i are given by $c_i = |m_i|/|m_i|^{max}$. For a constant external field $H = (0, 0, H)$ the model Hamiltonian may be written as

$$\hat{H} = -\frac{1}{2} \sum_{ij} J_{ij} S_i \cdot S_j c_i c_j - \sum_i D_i c_i (n_i \cdot S_i)^2 - g\mu_B H \sum_i S_i^z. \quad (12.3)$$

Here the first term describes the exchange interactions. The exchange couplings J_{ij} thereby depend both on the type of the coupled atoms and on the positions of these atoms. Even when only exchange couplings exclusively between atoms within some suitably defined nearest neighbour sphere are considered, the J_{ij} between the same type of atoms are spatially fluctuating quantities in an amorphous matrix because of the nonperiodic structure [12.3, 12.4]. The second term describes the influence of local uniaxial anisotropies, with D_i denoting the local anisotropy strengths and with the unit vectors n_i characterizing the local easy axes directions. In spatially disordered systems both D_i and n_i may exhibit spatial fluctuations. The third term describes the Zeeman coupling to the external field H. Starting from eq. (12.3) the following prototype classification can be made:

1. Systems with random exchange interactions which may arise from a random distribution of atoms with different magnetic moments or from a random distribution of exchange interactions J_{ij}.
   (a) Percolative systems: In the simplest case of a site-percolation problem such systems are binary alloys $A_x B_{1-x}$ with constant nearest neighbour ferromagnetic exchange interactions $J_{ij} = J_0 > 0$ and a spatially random distribution of magnetic A atoms and nonmagnetic B atoms on a regular lattice. In such a system clusters are defined as subgroups of A atoms which are connected via nearest neighbour paths. As x increases the average cluster size increases, and for the percolation concentration x_c there is for the first time an infinite cluster which spans the lattice from one side to the other. It is obvious that long-range ferromagnetic order can only appear for $x \geq x_c$ and for $T \leq T_c(x)$. For x, T close to x_c, T_c interesting tricritical effects occur. The simplest bond-percolation problem is given by a regular lattice for which all sites are occupied by magnetic atoms but for which a fraction $1 - x$ of the nearest neighbour exchange couplings is removed in a spatially random fashion. A cluster then is defined as a subgroup of atoms connected by nearest neighbour couplings. The site- and bond-problems may be modified by considering weakly magnetic B atoms (as in the case of amorphous FeNiB alloys) instead of nonmagnetic B atoms, or by considering strong and weak bonds instead of strong and zero bonds. For a review on percolation systems see [12.5].
   (b) Systems with spatially random fluctuations of the exchange couplings J_{ij} according to a continuous distribution function density

$$P(J_{ij}) = \frac{1}{\sqrt{2\pi}\delta J_0} \exp(-(J_{ij} - J_0)^2/2(\delta J_0)^2), \quad (12.4)$$

with $J_0 = \langle J_{ij} \rangle > 0$ and $\delta^2 = (\langle J_{ij}^2 \rangle - \langle J_{ij} \rangle^2)/\langle J_{ij} \rangle^2$. For small values of δ there are only a few antiferromagnetic couplings. Then the ferromagnetic ground state is maintained, and we can describe such a system essentially as a ferromagnet with spatially fluctuating ferromagnetic exchange couplings. Many of the amorphous ferromagnets are random exchange ferromagnets which exhibit both site disorder of type 1a (with a random distribution of atoms with large and small magnetic moments) and random bond disorder of type 1b induced by the nonperiodic amorphous structure. For increasing variance, δ, more and more antiferromagnetic couplings are added, and for some value of δ there is a transition from a ferromagnetic ground state to a spin glass state with no long-range order. For reviews on spin glasses see [12.6, 12.7].

2. Random anisotropy magnets. In the simplest case such systems exhibit ferromagnetic nearest neighbour couplings which are either constant (J_0) or randomly distributed around the mean value J_0, but with totally random orientations of the easy axis directions n_i and with constant local anisotropy strength D_0. The nature of the magnetic low-temperature phase then depends sensitively [12.8] on the ratio D_0/J_0, but for very large values of D_0/J_0 each magnetic moment can be directed almost perfectly along the local easy axis direction giving rise to a speromagnetic state which very much resembles that of an Ising spin glass. It has been shown by Hartree–Fock calculations [12.9] that macroscopically isotropic amorphous Fe-based ferromagnets may be realizations of random anisotropy systems with small D_0/J_0, and some amorphous rare earth magnets may represent speromagnetic systems [12.10]. In real materials there is always a nonvanishing volume average of the magnetic anisotropy ($\langle n_i \rangle \neq 0$), and therefore most of the Fe–Ni-based amorphous alloys exhibit a ferromagnetic low-temperature phase with only small spin cantings in reasonable external fields [12.9], see also Chapter 8.

Chapter 12 is mainly concerned with ferromagnetic random exchange systems which exhibit spatial fluctuations of the exchange interactions according to type 1a for concentrations far above the percolation limit and according to type 1b with small values of δ. In Section 12.6 spin glasses (systems of type 1b with large values of δ) and speromagnets (random anisotropy magnets of type 2 with large values of D_0/J_0) are considered.

12.2 Phase transition in random exchange ferromagnets

Before developing the micromagnetic theory of phase transitions in random exchange ferromagnets, some basic facts on the properties of phase transitions in these systems are reviewed.

12.2.1 Critical behaviour

It is well known that for a second-order phase transition at the critical temperature T_c the asymptotic critical behaviour (T very close to T_c, H close to zero) of thermodynamic quantities may be described by power laws, e.g.,

$$M_s(T) = \lim_{H \to 0} M(T, H) = B(-t)^\beta, \tag{12.5}$$

$$\chi_0(T) = \lim_{H \to 0} \chi(T, H) = \Gamma_\pm |t|^{-\gamma}, \tag{12.6}$$

$$M(T = T_c, H) = A_0 H^{1/\delta}, \tag{12.7}$$

with critical exponents β, γ, δ, and with the critical amplitudes $B, \Gamma_\pm$ (with $+$ sign for $T > T_c$ and $-$ sign for $T < T_c$), and A_0. Thereby, the famous universality concept holds [12.1], which states that (as long as there are no long-range interactions) the critical exponents, the critical amplitude ratios and the scaling functions [12.1] do not depend on the details of the interactions but just on two parameters: the dimension d of space and the number n of components of the spin variable (see Section 12.1). Originally, this universality concept was obtained for spatially homogeneous systems, and therefore the question arises whether the exchange fluctuations of random exchange ferromagnets will modify this universality concept. Both heuristic arguments [12.11] and renormalization group calculations (see, e.g., [12.12, 12.13]) predict that systems with short-range quenched randomness exhibit the same asymptotic critical behaviour as the corresponding homogeneous systems if the critical exponent α_p of the specific heat of the pure homogeneous system is negative. In contrast, for $\alpha_p > 0$ a crossover occurs from the pure fixed point of the renormalization group transformation to a new fixed point ('random fixed point') which determines the asymptotic critical behaviour. For three-dimensional Heisenberg ferromagnets ($\alpha_p < 0$) the prediction of this famous Harris criterion has been confirmed by studying the phase transitions in amorphous ferromagnets. A proper asymptotic analysis of the experimental data reveals (see, for instance, [12.14–12.17]) that the values of the asymptotic critical exponents of amorphous ferromagnets are indeed those of the homogeneous three-dimensional Heisenberg ferromagnet. The main difference between the homogeneous and the amorphous ferromagnet is that in the homogeneous ferromagnet all spins participate in the ferromagnetic–paramagnetic phase transition whereas in amorphous ferromagnets only a fraction of all the spins is involved, which becomes smaller when the critical concentration x_c for the appearance of long-range ferromagnetic order in alloys of type $Fe_x Ni_{80-x} P_{14} B_6$, for instance, is approached [12.14, 12.16]. The conjecture that for systems with $\alpha_p > 0$ spatial fluctuations will modify the critical behaviour was proved rigorously by Chayes *et al.* [12.18], and numerical values for the critical exponents of three-dimensional random exchange Ising ferromagnets (with $\alpha_p > 0$) were obtained by Monte Carlo simulations [12.19] and by Monte Carlo renormalization group calculations [12.20].

12.2.2 Crossover regime to the mean field behaviour

When leaving the asymptotic critical regime by increasing the variable $|t|$ while still being in a critical temperature range (in the sense that concepts like self-similarity of the spin system and hence scaling still hold), then two different situations must be distinguished:

1. There is only one fixed point of the renormalization group transformation which determines the critical behaviour. In this case there will be deviations from the asymptotic critical behaviour with increasing $|t|$ in the form of nonanalytical correction terms arising from irrelevant scaling fields of the renormalization group approach [12.1] and/or in form of analytical correction terms arising from the nonlinearities of the scaling fields [12.21], see also [12.22].
2. There is more than one fixed point which determines the critical behaviour. When increasing $|t|$, a crossover from behaviour according to the stable fixed point (which determines the asymptotic critical behaviour) to behaviour according to an unstable fixed point occurs. Such a possibility is outlined in [12.13], which yields for strongly disordered random ferromagnets a crossover from asymptotic critical behaviour to behaviour according to a tricritical fixed point with a large susceptibility exponent. As a result, the effective susceptibility exponent $\gamma(T)$ introduced by Kouvel and Fisher [12.23],

$$\gamma(T) = (T - T_c) \chi_0(T) \, d\chi_0^{-1}(T)/dT, \qquad (12.8)$$

exhibits a nonmonotonic temperature dependence with increasing $|t|$ (a hump).

Experimentally, both nonanalytical and analytical correction terms to the asymptotic critical behaviour were found [12.14, 12.15] for amorphous Heisenberg ferromagnets like Fe–Ni, Fe–Zr, and Fe–Co–Zr alloys. Furthermore, it turned out that a nonmonotonic temperature dependence of $\gamma(T)$ is a characteristic feature of all random spin systems, i.e., amorphous ferromagnets [12.15, 12.16, 12.24], random crystalline Heisenberg ferromagnets [12.24], random crystalline Ising ferromagnets [12.25], random classical Heisenberg models [12.26], random crystalline Heisenberg ferrimagnets [12.27], spin glasses [12.28] and systems with very strong random anisotropy [12.29] (where in the last two cases the effective exponent of the nonlinear susceptibility is addressed). In contrast, for homogeneous systems $\gamma(T)$ decreases monotonically from the asymptotic critical value to the mean field value. Figures 12.1 and 12.2 show two examples, namely experimental results on amorphous $Fe_{20}Ni_{56}B_{24}$ as compared to crystalline Ni and results of Monte Carlo simulations for the site-diluted three-dimensional Ising model as compared to the undiluted three-dimensional Ising model. In both cases the temperature range where the plot χ_0^{-1} vs. T is nonlinear is much larger for the disordered than for the ordered system, and $\gamma(T)$ exhibits a nonmonotonic temperature dependence for the

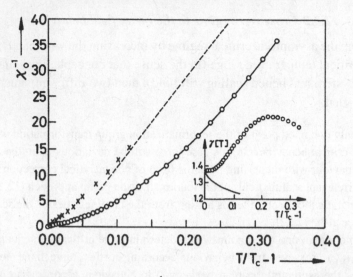

Fig. 12.1. Temperature dependence of $\chi_0^{-1}(T)$ for ($\times$) crystalline Ni [12.23] and ($\circ$) amorphous $Fe_{20}Ni_{56}B_{24}$ [12.30]. The data for Ni are multiplied by a constant factor so that the broken Curie–Weiss line is parallel to a line through the last three data points of the amorphous material. The inset shows $\gamma(T)$ (full curve: Ni, $\circ$ $Fe_{20}Ni_{56}B_{24}$).

disordered systems but decreases monotonically with increasing temperature for the ordered systems. By the Monte Carlo simulations it is demonstrated that these special temperature dependences of χ_0^{-1} and γ appear in systems for which there are definitely only atomic-scale spatially random fluctuations, so that structural inhomogeneities on a longer scale which may appear in real amorphous ferromagnets (see below) are not necessarily relevant for $\chi_0^{-1}(T)$ and $\gamma_{\text{eff}}(T)$. However, in the above discussed random spin systems the maximum of $\gamma_{\text{eff}}(T)$ in most cases appears for temperatures far outside the critical regime (typically at $t \approx$ some 10^{-1}). At these temperatures the thermal correlation length $\xi(T)$ in the spin system is already quite small (not much larger than the nearest-neighbour distance), so that concepts like self-similarity of the spin system on many length scales and scaling are no longer valid. It therefore does not make sense to describe the nonmonotonic temperature dependence of $\gamma(T)$ in terms of a critical crossover between different fixed points of the renormalization group transformation. In Section 12.4 a model is developed which explains the behaviour of $\gamma(T)$ by the interplay between the temperature-dependent noncritical thermal correlations in the spin system and the atomic-scale random exchange fluctuations. Although this model (which is the basis for the correlated molecular field theory) yields the nonmonotonic temperature dependence of $\gamma(T)$, it is not able to explain all the pecularities involved in the phase transition of amorphous ferromagnets, for instance, it cannot account for the fact that only a fraction of all the spins participates in the ferromagnetic–paramagnetic

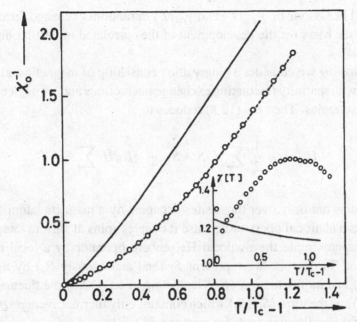

Fig. 12.2. Temperature dependence of $\chi_0^{-1}(T)$ for the three-dimensional simple-cubic Ising model without disorder [12.31] (full curve) and for the corresponding site-diluted system A_xB_{1-x} [12.25, 12.26] with $x = 0.3$, with $S_B/S_A = 0.2$. The data for the pure system are multiplied by a constant factor so that the high-temperature part of the full curve is parallel to a line through the last three data points of the diluted system. The broken curve is a numerically determined smoothing spline from which $\gamma_{eff}(T)$ of the inset has been calculated (open circles, note the large numerical fluctuations for $T \to T_c$). The full curve in the inset gives $\gamma_{eff}(T)$ for the undiluted system.

transition (Section 12.2.1). This latter observation found a straightforward but qualitative explanation [12.16] in terms of clusters of atoms which are embedded in the ferromagnetic matrix but which are magnetically decoupled from the matrix because of frustration zones that surround them and which originate from density fluctuations [12.32]. In our opinion the nonmonotonic behaviour of $\gamma(T)$ nevertheless is related to the atomic-scale fluctuations in the ferromagnetic matrix and does not result from the structure (i.e., shape and size) of these clusters, which certainly have a linear extension that is for $|t| \approx$ some 10^{-1} much larger than $\xi(T)$.

12.3 Molecular field theory and Landau–Ginzburg theory

This section describes the molecular field theory and the equivalent Landau–Ginzburg theory for which the correlations between the spin fluctuations at various sites are neglected. It will be shown that these theories are not able to reproduce the observed behaviour of $\chi_0^{-1}(T)$ and $\gamma_{\text{eff}}(T)$ in random exchange ferromagnets, but they are the basis for the development of the correlated molecular field theory (Section 12.4).

In the following we consider a binary alloy consisting of magnetic and nonmagnetic atoms with spatially fluctuating exchange interactions and we neglect the local uniaxial anisotropies. Then eq. (12.3) reduces to

$$\hat{H} = -\frac{1}{2} \sum_{ij} J_{ij} S_i \cdot S_j - g\mu_B H \sum_i S_i^z, \tag{12.9}$$

where the sums run only over those sites occupied by a magnetic atom. Note that $\hat{H}$ represents a nonlocal operator because it couples spins at various sites.

We now approximate the nonlocal Heisenberg operator by a local molecular field operator. To do this, we represent S_i (and analogously S_j) by its thermal average $\langle S_i \rangle$ and its fluctuation δS_i. Neglecting in eq. (12.9) the fluctuation term $\delta S_i(t)\,\delta S_j(t)$ and the term $\langle S_i \rangle \langle S_j \rangle$ which contains only thermal averages, we obtain the mean field Hamiltonian (see, for instance, [12.33])

$$H_0 = -\sum_i S_i^z (X_i + g\mu_B H), \tag{12.10}$$

with the spatially inhomogeneous mean field

$$X_i = \sum_j J_{ij} \langle S_j^z \rangle. \tag{12.11}$$

It has been shown by Wagner and Wohlfarth [12.34] and by Brauneck and Wagner [12.35] that this form of the mean field is also obtained by minimizing within the framework of the local mean field ansatz (12.10) a suitably defined variational functional with respect to X_i, where the functional represents an upper limit for the Gibbs free energy corresponding to the Heisenberg Hamiltonian (12.9). The mean field Hamiltonian is often written in the form

$$H_0 = -g\mu_B \sum_i S_i^z (H_i^{\text{mf}} + H), \tag{12.12}$$

with the molecular field

$$g\mu_B H_i^{\text{mf}} = \sum_j J_{ij} \langle S_j^z \rangle. \tag{12.13}$$

The inhomogeneous mean field Hamiltonian (12.10) was used to describe amorphous ferromagnets in a couple of papers (see, e.g., [12.34–12.37]), whereas Handrich [12.38] proposed a homogeneous mean field Hamiltonian for amorphous ferromagnets, replacing the quantities $\langle S_j^z \rangle$ in eq. (12.11) and eq. (12.13) by the volume average $\overline{\langle S_j^z \rangle}$.

For a homogeneous external magnetic field $\boldsymbol{H} = (0, 0, H)$ the mean field result $\langle \boldsymbol{S}_i \rangle = \langle S_i^z \rangle$ based on eq. (12.10) and eq. (12.11) is given [12.33] by

$$\langle S_i^z \rangle = S B_S \left[\frac{S}{kT} \sum_j J_{ij} \langle S_j^z \rangle + \frac{g \mu_0 H S}{kT} \right], \tag{12.14}$$

when we consider the quantum character of the spin system with spin quantum number S, where

$$B_S(x) = \frac{2S+1}{2S} \coth\left(\frac{2S+1}{2S} x \right) - \frac{1}{2S} \coth\left(\frac{1}{2S} x \right) \tag{12.15}$$

is the Brillouin function. Inverting eq. (12.14),

$$\frac{S}{kT} \sum_j J_{ij} \langle S_j^z \rangle + \frac{g \mu_0 H S}{kT} = B_S^{-1} \left(\frac{\langle S_i^z \rangle}{S} \right) \tag{12.16}$$

and evaluating the right hand side for $\langle S_i^z \rangle / S \ll 1$ (which holds for $T \le T_c$ and $T > T_c$) yields (see, e.g., [12.37])

$$B_S^{-1}(x) = C_1 x + C_2 x^3 + \cdots, \tag{12.17}$$

with

$$C_1 = 3 \frac{S^2}{S(S+1)}, \tag{12.18}$$

$$C_2 = \frac{9}{5} \frac{(2S)^4 \left[(2S+1)^4 - 1 \right]}{\left[(2S+1)^2 - 1 \right]^4}. \tag{12.19}$$

Going to a continuum approximation and introducing the magnetization

$$M(\boldsymbol{r}) = g \mu_B c(\boldsymbol{r}) \langle S^z(\boldsymbol{r}) \rangle \tag{12.20}$$

yields [12.37, 12.39]

$$\alpha(\boldsymbol{r}) M(\boldsymbol{r}) + B(\boldsymbol{r}) M^3(\boldsymbol{r}) = H + H_{\mathrm{mf}}(\boldsymbol{r}), \tag{12.21}$$

with

$$\alpha(r) = \frac{kT}{c(r)(g\mu_B S)^2}C_1; \qquad B(r) = \frac{kT}{c^3(r)(g\mu_B S)^4}C_2, \qquad (12.22)$$

and

$$H_{\mathrm{mf}}(r) = \left(\frac{1}{g\mu_B}\right)^2 \int J(r,r')\,M(r')\,\mathrm{d}^3r'. \qquad (12.23)$$

Assuming that the fluctuations of $M(r')$ vary on a much larger scale than the fluctuations of $J(r,r')$, $M(r')$ in eq. (12.23) may be expanded around $r' = r$, yielding

$$A(r)M(r) + B(r)M^3(r) - C_{mn}(r)\,\partial_m\partial_n M(r) - D_n(r)\,\partial_n M(r) = H, \qquad (12.24)$$

with

$$C_{mn}(r) = \frac{1}{2}\left(\frac{1}{g\mu_B}\right)^2 \int J(r,r')(r'-r)_m(r'-r)_n\,\mathrm{d}^3r', \qquad (12.25)$$

$$D_n(r) = \left(\frac{1}{g\mu_B}\right)^2 \int J(r,r')(r'-r)_n\,\mathrm{d}^3r', \qquad (12.26)$$

$$A(r) = \alpha(r) - V(r), \qquad (12.27)$$

$$V(r) = \left(\frac{1}{g\mu_B}\right)^2 \int J(r,r')\,\mathrm{d}^3r', \qquad (12.28)$$

and $\alpha(r)$, $B(r)$ given by eq. (12.22). For a random distribution of exchange interactions the off-diagonal elements of $C_{mn}(r)$ and the quantities $D_n(r)$ will be small [12.40] and they are often neglected for simplicity, yielding

$$A(r)M(r) + B(r)M^3(r) - C(r)\nabla^2 M(r) = H, \qquad (12.29)$$

with

$$C(r) = \frac{1}{6}\left(\frac{1}{g\mu_B}\right)^2 \int J(r, |r'-r|)(r'-r)^2\,\mathrm{d}^3r'. \qquad (12.30)$$

This equation has the form of a local Landau–Ginzburg equation [12.1]. The spatial fluctuations of $B(r)$ are thereby related to fluctuations of the concentration $c(r)$, the fluctuations of $C(r)$ are given by fluctuations of the exchange integrals, and the fluctuations of $A(r)$ originate from fluctuations of both the concentration and the exchange integral. Furthermore, it may be assumed that fluctuations of the concentration are more important than fluctuations of the exchange integrals, inserting $J(r,r') = J(|r'-r|)$ and yielding $C(r) = \mathrm{const.} = C$ and $V(r) = \mathrm{const.} = V$. The resulting equation is used in [12.34, 12.41–12.44]. Macroscopically heterogeneous materials correspond to long-wavelength fluctuations of $c(r)$, for example, and can be described by a theory neglecting the $\nabla^2 M$ term of eq. (12.29), see [12.41]. Finally, the homogeneous mean field Hamiltonian proposed by Handrich [12.38] corresponds to replacing $J(r, r') M(r')$ by $J(r, r') \overline{M(r')}$ in eq. (12.23), yielding

$$\alpha(r) M(r) + B(r) M^3(r) - V(r) \overline{M(r)} = H, \tag{12.31}$$

where the bar ($\overline{\quad}$) denotes the volume average.

For a homogeneous material with no spatial fluctuations and $c(r) = 1$, eq. (12.21) reduces to

$$A + BM^2 = H/M, \tag{12.32}$$

with

$$A = A'\left(\frac{T}{T_c^0} - 1\right), \tag{12.33}$$

$$A' = \left(\frac{1}{g\mu_B}\right)^2 \int J(|r - r'|)\, d^3r', \tag{12.34}$$

and the ferromagnetic Curie temperature T_c^0 of the homogeneous material,

$$kT_c^0 = \frac{1}{3}S(S+1) \int J(|r - r'|)\, d^3r'. \tag{12.35}$$

Equation (12.32) yields the mean field exponents for the homogeneous ferromagnet, $\gamma = 1$, $\beta = 1/2$, $\delta = 3$, i.e., the plot χ_0^{-1} vs. T is a straight line and the plot M^2 vs. H/M (Arrott plot [12.45]) is also a straight line which runs through the origin for $T = T_c^0$.

When introducing spatial inhomogeneities for $c(r)$ or $J(r, r')$, deviations from straight lines may occur for these plots, e.g., there will be curved Arrott plots. Figure 12.3 summarizes the results of the inhomogeneous and the homogeneous versions of the theories for $\chi_0^{-1}(T)$. As a result of the spatial fluctuations, the Curie temperature increases compared to T_c^0 and the curve $\chi_0^{-1}(T)$ develops a small curvature which is opposite in sign compared to the curvature of $\chi_0^{-1}(T)$ occuring in random exchange ferromagnets (Figs. 12.1 and 12.2). In contrast, the homogeneous mean field theory yields a linear plot for χ_0^{-1} vs. T, and the Curie temperature is that of a corresponding homogeneous ferromagnet with concentration $c = \overline{c(r)}$ and exchange couplings $\overline{J(r, r')}$ as long as there are no correlations between the fluctuations of $c(r)$ and $J(r, r')$ [12.37]. Obviously, neither the inhomogeneous nor the homogeneous theories are able to reproduce the strong upward curvature of $\chi_0^{-1}(T)$, which is characteristic for random exchange ferromagnets (Section 12.2.2).

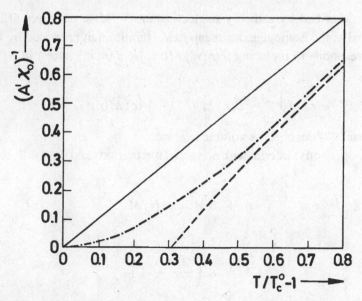

Fig. 12.3. Schematic plot of the results for $\chi_0^{-1}(T)$ for random exchange ferromagnets. The results were obtained by the inhomogeneous theories (dashed curve with very slight downward curvature), the homogeneous theory (full line), and by a theory which interpolates between the homogeneous theory for $T \to T_c$ and the inhomogeneous theory for high temperatures (Section 12.4).

12.4 Extended Landau–Ginzburg theory

The Landau–Ginzburg equation (12.24) which has been derived in Section 12.3 from the molecular field theory has two shortcomings:

1. It is an equation only for the z-component of the magnetization $M(r)$.
2. It neglects the influence of magnetostatic stray fields originating from the spatial fluctuations of $M(r)$ as well as the influence of the magnetoelastic coupling to internal stresses.

To extend the Landau–Ginzburg theory we first note that eq. (12.24) for $M(r) = M^z(r)$ may be obtained as a variational equation for the following Gibbs free energy functional near T_c

$$G = \int g(r)\, d^3r, \tag{12.36}$$

with

$$g(r) = \frac{1}{2}C_{mn}(r)\partial_m M(r)\partial_n M(r) + \frac{1}{2}A(r)M^2(r) + \frac{1}{4}B(r)M^4(r) - HM(r). \tag{12.37}$$

The above discussed effects may be incorporated [12.43] by starting from a modified Gibbs free energy density,

$$g(r) = \frac{1}{2} \sum_i C_{mn}(r) \partial_m M_i(r) \partial_n M_i(r) + \frac{1}{2} A(r) M^2(r) + \frac{1}{4} B(r) M^4(r) - H M(r)$$

$$- \frac{1}{2} H_s(r) \cdot M(r) - \frac{1}{2} \Lambda \sum_{i,j} M_i(r) M_j(r) \sigma_{ij}^{\text{def}}(r). \tag{12.38}$$

Here the fifth term denotes the magnetostatic stray field energy with the stray field $H_s(r)$ given by

$$H_s(r) = -\nabla U(r), \tag{12.39}$$

where $U(r)$ obeys Poisson's equation,

$$\Delta U(r) = 4\pi \rho(r), \tag{12.40}$$

with

$$\rho(r) = \text{div } M(r). \tag{12.41}$$

The last term in eq. (12.38) is the magnetoelastic coupling energy to internal stresses $\sigma^{\text{def}}(r)$, i.e., the analogue to eq. (2.76) for the case of macroscopically isotropic amorphous materials, and Λ is related to the saturation magnetostriction constant λ_s (Chapter 11) via

$$\lambda_s = \frac{1}{3} \Lambda M^2. \tag{12.42}$$

Minimizing the Gibbs free energy with respect to $M(r)$ yields

$$H_{\text{eff}, i} = 0, \tag{12.43}$$

with

$$H_{\text{eff}, i} = H_i - A(r) M_i(r) - B(r) M^2 M_i(r) + C_{mn}(r) \partial_m \partial_n M_i(r)$$

$$+ D_n(r) \partial_n M_i(r) + H_{s, i}(r) + \frac{1}{2} \Lambda \sum_j M_j(r) \sigma_{ij}^{\text{def}}(r) \tag{12.44}$$

and

$$D_n(r) = \partial_m C_{mn}(r). \tag{12.45}$$

Obviously, the variational equations for the Gibbs free energy near the phase transition may be written in the form of a vanishing effective field, $H_{\text{eff}} = 0$, which is different from the conventional micromagnetic equilibrium conditions which may be written in the form $H_{\text{eff}} \times M = 0$. This difference is due to the constraint

$|M^2| = $ const. which has to be taken into account for the derivation of the micromagnetic equations but which has to be released near the phase transition. The coupled system of eqs. (12.39)–(12.41) and (12.43)–(12.45) represents the basic equations of a generalized theory of second-order phase transitions of inhomogeneous ferromagnets by a micromagnetic approach.

As in Section 12.3, the off-diagonal elements of $C_{mn}(r)$ and the term containing $D_n(r)$ can be neglected [12.40]. With the resulting equations the influence of spatial variations of $A(r)$, $B(r)$ and of the internal stresses $\sigma_{ij}^{\text{def}}(r)$ as well as the influence of the magnetostatic stray fields associated with the spatial inhomogeneities on the Arrott plots and on $\chi_0^{-1}(T)$ have been investigated [12.43, 12.44]. As outlined in Section 12.3, fluctuations in the concentration of magnetic atoms yield fluctuations of $A(r)$ and $B(r)$. The spatially random fluctuations of $A(r)$ have the dominant effect and lead for all systems considered in the calculations to an increase of the ferromagnetic Curie temperature and a downward curvature of $\chi_0^{-1}(T)$, in agreement with the results of the inhomogeneous molecular field theory of Section 12.3 but in contrast to the experimental observation for amorphous ferromagnets. The medium-range microstructural internal stresses (see, e.g., Section 11.2) have only a very small effect (probably except for materials with very small exchange fields and very large magnetostriction), and the magnetostatic stray fields also have a negligible effect. Altogether, the situation is opposite to the one derived for the law of approach to saturation at low temperatures (Chapter 8). There, in most materials the spatially random fluctuations of the intrinsic material parameters were much less important than the effect of medium-range internal stresses. According to eqs. (12.27) and (12.33), the fluctuations of $A(r)$ may be interpreted as fluctuations of the local critical temperature, and this underpins their dominant influence on the magnetic properties near the phase transition.

12.5 Correlated molecular field theory

It has been shown in Section 12.3 that neither the homogeneous nor the inhomogeneous mean field theories are able to reproduce the experimentally observed upward curvature of $\chi_0^{-1}(T)$ in a wide temperature range which is characteristic for random exchange ferromagnets, whereas a theory interpolating between these two extreme situations does. In this section we introduce the correlated molecular field theory that achieves this interpolation.

12.5.1 Physical motivation

To illustrate the basic physical idea of the correlated molecular field theory we consider first the case of a crystalline Ising ferromagnet. At zero temperature all

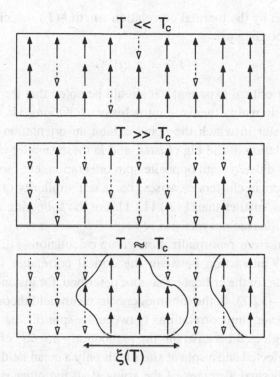

Fig. 12.4. Schematic representation of momentary spin configurations in an Ising ferromagnet at various temperatures.

spins are parallel because of the ferromagnetic exchange interactions. On increasing the temperature single spins are turned around because of the random thermal agitations of the heat bath, but at any time most of the spins are parallel because of the ferromagnetic exchange interactions (Fig. 12.4, top), i.e., the thermal average $\langle S_i \rangle$ of a spin at site i is nonzero (ferromagnetic regime), but there are thermal fluctuations $\delta S_i(t) = S_i(t) - \langle S_i \rangle$. For temperatures much larger than the Curie temperature, T_c, the thermal energy is much larger than the exchange interaction energy. Then the spins react more or less independently on the thermal agitations of the heat bath, and there are no correlations between the spin fluctuations $\delta S_i(t)$ and $\delta S_j(t)$ at different sites i and j (Fig. 12.4, middle). The thermal average $\langle S_i \rangle$ is zero (paramagnetic regime).

When approaching T_c there is a competition between the tendency to order spins because of the ferromagnetic exchange interactions and to disorder spins because of the random thermal agitations of the heat bath. As a consequence, the system creates large clusters of spins (Fig. 12.4, bottom) in each of which the ferromagnetic state is preserved for a short time. In this very simplified picture the spins in the clusters are strongly coupled together and react cooperatively. The mean size of the clusters is given by the thermal correlation length $\xi(T)$, which diverges when approaching T_c according to

$$\xi(T) = \xi_\pm |t|^{-\nu}, \tag{12.46}$$

where ν is another critical exponent. (It should be noted that the real situation is more complicated. In reality there are intracluster excitations, i.e., smaller clusters within the big cluster in which the spins exhibit an orientation opposite to the average spin orientation of the big cluster, and in these smaller clusters there are again even smaller clusters with opposite spin orientation, etc., so that the picture of clusters in clusters in clusters... arises, i.e., a self-similarity of the spin system on all length scales smaller than $\xi(T)$ [12.1]. For the following arguments these intracluster excitations are not relevant.)

It is well known from renormalization group calculations [12.1] that it is the nonlinear (quartic) interaction between the critical correlated spin fluctuations which is responsible for the critical behaviour described, for instance, by the power laws in eqs. (12.5)–(12.7). In the inhomogeneous mean field theories discussed in Section 12.3, however, any correlations between the spin fluctuations at different sites are totally neglected because the fluctuation term $\delta S_i \, \delta S_j$ of the Heisenberg Hamiltonian is neglected and a spin at site i feels only a mean field X_i (eq. (12.11)) produced by the thermal averages of the spins at all the other sites. As a result, mean field values for the critical exponents are obtained which deviate from the real values of the asymptotic critical exponents of the Heisenberg Hamiltonian. In the correlated molecular field theory the nonlinear interactions between the critical spin clusters are still neglected, i.e., the behaviour of a spin at site i will still be determined by a mean field produced by all the other spins so that the asymptotic critical behaviour will not be described correctly. However, the correlations in the spin system will be taken into account when formulating a more appropriate ansatz for the mean field.

For T approaching T_c the mean size of the clusters diverges, and then the system of Heisenberg spins as a whole reacts more or less cooperatively and does not feel anything from the spatially random fluctuations in a random exchange ferromagnet (the situation is different for Ising spins, see Section 12.2.1). Therefore, the main features of the phase transition in random exchange Heisenberg ferromagnets are very similar to those in homogeneous ferromagnets. In both cases there is a sharp phase transition, and the asymptotic critical exponents are the same for random and for homogeneous ferromagnets and are determined by the nonlinear interactions of the critical correlated spin fluctuations. On increasing T the mean size of the clusters decreases. Then the nonlinear interaction between the correlated spin fluctuations becomes negligibly small, and in homogeneous ferromagnets a more or less linear plot of χ_0^{-1} vs. T is observed. Nevertheless, there are still noncritical correlated spin fluctuations of smaller size in which the spins show a strong cooperative behaviour. These noncritical spin fluctuations must be taken into account in random exchange ferromagnets as long as the thermal correlation length $\xi(T)$ is larger than or comparable to the correlation length for the spatial fluctuations, i.e., of the fluctuations of the exchange couplings or of the concentration of magnetic atoms. The exchange fluctuations try to produce spatial fluctuations of the local spin values $\langle S^z(r)\rangle$ and hence spatial fluctuations of the magnetization (eq. (12.20)) and of the molecular field $H_{mf}(r)$ according to eq. (12.23). Because of the strong cooperative behaviour of the spins in the clusters, however, there are no fluctuations of $\langle S^z(r)\rangle$ with wavelengths shorter than the mean size of the clusters. For $T \to T_c$ the quantity $\xi(T)$ diverges, and then the local value $\langle S^z(r)\rangle$ is identical to the volume average $\langle S^z(r)\rangle$ as in the homogeneous mean field theory. For very large temperatures $\xi(T)$ is very small and then there are spatially random fluctuations of $\langle S^z(r)\rangle$ as in the inhomogeneous mean field theory. It is this transition between a homogeneous situation for $\langle S^z(r)\rangle$ at $T \to T_c$ and an inhomogeneous situation far away from T_c which will, within the framework of the correlated molecular field theory, reproduce the experimentally observed behaviour of random exchange ferromagnets in the crossover regime from the asymptotic critical behaviour to the mean field behaviour at high temperatures.

For a quantitative formulation, instead of the local mean field expression of eq. (12.23) we use a correlated molecular field defined by

$$H_{mf}(r)_{corr} = \left(\frac{1}{g\mu_B}\right)^2 \int J(r, r')\, M(r')_{corr}\, d^3r', \qquad (12.47)$$

with

$$M(r')_{corr} = g\mu_B c(r') \int f(r' - r'', \xi)\, S(r'')\, d^3r''. \qquad (12.48)$$

The magnetization $M(r) = g\mu_B c(r)S(r)$ is then obtained from the correlated molecular field equation

$$\alpha(r)M(r) + B(r)M^3(r) = H + H_{mf}(r)_{corr}, \qquad (12.49)$$

which replaces the inhomogeneous molecular field equation (12.21). Thereby the integral kernel $f(r' - r'', \xi)$ is the Ornstein–Zernicke function

$$f(r' - r'', \xi) = \frac{1}{4\pi\xi^2} \frac{\exp(-|r' - r''|/\xi)}{|r' - r''|}, \qquad (12.50)$$

which averages out the irrelevant fluctuations of $S(r'')$. For $T \to T_c$ this function varies only slowly in space because of the diverging $\xi(T)$, and then the atomic scale fluctuations of $S(r'')$ are averaged out. In contrast, for high temperatures (ξ small) the integral kernel $f(r' - r'', \xi)$ is well localized at r', yielding $M(r')_{corr} = g\mu_B c(r')S(r')$. For a discussion of the Gibbs free energy of the correlated molecular field theory, see [12.46].

For simplicity, we will confine ourselves in the following to model systems with $c(r) = $ const. but with spatially fluctuating exchange couplings $J(r, r')$. It has been shown in [12.47] that the qualitative results obtained from eq. (12.49) are similar for fluctuating exchange couplings and for fluctuating concentration $c(r)$. For $c(r) = $ const. and $T \to T_c$ the correlated magnetization is given by the volume average $\overline{M(r')}$ and the correlated molecular field equation approaches the homogeneous molecular field equation (12.31), whereas for very high temperatures the correlated magnetization is given by the local magnetization $M(r')$ and the inhomogeneous molecular field equation (12.21) is obtained. It should be noted that the real magnetization derived from eqs. (12.47)–(12.49) exhibits spatial fluctuations $m(r) = M(r) - \overline{M(r)}$ with $\overline{m^2(r)}/\overline{M}^2 \neq 0$ even for $T \to T_c$ (for which $M(r')_{corr}$ is given by $\overline{M(r')}$) because the correlated molecular field still exhibits the spatial fluctuations of $J(r, r')$. (The existence of magnetization fluctuations at $T \to T_c$ has been confirmed by measurements of the width of the Mössbauer lines for amorphous $Fe_{81}B_{13.5}Si_{3.5}C_2$, see [12.48] and references therein.) The correlated magnetization is just a mathematical tool to describe the smoothing effect of the spin clusters on the mean field experienced by the spins, and it should not be mixed up with the real magnetization. In the first version of the correlated molecular field theory [12.39] the correlated molecular field was defined as

$$H_{mf}(r)_{corr} = \int f(r - r', \xi) H_{mf}(r') \, d^3r', \qquad (12.51)$$

with the local molecular field $H_{mf}(r)$ given by eq. (12.23). Then all fluctuations of $H_{mf}(r)$ are averaged out for $T \to T_c$, resulting in $\overline{m^2}/\overline{M}^2 \to 0$.

It should be noted that the correlated molecular field in principle cannot be derived by the approach of Brauneck and Wagner [12.35], i.e., by a minimization of the mean field approximation for the Gibbs free energy, because this approximation excludes all correlation effects from the very beginning. The fact that the local mean field expression of eq. (12.11) minimizes the mean field approximation for the Gibbs free energy therefore does not mean [12.47] that it must be superior to the correlated molecular field expression of eq. (12.47), which includes in some approximation the effect of the spin correlations.

In the following subsection the zero-field susceptibility for the paramagnetic regime will be calculated for random exchange ferromagnets by the correlated molecular field theory. For an extension of the theory to the case of nonvanishing $\overline{M(r)}$, i.e., for a calculation of the temperature dependence of the spontaneous magnetization, of the Arrott plots and of the magnetization fluctuations $\overline{m^2(r)}/\overline{M}^2$ (relevant for the discussion of the widths of the Mössbauer lines [12.48]) we refer to [12.47].

12.5.2 Calculation of the paramagnetic zero-field susceptibility

The paramagnetic zero-field susceptibility $\chi_0(T) = \lim_{H \to 0} \chi(H, T > T_c)$ may be obtained from eq. (12.49) by neglecting the term $B(r)M^3(r)$ which is much smaller than $\alpha(r)M(r)$ for $M(r) \to 0$. For $c(r) = $ const. the remaining linear equation may then be written as (for an external magnetic field $H(r) = (0, 0, H(r))$)

$$\int d^3r' \, L(r, r') M(r') = H(r), \tag{12.52}$$

with the linear stochastic integral kernel

$$L(r, r') = \int d^3r'' \left[\alpha \delta(r - r') \delta(r - r'') - \left(\frac{1}{g \mu_B} \right)^2 J(r, r'') f(r'' - r') \right]. \tag{12.53}$$

Defining the Green's function via

$$\int d^3r' \, L(r, r') G(r', r'') = \delta(r - r''), \tag{12.54}$$

the solution of eq. (12.52) may be written in the form

$$M(r) = \int G(r, r') H(r') d^3r'. \tag{12.55}$$

Obviously, $G(r, r')$ plays the role of a position-dependent susceptibility $\chi_0(r, r')$. Equation (12.54) may be written in operator form as

$$LG = 1. \tag{12.56}$$

For spatially random fluctuations of $J(r, r'')$ the integral equation (12.54) cannot be solved exactly. To calculate the zero-field susceptibility, it suffices to determine the configuration-averaged Green's function $\overline{G}(|r' - r''|)$ of the stochastic but macroscopically isotropic and homogeneous system, for which we write

$$\int d^3r' \, L^{\text{eff}}(|r - r'|) \overline{G}(|r' - r''|) = \delta(r - r''), \tag{12.57}$$

with an effective integral kernel L^{eff} which has to be determined. For constant external field H the configurational averaged magnetization is obtained by integrating eq. (12.55) (assuming ergodicity of the system), yielding

$$\overline{M} = H \int \overline{G}(y) \, d^3y. \qquad (12.58)$$

The paramagnetic zero-field susceptibility is then given by

$$\chi_0 = \lim_{H \to 0} \overline{M}/H = \int \overline{G}(y) \, d^3y, \qquad (12.59)$$

or, by integrating eq. (12.57) over r'', by

$$\chi_0^{-1} = \int L^{\text{eff}}(y) \, d^3y. \qquad (12.60)$$

To determine $L^{\text{eff}}(|r - r'|)$ the integral kernel $L(r, r')$ is decomposed into a configuration-averaged part

$$L^0(r - r') = \int d^3r'' \left[\alpha \delta(r - r') \delta(r - r'') - \left(\frac{1}{g\mu_{\text{B}}} \right)^2 J_0(|\, r - r''|) f(r'' - r') \right], \qquad (12.61)$$

with $J_0(|r - r''|) = \overline{J(r, r'')}$, and a fluctuating part

$$\delta L(r, r') = - \left(\frac{1}{g\mu_{\text{B}}} \right)^2 \int j(r, r'') f(r'' - r') \, d^3r''. \qquad (12.62)$$

The Green's function $G^0(|r' - r''|)$ defined by

$$\int L^0(|r - r'|) \, G^0(|r' - r''|) \, d^3r' = \delta(r - r'') \qquad (12.63)$$

then has the meaning of a position-dependent susceptibility for the homogeneous and isotropic reference system defined by $L^0(|r - r'|)$. The zero-field susceptibility, $\chi_{0,\text{ref}}$, of the reference system is then given by

$$\chi_{0,\text{ref}} = \int G^0(y) \, d^3y, \qquad (12.64)$$

or by

$$\chi_{0,\text{ref}}^{-1} = \int L^0(y) \, d^3y. \qquad (12.65)$$

The configuration-averaged Green's operator $\overline{G}$ and the effective operator L^{eff} may be represented in terms of G^0, L^0 and δL by the expansions [12.49]

$$\overline{G} = G^0 - G^0 \, \overline{\delta L} \, G^0 + G^0 \, \overline{\delta L \, G^0 \, \delta L} \, G^0 - \cdots, \qquad (12.66)$$

$$L^{eff} = L^0 - \overline{\delta L\, G^0\, \delta L} + \overline{\delta L\, G^0\, \delta L\, G^0\, \delta L} + \cdots. \tag{12.67}$$

If we truncate the series of eq. (12.67) after the second-order term and calculate from the resulting L^{eff} the corresponding $\overline{G}$ we find the approximation

$$\overline{G} = G^0 + G^0\, \overline{\delta L\, G^0\, \delta L}\, \overline{G}, \tag{12.68}$$

which looks like the second-order approximation of the series of eq. (12.66) ($\overline{\delta L} = 0$) with the last G^0 replaced by $\overline{G}$. Obviously, truncating the two series at the same order yields different results, and the second-order result from eq. (12.67) looks superior to the second-order result from eq. (12.66) because it corresponds to a 'self-consistent' calculation of $\overline{G}$. In fact, it can be shown that truncating the series for L^{eff} at second order yields the same result as evaluating the infinite series of eq. (12.66) by applying the decoupling approximation introduced by Montgomery et al. [12.50] $\overline{(G^0\, \delta L)^{2n}} = \left(\overline{(G^0\, \delta L)^2}\right)^n$ (all odd moments assumed to be zero). We therefore prefer to use the second-order result for L^{eff} instead of the second-order result for $\overline{G}$ from eq. (12.66). It should be noted that the second-order approximation for L^{eff} is also equivalent to the result obtained by the method of moment equations [12.51] terminated by neglecting the third-order moments, which is used in [12.47] to solve the full nonlinear equation (12.49).

To evaluate the second-order term of eq. (12.67) the following ansatz for the two-point correlation function is used

$$\overline{\delta(r)\delta(r')} = \Delta^2 \exp(-(r - r')^2 / l_s^2), \tag{12.69}$$

with

$$\delta(r) = \int j(r, r')\, d^3r' \Big/ \int J_0(|r - r'|)\, d^3r'. \tag{12.70}$$

Here

$$\Delta = \sqrt{\overline{\delta^2(r)}} \tag{12.71}$$

describes the fluctuation strength and l_s is the structural correlation length. From eq. (12.60) the second-order result for the zero-field susceptibility is obtained,

$$\chi_0^{-1} = \chi_{0,ref}^{-1} - \frac{(A')^2}{(2\pi)^3} \int d^3k\, \Gamma(k)\chi_{0,ref}(k)f(k), \tag{12.72}$$

with

$$\chi_{0,ref}(k) = G^0(k) = \frac{1}{\alpha - \left(\dfrac{1}{g\mu_B}\right)^2 J_0(k)f(k)}, \tag{12.73}$$

where $J_0(k)$ is the Fourier transform of $J_0(|r - r'|)$,

$$f(k) = \frac{1}{1 + k^2 \xi^2}, \tag{12.74}$$

is the Fourier transform of the Ornstein–Zernicke function and $\Gamma(k)$ is the Fourier transform of the structural correlation function of eq. (12.69). Furthermore, a local expansion of $J_0(k)$ is used,

$$\left(\frac{1}{g\mu_B}\right)^2 J_0(k) = A' - Ck^2, \tag{12.75}$$

with A' and C given by eq. (12.34) and eq. (12.30), which is equivalent to expanding in eq. (12.47) $M(r')_{\text{corr}}$ around r, assuming that the fluctuations of $M(r')_{\text{corr}}$ vary on a much larger scale than those of $J(r, r')$, see Section 12.3. Equation (12.72) then yields

$$\chi_0^{-1} = \chi_{0,\,\text{ref}}^{-1} - (A')^2 \int d^3 k \, \Gamma(k) \frac{1}{\chi_{0,\,\text{ref}}^{-1} + C_{\text{eff}} k^2}, \tag{12.76}$$

with

$$C_{\text{eff}} = C + A' \xi^2 \frac{T}{T_c}. \tag{12.77}$$

The difference between the correlated and the inhomogeneous molecular field theory is given by the additional term in the stiffness constant C_{eff} of the correlated theory. For $T \to T_c$ the spin system as a whole reacts cooperatively and becomes totally stiff with respect to the spatial fluctuations $j(r, r')$ of the exchange couplings, which is reflected in the divergence of the effective stiffness C_{eff}. When neglecting this additional term, the result of the inhomogeneous molecular field theory is recovered.

Performing the integration in eq. (12.76) yields for the normalized quantities

$$\tilde{\chi}_0 = \chi_0 A', \tag{12.78}$$

$$\tilde{\xi} = \xi / \sqrt{C/A'}, \tag{12.79}$$

$$\Delta_c = \sqrt{C/A'} / \sqrt{l_s^2 / 2}, \tag{12.80}$$

$$t = T / T_c^0, \tag{12.81}$$

the relation

$$\tilde{\chi}_0^{-1} = \tilde{\chi}_{0,\,\text{ref}}^{-1} - \frac{\Delta^2}{\Delta_c^2} \frac{1}{1 + t\tilde{\xi}^2} \left[1 - \sqrt{\pi} p \, \exp(p^2) \, \text{erfc}(p) \right], \tag{12.82}$$

with

$$p = \frac{1}{\sqrt{2}\Delta_c} \sqrt{\frac{\tilde{\chi}_{0,\,\text{ref}}^{-1}}{1 + t\tilde{\xi}^2}}, \quad (12.83)$$

and with the complementary error function erfc(p). The correlation length ξ is calculated in a self-consistent way from the yet to be determined susceptibility χ_0 via

$$\xi = \sqrt{C\chi_0}, \qquad \tilde{\xi} = \sqrt{\tilde{\chi}_0}. \quad (12.84)$$

It then becomes obvious that $\tilde{\chi}_0$ is determined by just two material parameters, Δ and Δ_c, where Δ_c is composed of C, A' and the structural correlation length l_s. Assuming a very simple model for the exchange interaction,

$$J_0(|\boldsymbol{r} - \boldsymbol{r}'|) = \begin{cases} \text{const.}\dfrac{4\pi}{3}r_0^3 & \text{for } |\boldsymbol{r} - \boldsymbol{r}'| \leq r_0 \\ 0 & \text{for } |\boldsymbol{r} - \boldsymbol{r}'| > r_0, \end{cases} \quad (12.85)$$

yields from eqs. (12.30), (12.34) and (12.35) the relations

$$C = \frac{1}{10}\left(\frac{1}{g\,\mu_B}\right)^2 \cdot r_0^2 \, \text{const.}; \qquad A' = \left(\frac{1}{g\,\mu_B}\right)^2 \text{const.};$$

$$\frac{C}{A'} = \frac{r_0^2}{10}; \qquad\qquad \Delta_c = \sqrt{\frac{1}{5}}\frac{r_0}{l_s}. \quad (12.86)$$

In this model there are three characteristic lengths of the material, the range r_0 of exchange interactions, the structural correlation length l_s describing the range over which the spatial fluctuations of the material parameters are correlated, and the temperature dependent correlation length ξ which is related to the yet to be determined susceptibility $\tilde{\chi}_0$. The result for $\tilde{\chi}_0^{-1}$ of the correlated molecular field theory approaches the result for the homogeneous molecular field theory $\left(\tilde{\chi}_0^{-1} = \tilde{\chi}_{0,\,\text{ref}}^{-1}\right)$ for $\tilde{\xi} \to \infty$ according to $T \to T_c^0$ or for $\Delta_c \to \infty$ according to $r_0 \to \infty$ or $l_s \to \infty$, and it approaches the result for the inhomogeneous molecular field theory for $\xi \to 0$ according to $T \to \infty$ or for $\Delta_c \to 0$ according to $r_0 \to 0$ or $l_s \to \infty$.

The correlated molecular field theory takes into account the interplay between the noncritical correlations in the spin system and the structural fluctuations, but it neglects the nonlinear interactions between the spin clusters which determine the asymptotic critical behaviour, i.e., it is basically a mean field theory. It is therefore straightforward to use for $\tilde{\chi}_{0,\,\text{ref}}^{-1}$ the molecular field result for the homogeneous reference system, $\tilde{\chi}_{0,\,\text{ref}}^{-1} = \left(T/T_c^0 - 1\right)$. Then eq. (12.82) yields for $T \to T_c^0$ the mean field exponent $\gamma = 1$. To introduce the correct asymptotic critical exponent in a phenomenological approach [12.52], the molecular field result for $\chi_{0,\text{ref}}$ is replaced by the correct temperature dependence of the zero-field susceptibility in a homogeneous ferromagnet. It has been shown experimentally for the case of nickel [12.53] and theoretically [12.54] by the Padé approximant technique for many Ising and Heisenberg models on different lattices with various spin quantum numbers that a generalized Curie–Weiss law of the form

$$\tilde{\chi}_{0,\text{ref}}^{-1} = \frac{T}{T_c^0} \left(\frac{T - T_c^0}{T} \right)^{\gamma},$$

(12.87)

with the asymptotic critical exponent γ, describes the zero-field susceptibility of homogeneous ferromagnets very well from T_c^0 up to arbitrarily high temperatures. In the following, this ansatz will be used, with $\gamma = 1.38$ for three-dimensional Heisenberg ferromagnets. Figure 12.5 shows the temperature dependence of χ_0^{-1} for a fixed value of Δ_c and various values of the fluctuation strength Δ. Obviously, the exchange fluctuations produce a large temperature range of upward curvature of the plot χ_0^{-1} vs. T, in good agreement with the experimentally observed behaviour (Fig. 12.1). Figure 12.6 shows $\chi_0^{-1}(T)$ for $\Delta = 0.4$ and various values of Δ_c, exhibiting the expected behaviour (see above), i.e., for large values of Δ_c the result for the homogeneous reference system is obtained, whereas for small values of Δ_c the result of the inhomogeneous theory is reproduced (compare the dashed curve

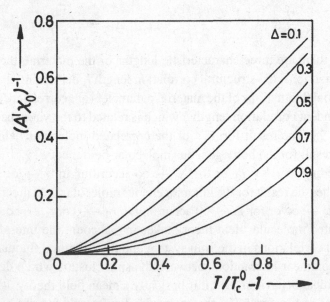

Fig. 12.5. Temperature dependence of χ_0^{-1} for a random exchange ferromagnet with material parameters A', C, d as in amorphous $Fe_{40}Ni_{40}P_{14}B_6$ [12.52] and $l_s = 2.5d$ (according to $\Delta_c = 0.3$) for various values of Δ.

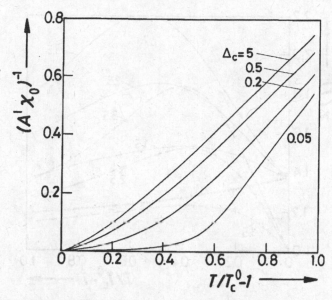

Fig. 12.6. Temperature dependence of χ_0^{-1} for a random exchange ferromagnet with $\Delta = 0.4$ and for various values of Δ_c.

in Fig. 12.3). Small values of Δ_c may be produced by rather large values of l_s, which describe macroscopically inhomogeneous systems for which the thermal correlation length $\xi(T)$ is smaller than l_s for nearly all temperatures, so that there is no interplay between the temperature dependent thermal correlations in the spin system and the structural randomness.

Figure 12.7 shows the Kouvel–Fisher exponent $\gamma(T)$ for $\Delta = 0.5$ and various values of l_s. The exponent exhibits the nonmonotonic temperature dependence characteristic for random exchange ferromagnets (Figs. 12.1 and 12.2).

Finally, Fig. 12.8 shows a fit of the theoretical curve to the experimental data for $Fe_{20}Ni_{56}B_{24}$ (Fig. 12.1), which is obtained for $\Delta = 0.15$ and $l_s = 5.7d$. This would mean that in $Fe_{20}Ni_{56}B_{24}$ the exchange interactions are correlated over a rather long range. However, one should caution that there is some arbitrariness in the ansatz of eq. (12.75) or of eq. (12.85) so that the actual values of the fit parameters should not be taken too literally.

Altogether, it can be concluded that the correlated molecular field theory is able to reproduce the characteristic features of the paramagnetic zero-field susceptibility of random exchange ferromagnets in the (noncritical) crossover regime from asymptotic critical behaviour to high-temperature behaviour and it allows discussion of the thermodynamic behaviour as function of two internal lengths, the range r_0 of the exchange interactions and the structural correlation length l_s.

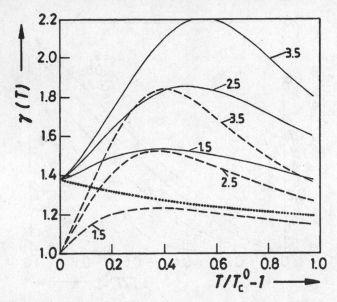

Fig. 12.7. Temperature dependence of $\gamma(T)$ for a random exchange ferromagnet with material parameters A', C, d as in amorphous $Fe_{40}Ni_{40}P_{14}B_6$ [12.52], $\Delta = 0.5$ and $l_s = pd$ for various values of p. The full lines are obtained when using eq. (12.87) for $\tilde{\chi}_{0\,\text{ref}}^{-1}$, whereas the dashed lines are for $\tilde{\chi}_{0\,\text{ref}}^{-1} = \left(T/T_c^0 - 1\right)$. The dotted line is $\gamma(T)$ for the reference system with $\tilde{\chi}_{0\,\text{ref}}^{-1}$ given by eq. (12.87).

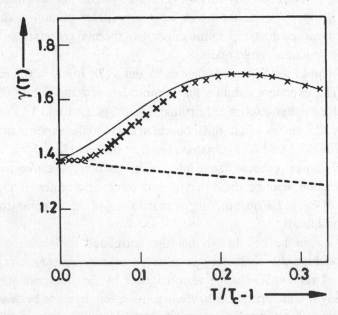

Fig. 12.8. Best fit of the theoretical curve (full line) to the experimental data (crosses) for amorphous $Fe_{20}Ni_{50}B_{24}$ [12.30]. The dashed line is $\gamma(T)$ for the reference system with $\tilde{\chi}_{0\,\text{ref}}^{-1}$ given by eq. (12.87).

12.6 Random ferrimagnets, spin glasses and random anisotropy magnets

The correlated molecular field theory has been extended to random ferrimagnets [12.55], spin glasses [12.28], and random anisotropy magnets [12.29]. In the simplest case, ferrimagnets are systems consisting of two sublattices A and B occupied by spins S^A and S^B and coupled by nearest neighbour exchange couplings $J_{ij} < 0$. In random ferrimagnets there are spatially random fluctuations of the exchange interactions originating from spatial fluctuations of the exchange couplings J_{ij} or from a random distribution of magnetic and nonmagnetic atoms on each of the two sublattices. The calculations predicted similar results as for the case of random ferromagnets, i.e., there is a strong upward curvature of the paramagnetic zero-field susceptibility $\chi_0^{-1}(T)$ in a large temperature range due to the exchange fluctuations, in constrast to the downward curvature of $\chi_0^{-''}(T)$ found in homogeneous ferrimagnets outside the critical regime. The predictions of the correlated molecular field theory and of Monte Carlo simulations [12.55] have been confirmed by measurements for structurally disordered crystalline ferrites $Zn_x Ni_{1-x} Fe_2 O_4$ [12.56].

For the rest of this chapter spin glasses and random anisotropy magnets with a speromagnetic state which resembles very much that of a spin glass (Section 12.1) are considered. These materials are characterized by a cusp in the susceptibility $\chi_0(T)$ at a freezing temperature T_f [12.6], but there is no spontaneous magnetization below T_f. On the other hand, the nonlinear zero-field susceptibility

$$\chi_2(T) = \lim_{H \to 0} \frac{\partial^3 M(T, H)}{\partial H^3}, \tag{12.88}$$

diverges for $T \to T_f$ [12.10, 12.57, 12.58]. Taking into account that for $T \to T_f$ the singular part of χ_2 may be written as [12.59]

$$\chi_2(T) \sim \frac{1}{T^3} \sum_{ij} \overline{\langle S_i \cdot S_j \rangle^2}, \tag{12.89}$$

this divergence shows that there must be strong correlations in the spin system at least during the time period of the measurement (with a diverging correlation length $\xi(T)$ for $T \to T_f$) like in a system with a second-order phase transition in thermal equilibrium. In the literature there has been a long discussion (see, e.g., [12.60]) whether the cusp in χ_0 and the divergence of χ_2 at T_f are signatures of a phase transition in thermodynamic equilibrium or whether they correspond to nonequilibrium phenomena. For instance, it has been argued that the ground states of the spin glasses and the random anisotropy magnets are highly degenerate, and that at finite temperatures there are always transitions between the various ground states so that no spin correlations will arise at all when averaging over an infinitely long time scale. On the other hand, one can argue from the divergence of χ_2 that the system is basically trapped in one of the various ground states during the time interval of the measurement (broken ergodicity) and that it exhibits thermal fluctuations around this ground state, whereby the ground states are characterized by spatially random orientations of 'frozen' spins. The following arguments are based on this assumption. The correlated spin fluctuations responsible for the divergence of χ_2 then have the form of spin clusters of linear extension $\xi(T)$ for which the local ground state configuration is momentary conserved, and there should again be an interplay between the temperature dependent thermal correlations in the spin system and the spatial structural fluctuations as in random exchange ferromagnets. This interplay, however, will probably not show up in the linear susceptibility χ_0, which – according to the fluctuation–dissipation relation – may be written as

$$\chi_0(T) \sim \frac{1}{T} \sum_{ij} \overline{\langle S_i \cdot S_j \rangle}. \qquad (12.90)$$

For spin glasses with a symmetric distribution of exchange couplings (corresponding to $J_0 = 0$ in eq. (12.4)) there is $\overline{\langle S_i \cdot S_j \rangle} \sim \delta_{ij}$ and hence $\chi_0(T) \sim 1/T$. In contrast, $\chi_2(T)$ is determined by the squares $\overline{\langle S_i \cdot S_j \rangle^2}$ which are influenced by the above discussed interplay. Thereby, these squares may be interpreted [12.61] as correlation functions for the Edwards–Anderson [12.62] spin glass order parameter $q = \overline{\langle S_i \rangle^2}$. As discussed by Binder [12.63] one can equally well define an order parameter $\psi = \overline{\langle \psi_i \rangle}$ with

$$\psi_i = \langle S_i \cdot e_i^n \rangle, \qquad (12.91)$$

where the unit vectors e_i^n describe the orientations of the spins S_i in the ground state n. One of these ground states may be stabilized by the application of a 'staggered' field $H_i^{n'} = He_i^{n'}$ whose direction varies from site to site. Of course such a field is not realizable in experiments, but nevertheless it is an appropriate quantity to discuss the critical behaviour of systems with nonparallel spin orientations [12.1]. Because we assume in the following that the system stays around a specific ground state for the duration of the measurement, the superscript n' will be omitted. For general reasons the order parameter ψ is an even more appropriate choice [12.61, 12.63] than the Edward–Anderson order parameter q [12.62]. The correlation length of the corresponding [12.61] correlation function $|\langle S_i \cdot S_j \rangle|$ again possibly diverges at T_f thus giving rise to a singular behaviour of the susceptibility $\chi_\psi(T)$ of the staggered magnetization [12.61],

$$\chi_\psi(T) \sim \frac{1}{T} \sum_{ij} \overline{|S_i \cdot S_j|}. \qquad (12.92)$$

Therefore χ_ψ is equally as appropriate to characterize the spin glass transition as the nonlinear susceptibility $\chi_2(T)$.

In the following, the correlated molecular field theory will be applied to calculate the temperature dependence of χ_ψ in the crossover regime from critical behaviour to high-temperature behaviour. Because of the great similarity of χ_ψ and χ_2 it is expected that the qualitative behaviour of χ_2 is similar to that of χ_ψ. Spin glasses and random anisotropy magnets with infinitely large local anisotropies are thereby treated on an equal footing. In both cases we assume that the spin system is kept around one specific ground state characterized by the set $\{e_i\}$ of unit vectors. For spin glasses the random directions $\{e_i\}$ originate from fluctuations of the J_{ij}, and the Hamiltonian including the Zeeman coupling to the staggered field may be written as

$$H - -\frac{1}{2}\sum_{ij} J_{ij} \mathbf{S}_i \cdot \mathbf{S}_j - H \sum_i (\mathbf{S}_i \cdot \mathbf{e}_i). \qquad (12.93)$$

For random anisotropy magnets with infinitely large D_i (eq. (12.3)) each spin $\mathbf{S}_i$ is parallel or antiparallel to the local easy axis direction $\mathbf{n}_i$, and in the ground state the directions are described by $\mathbf{e}_i = \pm \mathbf{n}_i$. By thermal excitations only the sign of $\mathbf{S}_i$ may be modified, $\mathbf{S}_i = \mathbf{e}_i S \sigma_i$ with $\sigma_i = \pm 1$, leaving the anisotropy energy (second term in eq. (12.3)) unchanged so that we can omit it. The Hamiltonian including the Zeeman coupling to the staggered field may then be written again in the form of eq. (12.93), with

$$\mathbf{S}_i = \mathbf{e}_i \cdot S \sigma_i, \qquad \sigma_i = \pm 1. \qquad (12.94)$$

To calculate the ψ_i we start from the molecular field equation (12.16) and replace the constant field $\mathbf{H} = (0, 0, H)$ by the staggered field $\mathbf{H}_i$. Evaluating the right hand side of this equation to linear order (for the example $S = 1/2$), multiplying the resulting equation by $\mathbf{e}_i$ and introducing the (normalized) staggered magnetization $\psi_i = \langle \mathbf{S}_i \rangle \cdot \mathbf{e}_i / S$ yields

$$kT\psi_i = H + \sum_j J_{ij}^{\text{eff}} \psi_j, \qquad (12.95)$$

with

$$J_{ij}^{\text{eff}} = S^2 J \mathbf{e}_i \cdot \mathbf{e}_j. \qquad (12.96)$$

Obviously, we have transformed the very complicated ground state of the spin glass or of the random anisotropy magnet (in terms of $\langle \mathbf{S}_i \rangle$) to a 'ferromagnetic' ground state (in terms of ψ_i) by using renormalized effective exchange couplings J_{ij}^{eff} which contain all the information about the ground state, and by simple physical arguments a reasonable ansatz for the distribution function density $P(J_{ij}^{\text{eff}})$ may be obtained [12.28, 12.29].

In a continuum version eq. (12.95) reads

$$kT\,\psi(r) = H + H_{\mathrm{mf}}(r); \qquad H_{\mathrm{mf}}(r) = \int J^{\mathrm{eff}}(r,r')\,\psi(r')\,\mathrm{d}^3r'. \qquad (12.97)$$

Formally, this equation is totally equivalent to the linear part of eq. (12.49) for a random exchange ferromagnet, and again it may be generalized to a correlated molecular field equation by replacing $H_{\mathrm{mf}}(r)$ by

$$H_{\mathrm{mf}}^{\mathrm{corr}}(r) = \int J^{\mathrm{eff}}(r,r')\,\psi(r')^{\mathrm{corr}}\,\mathrm{d}^3r', \qquad (12.98)$$

with

$$\psi(r')^{\mathrm{corr}} = \int f(r' - r'', \xi)\,\psi(r'')\,\mathrm{d}^3r'' \qquad (12.99)$$

and the Ornstein–Zernicke function $f(r' - r'', \xi)$ according to eq. (12.50).

The staggered zero-field susceptibility $\chi_\psi = \lim_{T\to 0} \partial\overline{\psi}/\partial H$ may then be calculated on the lines described in Section 12.5.2. The corresponding Kouvel–Fisher exponent $\gamma_\psi(T) = (T - T_{\mathrm{f}})\,\chi_\psi(T)\,\mathrm{d}\chi_\psi^{-1}(T)/\mathrm{d}T$ again exhibits a nonmonotonic temperature dependence [12.28, 12.29], in close analogy to the exponent $\gamma(T)$ for random exchange ferromagnets.

Experimentally, different types of results were obtained for $\chi_2(T)$. Omari *et al.* [12.58] managed to fit a power law of the form

$$\chi_2^{-1}(T) \sim \left(\frac{T - T_{\mathrm{f}}}{T}\right)^{\gamma_{\mathrm{s}}}; \qquad \gamma_{\mathrm{s}} = 3.65 \pm 0.35, \qquad (12.100)$$

to their data for CuMn, which is analogous to the generalized Curie–Weiss law for homogeneous ferromagnets (eq. (12.87)), and which would correspond to a monotonic $\gamma_\psi(T)$. In contrast, Monod and Bouchiat [12.64] describe their data for AgMn by assuming a temperature dependence of γ_{s}. Altogether, the experimental situation for the nonlinear susceptibility of spin glasses is not as clear as that for the linear susceptibility of random exchange ferromagnets.

12.7 Dynamic correlated molecular field theory

To describe dynamic phenomena at the phase transition, the static molecular field equation eq. (12.14) is replaced by the dynamic analogue [12.65]

$$\gamma\,\frac{\mathrm{d}}{\mathrm{d}t}\langle S_i^z\rangle = -\langle S_i^z\rangle + S B_S\left[\frac{\gamma\mu_{\mathrm{B}}S}{kT}\left(H_{\mathrm{mf}}^i + H(t)\right)\right], \qquad (12.101)$$

with a time-dependent external magnetic field $H(t)$ and a constant γ which has the dimension of time and which determines the time scale for dynamic processes.

Taking into account only the linear term in the Taylor expansion of B_S and going to a continuum description ($H_{mf}^i \to H_{mf}(r)$, $\langle S_i^z \rangle \to S(r)$, $M(r) = g\mu_B S(r)$, for simplicity a system consisting only of magnetic atoms of one kind but with spatial fluctuations of the exchange interactions is considered) yields

$$\alpha\gamma \frac{d}{dt} M(r, t) = -\alpha M(r, t) + H_{mf}(r, t) + H(t), \qquad (12.102)$$

with α given by eq. (12.22) and with the molecular field

$$H_{mf}(r, t) = \left(\frac{1}{g\mu_B}\right)^2 \int J(r, r') M(r', t)\, d^3r'. \qquad (12.103)$$

To generalize eq. (12.103) it should be recalled that as an effect of the strong correlations of the spin fluctuations near T_c there are clusters of spins in which a net fraction of the spins are aligned and which react more or less cooperatively, i.e., the clusters as a whole change orientation because of the thermal agitations of the heat bath. Because the size of the clusters diverges when approaching T_c, and because the thermal agitations flip the spins in these clusters one after the other, it takes more time to turn around large spin clusters than to turn around small spin clusters, and therefore the characteristic time required to turn around the clusters diverges when approaching T_c [12.1].

Suppose now that we consider a spin system exposed to a homogeneous magnetic field at $T > T_c$, which exhibits a nonvanishing volume average $\overline{M(r)}$ of the magnetization. Switching off the magnetic field at $t = 0$ the system will relax to the new equilibrium state $\overline{M(r)} = 0$. It becomes obvious from the above discussion that the relaxation time will be determined by the characteristic time scale for the spin clusters, because the relaxation corresponds to a redistribution of the spin clusters, whereas the single-spin dynamics is not relevant.

For a dynamic generalization of the correlated molecular field theory [12.66, 12.67] a time-dependent correlated molecular field is introduced where spatial fluctuations with wavelengths smaller than $\xi(T)$ and temporal fluctuations with frequencies higher than the inverse characteristic time of the clusters are suppressed,

$$H_{mf}(r, t) = \left(\frac{1}{g\mu_B}\right)^2 \int J(r, r') M(r', t)_{corr}\, d^3r', \qquad (12.104)$$

$$M(r', t)_{corr} = \int_{t-\infty}^{t+\infty} \int g(r' - r'', t - t') M(r'', t')\, d^3r''\, dt'. \qquad (12.105)$$

Here the integral kernel $g(r' - r'', t - t')$ averages out all the irrelevant fluctuations of the magnetization, and in $M(r', t)_{corr}$ all the information about the high-frequency dynamics is lost. Equations (12.104) and (12.105) represent a noncausal ansatz for the molecular field (which does not constitute a problem because the correlated molecular field is just a mathematical tool to take into account the effect of spin correlations within the framework of a mean field theory). Therefore one must bear in mind that the dynamic susceptibility calculated by this ansatz (see below) has a physical meaning only for frequencies much smaller than the inverse characteristic lifetime of the clusters, and that the Kramers–Kronig relations do not hold. For the Fourier transform $g(k, \omega)$ of the integral kernel the real part of the dynamic susceptibility of the conventional molecular field theory [12.65] according to eqs. (12.102) and (12.103) is used:

$$g(k, \omega) = A \operatorname{Re}[\chi(k, \omega)], \tag{12.106}$$

$$\chi(k, \omega) = \frac{1}{-i\omega\alpha\gamma + A + Ck^2}, \tag{12.107}$$

with A given by eq. (12.33) and $C = \overline{C(r)}$ with $C(r)$ given by eq. (12.30). The static limit ($\omega = 0$) yields $g(k, 0) = f(k) = 1/(1 + k^2\tilde{\xi}^2)$ with $\xi = \sqrt{C/A}$, i.e., the static correlated molecular field theory is regained.

Equation (12.102) with eqs. (12.104)–(12.107) represents a linear integrodifferential equation with spatially random fluctuations of the exchange couplings $J(r, r')$ which may be treated by the same method as described in Section 12.5.2. Then the following implicit equation can be derived [12.66, 12.67] for the relaxation time τ which determines the exponential decay of the volume-averaged magnetization when switching off the external field (see above)

$$-\alpha\tau^{-1} = -\alpha + A'g(0, \tau^{-1}) + (A')^2 g(0, \tau^{-1}) \left(\frac{1}{2\pi}\right)^3$$

$$\times \int d^3k\, \Gamma(k)\, \tilde{\chi}(k, \tau^{-1})\, g(k, \tau^{-1}), \tag{12.108}$$

which is the counterpart to eq. (12.76) for the static susceptibility. The quantity $\Gamma(k)$ is the Fourier transform of the structural correlation function for the exchange fluctuations, eq. (12.69), and

$$\tilde{\chi}(k, \tau^{-1}) = \frac{1}{-\alpha\tau^{-1} + \alpha - \left(\dfrac{1}{\gamma\mu_{\mathrm{B}}}\right)^2 g(k, \tau^{-1})\, J_0(k)} \tag{12.109}$$

is the above discussed noncausal dynamic susceptibility for $\omega\gamma = -i/\tau$.

For homogeneous systems with no fluctuations of the exchange interactions eq. (12.108) with eq. (12.106) and eq. (12.107) yields

$$\frac{\alpha^3}{A^2}(\tau^{-1})^3 - \frac{\alpha^3}{A^2}(\tau^{-1})^2 - \alpha\tau^{-1} + A = 0. \tag{12.110}$$

The physically relevant solution of this equation yields for $T \to T_c$ the asymptotic behaviour

$$\tau^{-1} = \left(\frac{A}{\alpha}\right)^{3/2} = \left(\frac{T - T_c}{T}\right)^{\Delta_{MM}}, \tag{12.111}$$

with $\Delta_{MM} = 3/2$. Obviously, the relaxation time τ becomes infinitely large for T approaching T_c (critical slowing down). The conventional molecular field theory based on eqs. (12.102) and (12.103) gives $\Delta_{MM} = 1$, i.e., the results of the dynamic molecular field theory differ from those of the conventional theory for homogeneous systems. For comparison, a high-temperature series expansion for the Ising model on a simple cubic lattice [12.68] yields $\Delta_{MM} = 1.4$ which is close to our result $\Delta_{MM} = 1.5$.

For random exchange ferromagnets the dynamic correlated molecular field theory predicts that the asymptotic exponent Δ_{MM} is not influenced by the spatial fluctuations of the exchange couplings. By the dynamic renormalization group theory [12.69] it has been shown that the critical relaxation dynamics should indeed not be affected by isotropic exchange fluctuations for three-dimensional Heisenberg ferromagnets, whereas the critical slowing down is enhanced by the spatial fluctuations for Ising systems. For the crossover regime from critical behaviour to high-temperature behaviour the correlated molecular field theory yields an enhancement of the relaxation time due to the spatial fluctuations [12.66].

References

[12.1] Ma, S.-K., 1976, *Modern Theory of Critical Phenomena* (Benjamin, Reading, Massachusetts).
[12.2] Liebs, M., and Fähnle, M., 1996, *J. Phys. Condens. Matter* **8**, 3207.
[12.3] Handrich, K., and Kobe, S., 1980, *Amorphe Ferro- und Ferrimagnetika* (Akademie-Verlag, Berlin).
[12.4] Kaneyoushi, T., 1992, *Introduction to Amorphous Magnets* (World Scientific, Singapore).
[12.5] Stauffer, D., 1985, *Introduction to Percolation Theory* (Taylor & Francis, London–Philadelphia).
[12.6] Fischer, K.H., 1983, *Phys. Stat. Sol. (B)* **116**, 357.
[12.7] Binder, K., and Stauffer, D., 1979, *Monte Carlo Methods in Statistical Physics* (Ed. K. Binder, Springer, Heidelberg).
[12.8] Chudnovsky, E.M., Saslow, W.M., and Sarota, R.A., 1986, *Phys. Rev. B* **33**, 251.
[12.9] Elsässer, C., Fähnle, M., Brandt, E.-H., and Böhm, M.C., 1988, *J. Phys. F: Met. Phys.* **18**, 2463.

[12.10] Sellmyer, D.J., and Nafis, S., 1985, *J. Appl. Phys.* **57**, 3584.
[12.11] Harris, A.B., 1974, *J. Phys. C: Solid State Phys.* **7**, 1671.
[12.12] Weinrib, A., and Halperin, B.I., 1983, *Phys. Rev. B* **27**, 413.
[12.13] Heuer, H.-O., Wagner, D., 1989, *Phys. Rev. B* **40**, 2502.
[12.14] Sambasiva Rao, M., and Kaul, S.N., 1995, *J. Magn. Magn. Mater.* **147**, 149.
[12.15] Babu, P.D., and Kaul, S.N., 1997, *J. Phys. Condens. Matter* **9**, 7189.
[12.16] Kaul, S.N., 1985, *J. Magn. Magn. Mater.* **53**, 5.
[12.17] Fähnle, M., Kellner, W.-U., and Kronmüller, H., 1987, *Phys. Rev. B* **35**, 3640.
[12.18] Chayes, J.T., Chayes, L., Fisher, D.S., and Spencer, T., 1986, *Phys. Rev. Lett.* **57**, 2999.
[12.19] Braun, P., Staaden, U., Holey, T., and Fähnle, M., 1989, *Int. J. Mod. Phys.* **B3**, 1343.
[12.20] Fähnle, M., Holey, T., Staaden, U., and Braun, P., 1990. In *Festkörperprobleme, Advances in Solid State Physics* 30, (Ed. U. Rössler, Vieweg & Sohn, Braunschweig, Wiesbaden) p. 425.
[12.21] Aharony, A., and Fisher, M.E., 1983, *Phys. Rev. B* **27**, 4394.
[12.22] Kaul, S.N., 1994, *Phase Transitions* **47**, 23.
[12.23] Kouvel, J.S., and M.E. Fisher, 1964, *Phys. Rev.* **136**, A 1626.
[12.24] Kellner, W.-U., Fähnle, M., Kronmüller, H., and Kaul, S.N., 1987, *Phys. Stat. Sol. (B)* **144**, 397.
[12.25] Fähnle, M., 1983, *J. Phys. C.: Solid State Phys.* **16**, L 819.
[12.26] Fähnle, M., 1985, *J. Phys. C.: Solid State Phys.* **18**, 181.
[12.27] Haug, M., Fähnle, M., Kronmüller, H., and Haberey, F., 1987, *J. Magn. Magn. Mater.* **69**, 163.
[12.28] Fähnle, M., and Egami, T., 1982, *J. Appl. Phys.* **53**, 7693.
[12.29] Fähnle, M., 1985, *Solid State Commun.* **55**, 743.
[12.30] Fähnle, M., Herzer, G., Kronmüller, H., Mayer, R., Saile, M., and Egami, T., 1983, *J. Magn. Magn. Mater.* **38**, 240.
[12.31] Baker, G.A., 1961, *Phys. Rev.* **124**, 768.
[12.32] Kaul, S.N., Siruguri, V., and Chandra, G., 1992, *Phys. Rev. B* **45**, 12343.
[12.33] Gebhardt, W., and Krey, U., 1980, *Phasenübergänge und kritische Phänomene* (Vieweg & Sohn, Braunschweig, Wiesbaden).
[12.34] Wagner, D., and Wohlfarth, E.P., 1979, *J. Phys. F: Met. Phys.* **9**, 717.
[12.35] Brauneck, W., and Wagner, D., 1984, *J. Magn. Magn. Mater.* **44**, 20.
[12.36] Kobe, S., 1970, *Phys. Stat. Sol.* **41**, K13.
[12.37] Fähnle, M., 1980, *Phys. Stat. Sol. (B)* **99**, 547.
[12.38] Handrich, K., 1969, *Phys. Stat. Sol.* **32**, K55.
[12.39] Herzer, G., Fähnle, M., Egami, T., and Kronmüller, H., 1980, *Phys. Stat. Sol. (B)* **101**, 713.
[12.40] Henderson, R.G., and de Graaf, A.M., 1973, *Amorphous Magnetism* (Ed. H.O. Hooper and A.M. de Graaf, Plenum Press, New York–London) p. 331.
[12.41] Shtrikman, S., and Wohlfarth, E.P., 1972, *Physica* **60**, 427.
[12.42] Yamada, H., and Wohlfarth, E.P., 1975, *Phys. Lett.* **A51**, 65.
[12.43] Kronmüller, H., and Fähnle, M., 1980, *Phys. Stat. Sol. (B)* **97**, 513.
[12.44] Fähnle, M., and Kronmüller, H., 1980, *Phys. Stat. Sol. (B)* **98**, 219.
[12.45] Arrott, A., 1957, *Phys. Rev.* **108**, 1394.
[12.46] Fähnle, M., 2000, *J. Magn. Magn. Mater.* **210**, L1.
[12.47] Holey, T., and Fähnle, M., 1987, *Phys. Stat. Sol. (B)* **141**, 253.
[12.48] Fähnle, M., and Herzer, G., 1986, *Solid State Commun.* **57**, 449.
[12.49] Kröner, E., and Koch, H., 1976, *SM Archives* **1**, 183.

[12.50] Montgomery, C.G., Krugler, J.L., and Stubbs, R.M., 1970, *Phys. Rev. Lett.* **25**, 669.

[12.51] Beran, M.J., 1968, *Statistical Continuum Theories* (Interscience Publishers, New York).

[12.52] Fähnle, M., and Herzer, G., 1984, *J. Magn. Magn. Mater.* **44**, 274.

[12.53] Souletie, J., and Tholence, J.L., 1983, *Solid State Commun.* **48** 407.

[12.54] Fähnle, M., and Souletie, J., 1984, *J. Phys.* C **17**, L 469.

[12.55] Fähnle, M., and Haug, M., 1987, *Phys. Stat. Sol. (B)* **140**, 569.

[12.56] Haug, M., Fähnle, M., Kronmüller, H., and Haberey, F., 1987, *Phys. Stat. Sol. (B)* **144**, 411.

[12.57] Barbara, B., Malozemoff, A.P., and Imry, Y., 1981, *Phy. Rev. Lett.* **47**, 1852.

[12.58] Omari, R., Préjean, J.J., and Souletie, J., 1983, *J. Physique* **44**, 1069.

[12.59] Ueno, Y., and Oguchi, T., 1980, *Prog. Theor. Phys.* **63**, 342.

[12.60] Binder, K., and Kinzel, W., 1981, *Lecture Notes Phys.* **149**, 124.

[12.61] Aharony, A., and Imry, Y., 1976, *Solid State Commun.* **20**, 899.

[12.62] Edwards, S.F., and Anderson, P.W., 1975, *J. Phys.* F **5**, 965.

[12.63] Binder, K., 1977, *Z. Phys.* B **26**, 339.

[12.64] Monod, P., and Bouchiat, H., 1982, *J. Physique Lett.* **43**, L 45.

[12.65] Suzuki, M., and Kubo, R., 1968, *J. Phys. Soc. Japan* **24**, 51.

[12.66] Fähnle, M., 1981, *Phys. Stat. Sol. (B)* **106**, 519.

[12.67] Fähnle, M., Herzer, G., Egami, T., and Kronmüller, H., 1982, *J. Appl. Phys.* **53**, 2326.

[12.68] Yahata, H., 1971, *J. Phys. Soc. Japan* **30**, 657.

[12.69] Krey, U., 1977, *Z. Physik* B **26**, 355.

13

Computational micromagnetism of thin platelets and small particles

13.1 Introduction

By means of the continuum theory of micromagnetism only a few problems can be solved rigorously. Well-known examples are planar domain walls, the nucleation fields of cylindrical domain particles, the law of approach to ferromagnetic saturation and weak stripe domains in thin films. However, all problems dealing with the effects of the microstructure have to be solved by approximate methods without a rigorous solution of the nonlinear micromagnetic equation. Also domain patterns and dynamic magnetization processes as strongly nonlinear problems cannot be determined by analytical micromagnetic solutions. A further very extensive research field up to now is the spin configurations in thin ferromagnetic films and small particles. Here the first numerical methods have been applied by Holz [13.1], Hubert [13.2], Brown and La Bonte [13.3], La Bonte [13.4] and Holz and Hubert [13.5]. Holz and Hubert used the Ritz method, minimizing the total magnetic Gibbs free energy with respect to the free parameters of a functional ansatz for the distribution of the magnetization. Since the Ritz technique in particular does not allow a correct treatment of the long-range dipolar fields (stray fields) other methods have been developed. Here the finite element method (FEM) [13.6–13.14] and the fast Fourier transform (FFT) [13.15–13.18] have to be mentioned as the most efficient techniques. Also the method of finite differences (FDM) [13.19–13.26] has been successfully applied to determine the distribution of the magnetization in small particles. It is obvious that the latter two methods are restricted to geometries of cubes, rectangular platelets or more generally to parallelepipeds, whereas the FEM also allows the treatment of ensembles of polyhedral grains and arbitrary shapes of samples. One of the very crucial problems in all computational micromagnetic calculations is the treatment of the magnetic stray field resulting from either surface charges $\sigma = \boldsymbol{n} \cdot \boldsymbol{J}_s$, or volume charges $\rho = -(1/\mu_0)\,\mathrm{div}\,\boldsymbol{J}_s$. These magnetic charges give rise to a nonlocal stray field, the correct treatment of

Table 13.1. *Material parameters and the exchange lengths l_s and l_K of $Nd_2Fe_{14}B$ and permalloy $Ni_{80}Fe_{20}$.*

Material	J_s [T]	K_1 [10^6 J/m]	A [pJ/m]	l_s [nm]	l_K [nm]
$Nd_2Fe_{14}B$	1.61	4.3	7.3	2.8	1.4
$Ni_{80}B_{20}$	1.00	0.05	13	5.7	161

which is especially important in the case of soft magnetic materials with $Q \ll 1$. In hard magnetic materials the dipolar interactions are less important because in this case the magnetocrystalline energy corresponds to the dominant energy term. As outlined in Chapter 4 the extension of spin inhomogeneities is governed by the so-called exchange lengths

$$l_K = \left(\frac{A}{K_1}\right)^{1/2}, \quad l_s = \left(\frac{2A}{M_s J_s}\right)^{1/2}, \quad l_H = \frac{2A}{H_{ext} J_s}. \quad (13.1)$$

As characteristic examples, Table 13.1 shows numerical values of the exchange lengths for permalloy and $Nd_2Fe_{14}B$.

The smallest exchange length always determines the extension of spin inhomogeneities and, as a consequence, either the energy terms, K_1, $M_s J_s/2$, or $H_{ext} J_s$ contribute the largest amount to the total energy. In the case of soft magnetic materials the spin arrangements are characterized by reduced volume and surface charges, i.e., the spin arrangements are more or less divergence free, $\text{div } M_s = 0$, and in a first approximation may be derived from a vector potential, $J_s = \text{rot } A$. On the other hand, in hard magnetic materials the leading energy term corresponds to the magnetocrystalline energy. The numerical treatment of magnetic states and magnetization processes in modern nanocrystalline materials has been shown to be a rather effective method to determine the hysteresis loop and its characteristic parameters. In the following, a review will be given on the present state of computational micromagnetism and its applications to magnetic configurations and magnetization processes in small particles, thin platelets and assemblies of nanocrystalline grains.

13.2 Applications of the finite difference method

The numerical solution of the micromagnetic equilibrium conditions requires a discretization of the functions $J_s(r)$ or $M_s(r)$, which should establish an energy minimum. Here one may differentiate between the static and the dynamic equilibrium equations given in Chapters 3 and 14. A mixed form of discretization has

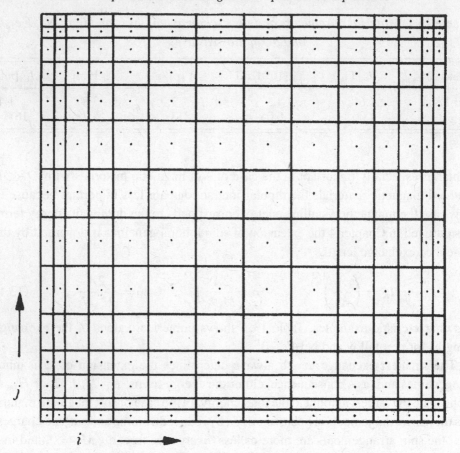

Fig. 13.1. Grid with unequal spacings. Refinement of the mesh at the corners takes care of the large demagnetization fields [13.20].

been used by La Bonte [13.4] and Chantrell *et al.* [13.27–13.30], where the local magnetic torque vanishes and where the equilibrium state is found by allowing the precessional movement of the magnetization. In the case of the finite difference method (FDM) the space is discretized by rectangular blocks in three dimensions or by rectangles in two dimensions. Figure 13.1 shows schematically the discretization of a two-dimensional function $F(x_i, y_i)$ according to the FDM. Within each element the magnetization is assumed to be homogeneous. The stray field energy therefore results from the surface charges at the element's surfaces and the exchange energy is due to the change of the easy directions between neighbouring elements. Near corners and edges in general large changes of the direction of the magnetization have to be expected which requires a reduction of the cell sizes. The dipolar energy has to be determined as the sum of all contributions of the surface charges of all elements as demonstrated by Fig. 13.2. Figure 13.2 shows a rectangular homogeneously

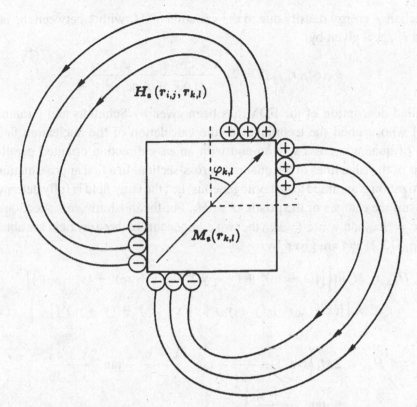

Fig. 13.2. Rectangular element at the lattice point $r_{k,l}$ with homogeneous magnetization $M_s(r_{k,l})$ of direction $\varphi_{k,l}$. The surface charges cause a magnetic stray field $H_s(r_{i,j}, r_{k,l})$. [13.20].

magnetized element in the x, y plane at the lattice points $r_{k,l} = (x_k, x_l)$ with arbitrary direction of M_s, where $\varphi_{k,l}$ is the angle of $M_s(r_{k,l})$ with respect to the y-axis. ΔX_k and ΔY_l are the lengths of the elements in the x and y directions, respectively. The surface charges at $r_{k,l}$ cause a magnetic stray field, $H_s(r_{i,j}, r_{k,l})$, to be determined from the potential according to $H_s(r_{i,j}, r_{k,l}) = -\nabla U(r_{i,j}, r_{k,l})$:

$$U(r_{i,j}, r_{k,l}) = \frac{1}{2} \int\limits_{S_{k,l}} \frac{M_s(r'_{k,l})\, \mathrm{d}S(r'_{k,l})}{|r_{i,j} - r'_{k,l}|}, \qquad (13.2)$$

where the integration extends over the surface $S_{k,l}$ of the element $r_{k,l}$ and the factor $1/2$ takes care of the fact that each surface $S_{k,l}$ belongs to two elements. Thus the local magnetic field $H_s(r_{i,j})$ at the grid point $r_{i,j}$ is given by

$$H_s(r_{i,j}) = \sum_{k=1}^{n_x} \sum_{l=1}^{n_y} H_s(r_{i,j}, r_{k,l}). \qquad (13.3)$$

The exchange energy density due to the variation of M_s with x between the points $r_{i,j}$ and $r_{i,j+1}$ is given by

$$\phi_A(r_{i,j}, r_{i,j+1}) = 2A \frac{1 - \cos(\varphi_{i,j} - \varphi_{i+1,j})}{\frac{1}{4}(\Delta x_i + \Delta x_{i+1})^2}. \tag{13.4}$$

A detailed description of the FDM has been given by Schmidts and Kronmüller [13.22] who applied the technique for the calculation of the nucleation field of prisms of quadratic cross-section and with an easy direction oriented parallel to one pair of the edge lines of the quadratic cross-section. In a first approximation the magnetization is assumed to be homogeneous, i.e., the stray field is fully determined by the surface charges of the prism, $\sigma = M_s$. For the quadratic cross-section with edges $x = \pm x_0$ and $y = z = \pm z_0$ the two components of the stray field are obtained from eq. (13.2) and are given by

$$H_{s,x} = M_s \ln \Big\{ \big[\{(x - x_0)^2 + (y - y_0)^2\} \cdot \{(x + x_0)^2 + (y + y_0)^2\} \big]$$
$$\times \big[\{(x - x_0)^2 + (y + y_0)^2\} \cdot \{(x - x_0)^2 + (y - y_0)^2\} \big]^{-1} \Big\}, \tag{13.5}$$

$$H_{s,y} = 2M_s \Big\{ \tan^{-1} \frac{x - x_0}{y - y_0} - \tan^{-1} \frac{x - x_0}{y + y_0} - \tan^{-1} \frac{x + x_0}{y - y_0}$$
$$+ \tan^{-1} \frac{x + x_0}{y + y_0} \Big\}. \tag{13.6}$$

From eqs. (13.5) and (13.6) it may be concluded that at each corner the magnetic stray field $H_{s,x}$ has a logarithmic singularity. Naturally, this singularity is a consequence of the micromagnetic continuum approach which would be avoided if we consider discrete atomic moments. Due to the singularity the magnetization at the corner is oriented parallel to the diagonal of the square. For a second approximation the magnetic stray field of the first approximation is used to calculate a new distribution of $M_s(r_{i,j})$. Here for each element the micromagnetic equilibrium condition,

$$2A \, \Delta\varphi_{i,j} + (H_{\text{ext}} + H_{s,j}(r_{i,j})) \sin(\psi_{i,j} - \varphi_{i,j}) - \frac{\partial \phi_K(\varphi_{i,j})}{\partial \varphi_{i,j}} = 0, \tag{13.7}$$

has to be solved, where $H_{s,j}(r_{i,j})$ is given by eqs. (13.5) and (13.6) and $\psi_{i,j}$ denotes the angle between the y-axis and the total field $H_t = H_{\text{ext}} + H_s$. In eq. (13.7) the exchange term $2A\Delta\varphi_{i,j}$ is given by

$$2A\Delta\varphi_{i,j} = 2A \Big\{ B_{i,j}^{(1)}(\varphi_{i,j} - \varphi_{i+1,j}) + B_{i,j}^{(2)}(\varphi_{i,j} - \varphi_{i-1,j})$$
$$+ B_{i,j}^{(3)}(\varphi_{i,j} - \varphi_{i,j+1}) + B_{i,j}^{(4)}(\varphi_{i,j} - \varphi_{i,j-1}) \Big\} \tag{13.8}$$

and the coefficients $B_{i,j}^n$ are written

$$B_{i,j}^{(1)} = -8/\{\Delta x_{i+1} + \Delta x_i)(\Delta x_{i+1} + 2\Delta x_i + \Delta x_{i-1})\},$$
$$B_{i,j}^{(2)} = -8\{(\Delta x_{i-1} + \Delta x_i)(\Delta x_{i+1} + 2\Delta x_i + \Delta x_{i-1})\},$$
$$B_{i,j}^{(3)} = -8/\{(\Delta y_{j+1} + \Delta y_j)(\Delta y_{j+1} + 2\Delta y_j + \Delta y_{j-1})\},$$
$$B_{i,j}^{(4)} = -8\{(\Delta y_{j-1} + \Delta y_j)(\Delta y_{j+1} + 2\Delta y_j + \Delta y_{j-1})\}. \tag{13.9}$$

If H_t is known for each lattice point then eq. (13.7) may be solved for each element in order to derive the new magnetization $M_s(r_{i,j})$. This procedure is continued until the difference between the input state and the output state falls below a fixed tolerance.

Due to the inhomogeneous stray field H_s, the magnetization M_s in prisms and all specimens with nonellipsoidal shape always remains inhomogeneous unless H_{ext} approaches infinity. The magnetic states corresponding to the demagnetization curve change reversibly with increasing field applied inversely to the original orientation of M_s. Figures 13.3 and 13.4 present the distribution of M_s and H_s for the ground state of $H_{ext} = 0$ and the so-called undercritical state just below the nucleation field at 4.35 T for a prism of the hard magnetic material $Nd_2Fe_{14}B$ ($K_1 = 4.3 \cdot 10^6$ J/m^3, $K_2 = 0.65 \cdot 10^6$ J/m^3, $J_s = 1.61$ T, $A = 7.7 \cdot 10^{-12}$ J/m).

It is of interest that in the case of hard magnetic particles with nonellipsoidal shape the hysteresis loop still shows rectangular behaviour with only a negligible reversible decrease of M due to the rotations at the edges. Hysteresis loops have been determined as a function of magnetic fields applied at an angle, Ψ_0, with

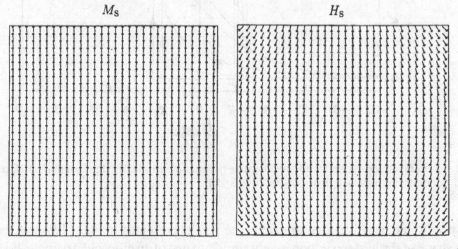

M_s H_s

Fig. 13.3. Remanent state of a prismatic particle of edge length $d = 60$ nm and the material parameters of $Nd_2Fe_{14}B$. Left, M_s; right, H_s [13.20].

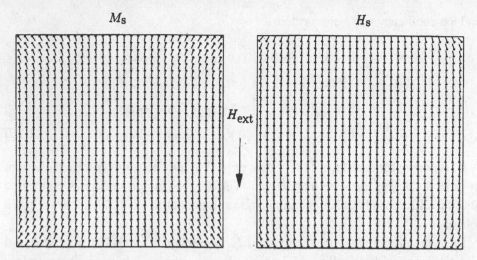

Fig. 13.4. Undercritical state of a prismatic particle of edge length $d = 60$ nm and the material parameters of $Nd_2Fe_{14}B$. Left, M_s; right, H_s [13.20]. Applied field: $H_{ext} = 4.35 \cdot 10^6$ A/m [13.20].

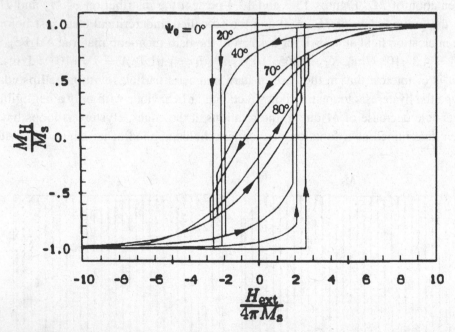

Fig. 13.5. Hysteresis loop for the rectangular prismatic particle of edge length $d = 10$ μm and the material parameters of $Nd_2Fe_{14}B$ [13.20].

respect to the negative easy direction. The results presented in Fig. 13.5 show that the shape of the hysteresis loop is very similar to that of an ellipsoidal particle, however, with a reduced nucleation field. Also the angular dependence of the nucleation field of the prism, shown in Fig. 13.6, reveals a similar behaviour to the conventional

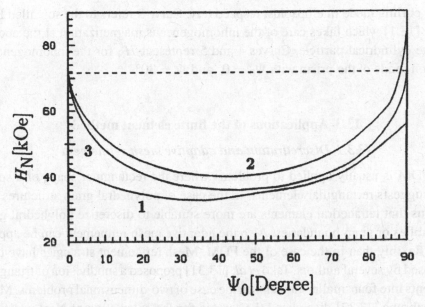

Fig. 13.6. Nucleation field of $Nd_2Fe_{14}B$ for the rectangular prismatic particle for various angles Ψ_0 of the applied field with respect to the negative c-axis (1); nucleation field of an ellipsoidal particle taking into account K_2 (2); H_N of the Stoner–Wohlfarth particle (3).

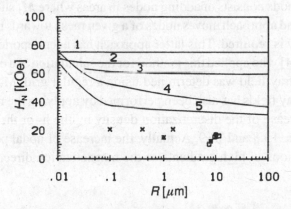

Fig. 13.7. Nucleation field H_N versus particle diameter R for various models. (1) Coherent rotation mode, (2) curling mode, (3) Holz mode. Parallelepiped with (4) $\Psi_0 = 0°$, (5) $\Psi_0 = 30°$. Measured coercivities: × Grönefeld, M., and Kronmüller, H., 1989, *J. Magn. Magn. Mater.* **80**, 223; ∗ Durst, K.-D., and Kronmüller, H., 1986, *J. Magn. Magn. Mater.* **59**, 86; □ Kiss, L.F., Martinek, G., Forkl, A., and Kronmüller, H., 1989, *Phys. Stat. Sol. (A)* **114**, 685.

Stoner–Wohlfarth particle (compare Fig. 6.19), however, with a considerable reduction of the nucleation field for $\Psi_0 = 0$. Figure 13.7 presents the nucleation field of an ideally oriented particle ($\Psi_0 = 0$) as a function of the particle size, R. The results obtained by various other models are included and compared with the experimental results. Curves 1 and 2 correspond to the coherent rotation mode and to the curling mode in ellipsoids, respectively. Curve 3 refers to the so-called Holz mode [13.1] which takes care of the inhomogeneous magnetization at the ends of a finite cylindrical particle. Curves 4 and 5 represent H_N for the inhomogeneous rotation mode of the prism with $\Psi_0 = 0°$ and $\Psi = 30°$.

13.3 Applications of the finite element method

13.3.1 Discretization and adaptive mesh refinement

The FDM is usually applied to problems where the rectangular shape of a specimen suggests rectangular elements. In the case of polyhedral grain structures it is obvious that tetrahedral elements are more suitable to discretize polyhedral grain assemblies or grain boundaries. Also the adaptive mesh refinement can be applied more flexibly than in the case of the FDM. Mesh refinement strategies have been proposed by several authors. Tako *et al.* [13.31] proposed a subdivision of triangular elements into four smaller elements in the case of two-dimensional problems. Miltat and Labrune [13.32] discretized the sample for the calculation of Néel walls into rectangular prisms of variable size with small elements in the centre of the wall and increasingly larger elements according to the distance from the wall centre. In the following the so-called h- and r-type refinements are applied [13.33–13.35]. The first of these methods consists of adding nodes in areas where M_s shows large gradients and the second approach moves nodes of a given mesh towards regions in which a higher accuracy is required. This latter approach has been reported by Lewis and Della Torre [13.34] who applied this method for the calculation of a one-dimensional wall, while the stray field was determined using a regular grid. More recently, the calculation of stray fields has also been performed by applying the mesh adaptation [13.11]. The increase of the discretization density by the h- or the r-refinement is visualized in Figs. 13.8 and 13.9. Actually, the increase of nodal point densities is performed in regions of high gradients of the magnetization direction, whereas in

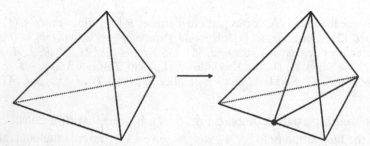

Fig. 13.8. Bisection of a finite element tetrahedron according to the so-called h-refinement [13.33]. Bisection in one element requires bisection also in the neighbouring grain.

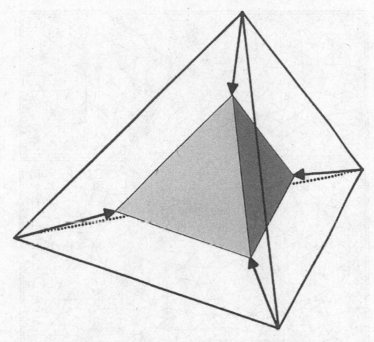

Fig. 13.9. Shrinking of a tetrahedron according to the r-refinement [13.35]. The nodes are shifted uniformly along the diagonals.

regions of small gradients the nodal point density may be diminished. In the case of the r-refinement the factor by which the element shrinks has to be determined by means of the maximum angle $\Delta \varphi_{max}$ found between two magnetization vectors $M_{s,i}$ and $M_{s,j}$ at the nodes i and j, $\Delta \varphi_{max} = \max_{i,j}\{| \arccos(M_{s,i} \cdot M_{s,j}/M_s^2)|\}$. A possible shrinking function may then be

$$f(\Delta \varphi_{max}) = 1 - \frac{1}{2}\sqrt{\Delta \varphi_{max}/\pi}. \qquad (13.10)$$

The element then remains unchanged for $\Delta \varphi_{max} = 0$ and for $\Delta \varphi_{max} = \pi$ reduces the volume of the element by a factor of $1/8$. Naturally, there exist more constraints for the shifting of nodal points. For example, nodes at the surface must remain on the surface. Nodes at corners must not be moved and nodes on edges may move only along the edges. Tunneling of a node through the opposite plane of the tetrahedra should be avoided. Similarly, ill-shaped elements should be eliminated by defining a maximum finite angle between the edges and the planes. A threshold for the minimum volume of an element may be chosen which may be of the order of the cube of the smallest exchange length l_s^3 or l_K^3.

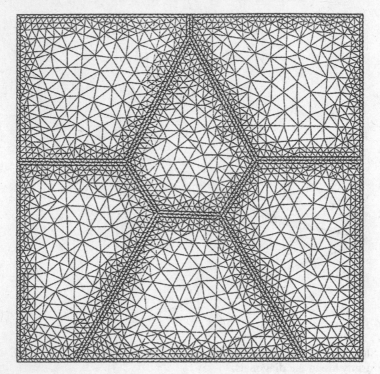

Fig. 13.10. Finite element mesh showing the refinement of the mesh length near grain boundaries [13.68].

After each construction of a new mesh the magnetization has to be calculated from the minimization of the magnetic Gibbs free energy performed by means of the conjugate gradient technique. This procedure is repeated until the change of the total Gibbs free energy approaches a fixed limit, e.g., 1/1000 of the absolute value. As an example, Fig. 13.10 shows the finite element mesh with adaptive element refinement near corners, edges and grain boundaries.

13.3.2 Discretization of the Gibbs free energy used for computational micromagnetism

Neglecting the magnetoelastic coupling energy, the magnetic Gibbs free energy for a uniaxial crystal is a sum of four terms

$$\phi_t = \int \left[A \sum_i \{\nabla \gamma_i(\vartheta, \varphi)\}^2 + K_1 \sin^2 \alpha(\vartheta, \varphi) - \frac{1}{2} \boldsymbol{H}_s \cdot \boldsymbol{J}_s - \boldsymbol{J}_s \cdot \boldsymbol{H}_{\text{ext}} \right] dV,$$

(13.11)

where the individual energy terms are due to the exchange, the magnetocrystalline, the dipolar and the magnetostatic energy, and $\alpha(\vartheta, \varphi)$ denotes the angle between J_s and the easy direction. $\vartheta(r)$ and $\varphi(r)$ are the space-dependent polar and the azimuthal angles of $J_s(r)$ in a spherical coordinate system. Usually the stray field H_s complicates the determination of the energy minimum of ϕ_t. Sometimes it it suggested that the stray field term can be neglected in the case of hard magnetic materials where the exchange length l_K determines the distribution of J_s [13.36]. Here it has to be taken into account, however, that l_s in general is rather small, e.g., 2.7 nm in the case of $Nd_2Fe_{14}B$, a value which is comparable to $l_K = 1.3$ nm. Stray field terms therefore should not be neglected, in particular because the demagnetization process starts at corners and edges where the demagnetization fields are very large and of the order of the magnetization M_s.

For minimizing the Gibbs free energy Brown [2.52] proposed two functionals which present upper and lower bounds for the magnetostatic energy. These functionals are defined either by the scalar potential $U(r)$ or the magnetic vector potential $A(r)$. The magnetostatic energy ϕ_s is approximated by the functionals

$$\phi_s \geq W_1(U) = \int_{V_m} (\nabla U \cdot J)\, dV - \frac{\mu_0}{2} \int_{V_t} (\nabla U)^2\, dV, \qquad (13.12)$$

or

$$\phi_s \leq W_2(A) = \frac{1}{2\mu_0} \int_{V_t} (\text{rot}\, A - J)^2\, dV = \frac{1}{2\mu_0} \int_{V_t} (B - J)^2\, dV, \qquad (13.13)$$

where the integrals extend either over the magnet's volume V_m or the total volume V_t. Maximizing $W_1(U)$ gives the exact stray field energy ϕ_s, whereas in the case of $W_2(U)$ minimizing with respect to A and J gives the exact stray field energy. In the case of $W_1(U)$ the maximization with increasing numbers of iterations leads to a fluctuation of the variables ϑ and φ as shown in Fig. 13.11a, whereas in the case of the minimization of $W_2(A)$ the true equilibrium values of ϑ and φ are approached monotonously as demonstrated by Fig. 13.11a. Accordingly, a stable solution is found much faster using the functional $W_2(A)$. In order to test the precision obtained by using the functional W_1 or W_2 the magnetostatic energy of a homogenenous magnetized ($\phi_{ex} = 0$, $\phi_K = 0$) quadrat of edge length 1 μm and thickness 20 nm has been calculated for different numbers of finite elements in inner and outer space [13.38]. These numerical results have been compared with the precise analytical result, $(1/2)N\mu_0 M_s^2$, using Aharoni's calculations [13.37] for the effective demagnetization factor, N, of parallelepipeds. According to Fig. 13.11b the scalar potential underestimates the magnetostatic energy

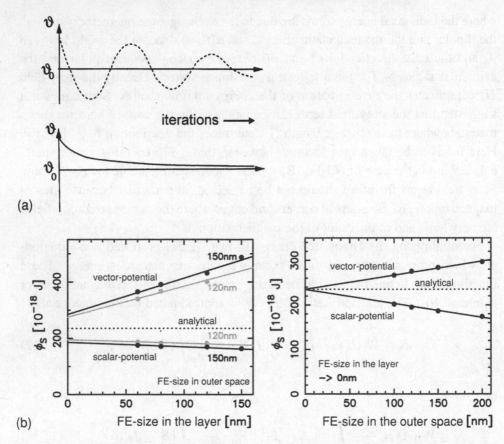

Fig. 13.11. (a) Oscillating approach and monotonous approach to the equilibrium value of the total energy for $W_1(U)$ and $W_2(A)$, respectively. (b) Test of precision of the scalar and the vector potential for the case of a homogeneously magnetized quadratic platelet 1 μm × 1 μm × 20 nm as a function of the number of elements in the outer and inner space and comparison with the precise analytical result [13.38].

throughout, whereas the vector potential overestimates it. An extrapolation of both straight lines to an infinite number of elements leads to the correct result. Here it should be noted that such an extrapolation has to be performed for each thickness of the platelet.

Using $W_1(U)$, the corresponding Euler–Lagrange equations are those of the magnetostatic boundary problem: $\Delta U = 0$ outside the magnetic particle and $\Delta U = \mathrm{div}\, \boldsymbol{M}_{\mathrm{s}}$ inside the magnetic particle.

At the grain boundaries between grains i and j the following boundary conditions hold for the scalar potential U:

$$U^i = U^j, \qquad (\nabla U^i - \nabla U^j) \cdot \boldsymbol{n} = (\boldsymbol{M}_{\mathrm{s}}^i - \boldsymbol{M}_{\mathrm{s}}^j) \cdot \boldsymbol{n}, \qquad (13.14)$$

where n corresponds to the normal vector of the grain boundary pointing from grain i to grain j. Using the functional $W_2(A)$, the Euler–Lagrange equations are $\nabla \times \nabla \times A = 0$ outside the particle and $\nabla \times \nabla \times A = \nabla \times M_s$ inside the particle.

The grain boundary conditions for the vector potential A are

$$(\nabla \times A^i - \nabla \times A^j) \cdot n = 0, \qquad (\nabla \times A^i - \nabla \times A^j) \times n = J_s \times n. \quad (13.15)$$

Equations (13.15) show that the tangential component of A at the boundary is continuous, whereas the normal component reveals a discontinuous jump. It should be noted, however, that for the calculation of the spin distribution the vector potential itself is meaningless because in the B-field (rot A) its boundary behaviour is reproduced correctly. Similarly the vector potential has also been shown to be irrelevant by Demerdash *et al.* [13.39]. In order to calculate the equilibrium distribution of J_s eq. (13.12) or eq. (13.13) has to be solved simultaneously with minimization of eq. (13.11) where the stray field energy is replaced either by W_1 or W_2. Naturally, it is an advantage that in the case where W_2 is used both terms ϕ_t as well as W_2 have to be minimized. The discretization of the Gibbs free energy density, g_t, proceeds according to the following scheme:

$$\phi_t(f(r)) = \int_{V_t} g_t[f(r)] \, dV, \quad (13.16)$$

with

$$f(r) = F\big(\vartheta(r), \varphi(r), A_x(r), A_y(r), A_z(r)\big).$$

$\phi_t(f(r))$ is now given as a sum over all contributions from the elements E,

$$\phi_t(f(r)) = \sum_{e=1}^{E} \int_{V_e} g_t[f^e(r)] \, dV. \quad (13.17)$$

The function $f^e(r)$ within each element is linearly interpolated between the four nodes of each tetrahedron

$$f^e(r) = \sum_{\alpha=1}^{4} f_\alpha^e N_\alpha^e(r), \quad (13.18)$$

with

$$N_\alpha^e(x, y, z) = a_\alpha + b_\alpha x + c_\alpha y + d_\alpha z, \quad (13.19)$$

where the coefficients $a_\alpha - d_\alpha$ depend on the shape of the element and f_α^e denotes a constant for each element. The linear interpolation of the variables ϑ, φ and A is especially important for the exchange energy, which otherwise is overestimated.

The integral of $W_2(A)$ extends over the whole space. Therefore, the ferromagnetic particle as well as the external space has to be distributed into finite elements. In the literature there are a number of proposals on how to treat this open boundary problem. In order to reduce the infinite external space and to avoid cut-off errors a spatial transformation is applied which changes the open boundary problem into a closed boundary problem [13.9, 13.39]. The transformation in the case of a parallelepiped is demonstrated by Fig. 13.12. Here the external space is mapped into a shell which encapsulates the ferromagnetic particle. The mapping transformation for one of the six outer regions shown for three dimensions in Fig. 13.12 is

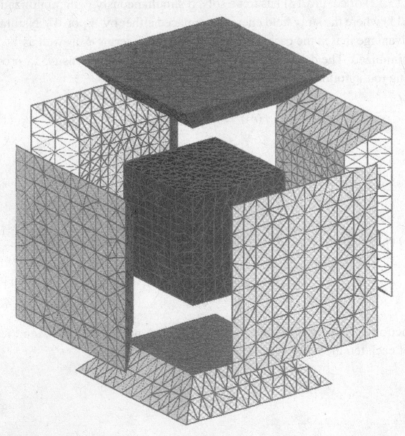

Fig. 13.12. Exploded view of the different domains considered in the computational region. The unbounded area outside the magnetic cube in the middle is mapped into six finite domains surrounding the ferromagnetic cube [13.40].

performed by the following equations:

$$X = \frac{Y}{y}x,$$

$$Y = y_b - (y_b - y_a)\left(\frac{y_a}{y}\right)^\alpha,$$

$$Z = \frac{Y}{y}z, \tag{13.20}$$

where X, Y, Z are the mapped coordinates of the original space (x, y, z) and (x_a, y_a) and (x_b, y_b) are the corner points of the mapping region. If (x, y) belongs to one of the other regions the new transformation equations are obtained by cyclic exchange in eq. (13.20).

13.3.3 Magnetic structures and magnetization processes in thin platelets

The discovery of the giant magnetic resistance and the spin-dependent tunnelling effect in thin films systems [13.41–13.43] has led to the development of new magnetic recording devices. The understanding of magnetic structures and magnetization processes in thin platelets and small particles has become a prerequisite for the successful development of high-density recording systems. After the achievement of recording densities of 1 Gb/inch², future recording densities of 1 Tb/inch² are envisaged. The basic properties to be understood are the hysteresis loop and its characteristic properties such as the coercive field, the remanent state, and the switching time, as well as the dependence of these properties on magnetic material parameters and the shape and dimensions of the particles. The finite element method is a rather effective technique to determine the magnetic configurations and the magnetization process. The arrangement of the magnetization in a thin magnetically uniaxial platelet of geometry 1×2 μm and a thickness of 20 nm has been the subject of the so-called μMAG standard problem No. 1 [13.44]. For the numerical calculations the material parameters of permalloy have been used: $J_s = 1.04$ T, $K_1 = 500$ J/m³, $A = 1.3 \cdot 10^{-11}$ J/m. The FEM calculations show that two characteristic types of domain patterns may exist at zero magnetic field. Figure 13.13 schematically shows the corresponding high-remanence (type I) and low-remanence (type II) states [13.45–13.47]. Low-remanence states are obtained if the starting structure for the simulation corresponds to a two-domain state with domains magnetized antiparallel to each other. The high-remanence states of Fig. 13.14 are formed if the starting conditions correspond to a single domain state. In this case we deal either with the so-called C- or the S-state. In both cases so-called quasi-closure domains form at the edges of the platelet within a region of width $l_K = 161$ nm as shown in Fig. 13.14. The C-state forms if the initial state is symmetrical to the long symmetry

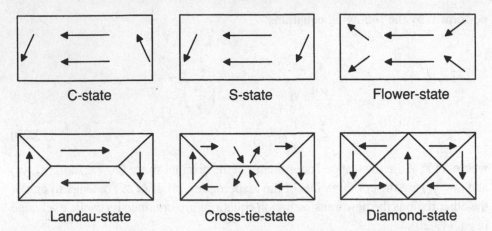

C-state   S-state   Flower-state

Landau-state   Cross-tie-state Diamond-state

Fig. 13.13. Schematic representation of domain configurations in a thin rectangular platelet of dimensions 1 μm × 2 μm × 0.02 μm.

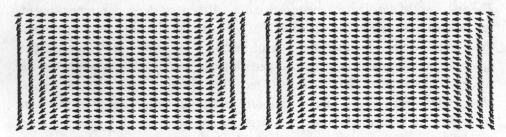

Fig. 13.14. High-remanence states of S- and C-configurations as obtained from the high-field saturated state [13.45].

axis whereas the S-state forms if the symmetry is broken by a small rotation of the applied field of the long symmetry axis. Both configurations have nearly the same free enthalpy, which mainly results from the stray field energy (Table 13.2). The so-called flower state has a large free energy because of its increased exchange energy.

The low-remanence type II configurations derive from the conventional Landau pattern. Figure 13.15 shows a Landau-type structure based on symmetrical Néel walls which exist for thickness $D < 10l_s = 57$ nm [13.48]. For thicknesses $D > 10l_s$ the Landau pattern is formed by Bloch walls. In the case of a 20 nm platelet we deal with Néel walls and on the right side between the central 180°-wall and the two 90°-walls at the closure domains a vortex-like configuration exists establishing an asymmetric structure. Here it should be noted that Arrott and Templeton [13.48] made the following remark: 'The simplest problem in continuum

Table 13.2. *Total energies obtained by the FEM method for the different types of spin configurations in a permalloy platelet of dimensions 1 μm × 2 μm × 20 nm (μMAG standard problem No. 1) according to [13.45].*

Structure	ϕ_t [10^{-18} J]
Landau + cross-tie wall	79.616
Diamond	79.903
Landau vortex centred	84.823
Landau vortex asymmetrical	88.733
S-state	169.897
C-state	169.890

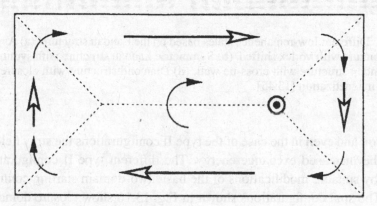

Fig. 13.15. Schematic asymmetrical Landau structure of Néel walls with vortex-like structure on the right side according to [13.48].

magnetism in a singly connected finite body is the Landau structure . . . but it remains an unsolved problem to describe it in all its details.'

Figure 13.16 shows type II configurations which are denoted as follows: asymmetrical Landau structure (vortex shifted), symmetrical Landau structure (vortex in the centre), Landau structure with cross-tie wall, and the diamond structure composed of seven domains. Table 13.2 summarizes the total free enthalpies of six different structures. Obviously the Landau structure with cross-tie wall is the energetically most favourable configuration because the 180°-Néel wall is replaced by 90°-Néel walls, the energy of which is only 0.32 of 180°-Néel walls as shown in Chapter 4. For all configurations the stray field energy is the dominating energy

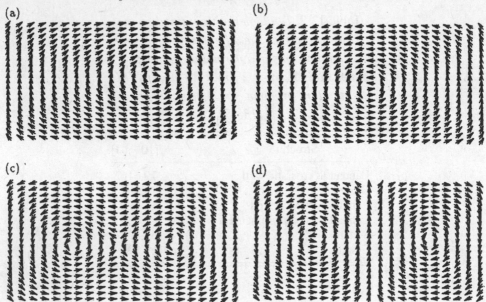

Fig. 13.16. Different low-remanence states based on the Landau structure. (a) Asymmetric Landau structure with vortex shifted. (b) Symmetric Landau structure with symmetric vortex. (c) Landau structure with cross-tie wall. (d) Diamond structure with closure domains of parallel magnetization [13.45].

contribution and even in the case of the type II configurations the stray field energy exceeds the increased exchange energy. The different type II configurations are obtained by suitable modifications of the basic two-domain starting configuration [13.45]. The spin configurations shown in Fig. 13.16 show closure domains with parallel or antiparallel magnetization. The transition between these two types of configurations is prevented because of large energy barriers between these different configurations. Therefore the transition into the configuration with the lowest Gibbs free energy, i.e., the configuration with cross-tie wall, cannot be obtained from the S-state or the diamond state under quantitative conditions. Naturally, in real materials with defects a large number of nucleation centres exist with locally reduced energy barriers, which promote the transition into the lowest energy configurations. In the following we consider the magnetization processes of type I and type II configurations.

Magnetization process of the S-state

The magnetization process of the S-state is presented in Fig. 13.17 under the action of a magnetic field applied antiparallel to the magnetization. Reducing the applied

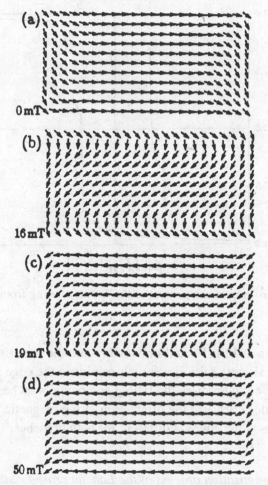

Fig. 13.17. Magnetization process of the S-state for a field applied antiparallel to the long edge. The field values are given as positive. (a) $\mu_0 H_{ext} = 0$ mT, (b) $\mu_0 H_{ext} = 16$ mT, (c) $\mu_0 H_{ext} = 19$ mT, (d) $\mu_0 H_{ext} = 50$ mT [13.45].

field from +50 mT in steps of 1 mT into the opposite direction of −50 mT we observe a reversible rotation of J_s at the long edges of the platelet. At an opposite field of 16 mT the magnetization in the centre of the platelet rotates reversibly into the direction of the applied field, whereas the magnetization along the long edges still resembles the configuration of the remanent state at $H = 0$ because of the stray field barrier to be overcome by rotation. Finally, at 19 mT the magnetization at the upper edge orients irreversibly parallel to the applied field and the same happens at a slightly higher field at the lower edge. At 50 mT the magnetization has converted completely into the inverse S-shape structure. The corresponding hysteresis loop shown in Fig. 13.18 reveals a big step in magnetization when the centre of the

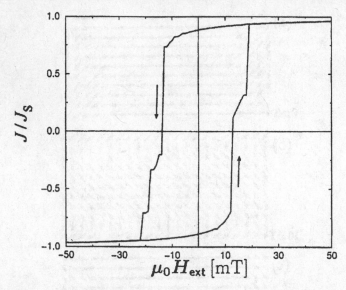

Fig. 13.18. Hysteresis loop of the platelet in the S-state starting from the saturated state [13.45].

S-type configuration at 18 mT rotates reversibly into the opposite direction. Since the applied field is slightly misaligned with respect to the edge length in order to stabilize the S-type configuration, the falling and rising branches of the hysteresis show small differences because the orientation of the magnetic field with respect to the quasidomains at the edges is different for both branches.

Magnetization process of the Landau cross-tie state

In the case of the low remanence type II configurations the magnetization process takes place by domain wall displacements and deformations of domain patterns. The magnetization process is shown in Fig. 13.19 for a permalloy platelet of dimensions $1 \times 3.2 \ \mu m$ and a thickness of 80 nm [13.45, 13.47], and a field applied parallel to the long edge. Up to fields of 21 mT the upper domain with magnetization parallel to the field grows reversibly by shifting the centre of the cross-tie wall to the lower edge of the platelet. This process is accompanied by a bowing of the central domain wall and a shearing of the closure domains indicated by the shadowed regions. The two swirls at the central corners of the closure domains move towards the lower corners of the platelet and they disappear at 28 mT where the platelet is magnetized into the C-type configuration. This transition is an allowed one because this is a transition between configurations with closure domains magnetized antiparallel to each other. It is of interest to note that the magnetic structure found in Fig. 13.19

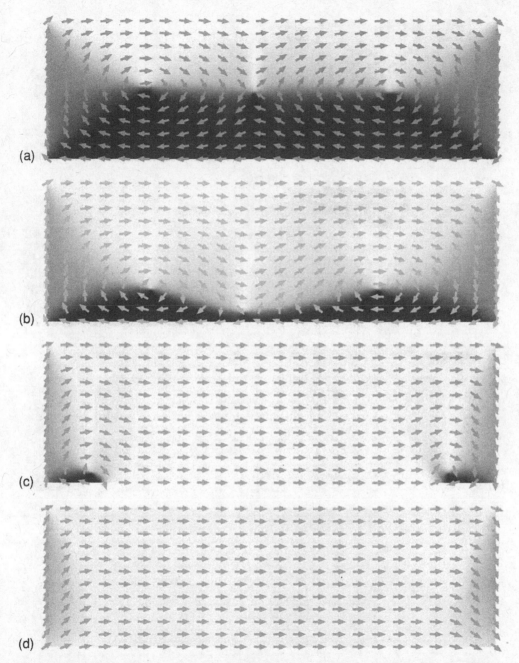

Fig. 13.19. Magnetization process of the four-domain Landau structure with symmetric cross-tie wall. The magnetic field is applied parallel to the long edge from left to right: (a) $\mu_0 H_{ext} = 0$ mT, (b) $\mu_0 H_{ext} = 20.5$ mT, (c) $\mu_0 H_{ext} = 21.5$ mT, (d) $\mu_0 H_{ext} = 28$ mT [13.45].

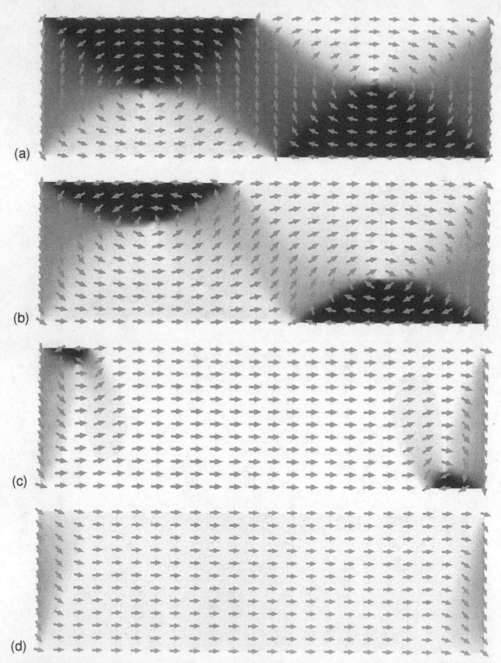

Fig. 13.20. Magnetization process of the diamond structure with seven domains for a platelet of dimensions $1\,\mu m \times 3.2\,\mu m \times 0.08\,\mu m$. The field is applied parallel to the long edge and oriented from left to right: (a) $\mu_0 H_{ext} = 0$ mT, (b) $\mu_0 H_{ext} = 24$ mT, (c) $\mu_0 H_{ext} = 25.5$ mT, (d) $\mu_0 H_{ext} = 31$ mT [13.45].

has been detected previously by Van den Berg and Vatvani [13.49] by the Bitter technique in a platelet of dimensions 15×50 μm.

Magnetization process of the diamond structure

The magnetization process of the diamond structure is shown in Fig. 13.20. The combined magnetization process takes place by a reversible rotation of M_s within the central diamond and growth of these longitudinal domains with M_s parallel to the applied field. The closure domains are deformed by a shearing process. These processes take place reversibly up to a magnetic field of 25 mT. The magnetization proceeds irreversibly if the neighbouring two 90°-walls of the central diamond touch each other and annihilate. After this process, at 25.5 mT the two swirls are located at diagonal corners as shown in Fig. 13.20c. Above fields of 31 mT, in Fig. 13.20d the swirls have vanished and the S-type configuration is found as expected for transitions within spin structures with closure domains magnetized parallel to each other.

It is of interest to note that the magnetization process of the cross-tie and the diamond structure proceeds reversibly up to fields slightly smaller than the average demagnetization field, $\mu_0 H_d = N \cdot J = 28.5$ mT, of the saturated specimen ($N =$ demagnetization factor according to Aharoni [13.37]). The quasi-saturation fields are smaller than H_d because the ever-present closure domains reduce the surface charges in comparison to the fully homogeneously magnetized platelet.

Critical thickness for high-remanence single domain configurations

According to Table 13.2 the multidomain states for platelets of thickness 20 nm have the lowest Gibbs free energies. This is a consequence of the high stray field energy which results from the surface charges. At lower thickness of the platelet, however, the magnetostatic energy decreases more rapidly than the wall energies and at a certain critical thickness shown in Fig. 13.21 the total energy of the quasi-homogeneous state of a quadrat of edge length 1 μm (C-state) becomes lower than that of the four domain states. The numerical calculations show that the transition for permalloy takes place at 1.5 nm, for α-Fe at $\sim$2.5 nm and for Co at $\sim$4 nm [13.38]. These results are fundamental for the development of thin film elements in magnetic recording technology.

For disc-shaped particles the critical thickness has been determined by Cowburn *et al.* [13.50]. Using a high-sensitivity magneto-optical method it was shown that for discs of diameter 20 nm the transition to a single domain state takes place at a thickness of 6 nm. For discs of diameter 100 nm a critical thickness of 15 nm was found.

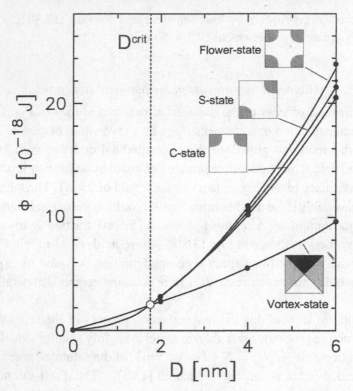

Fig. 13.21. Total Gibbs free energy of a quadratic platelet as a function of thickness for different configurations (flower, C-, S- and four-domain state): (————) numerical [13.38].

13.3.4 Magnetic structures and magnetization processes in small particles

The magnetic structures and magnetization processes of single domain ellipsoidal, spherical and cylindrical particles have been described in Chapter 6 and Section 13.3.3. The magnetic ground states of these particles below the critical radii correspond to a homogeneous magnetization, whereas in particles with nonellipsoidal shape the ground states become inhomogeneous due to inhomogeneous stray fields at edges and corners. Numerical calculations have been performed for isolated cubes and spherical particles embedded into a hard magnetic matrix.

Cubic, free particles

The magnetic ground states of cubic particles have been calculated for magnetically soft and magnetically hard properties. When starting from the homogeneous arrangement for a material with $Q = 0.1$ at small edge lengths, according to Fig. 13.22, the expected so-called flower state exists with symmetric inhomogeneities of M_s near the edges and corners. These tiltings have been studied by

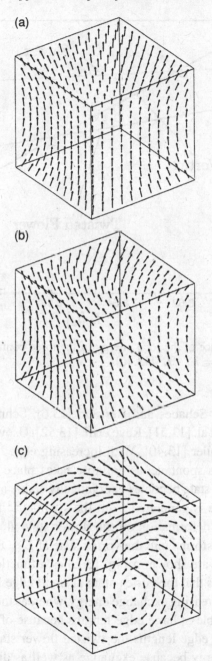

Fig. 13.22. (a) Three-dimensional representation of the flower state for small edge lengths $L < 8.6\,l_\mathrm{s}$. (b) Magnetization structure of the twisted flower state. (c) Vortex state obtained by using a two-domain state for the starting conditions [13.40].

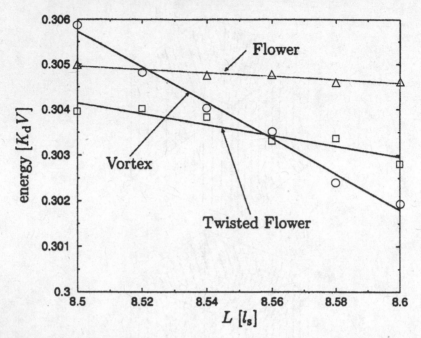

Fig. 13.23. Size dependence of the total energy of the three configurations of Fig. 13.22. Energy in units of $K_u \cdot V = K_u \cdot L^3$ [13.40].

several authors such as Schabes and Bertram [13.6], Schmidts and Kronmüller [13.20, 13.21], Schrefl et al. [13.51], Rave *et al.* [13.52], Usov and Peschang [13.53] and Hertel and Kronmüller [13.40]. With increasing edge length the flower state becomes unstable and a spontaneous collapse takes place at $L \approx 8.0 l_s$ where a so-called twisted flower state becomes energetically stable as shown in Fig. 13.22. The twisted flower state is generated from the symmetric flower state by super-imposing a vortex-type deformation on the flower state along the easy axis. The twisted flower state transforms into the vortex state at $L = 8.57 l_s$. The energies of the three configurations are shown in Fig. 13.23 as a function of the edge length. Here it becomes obvious that the vortex state at small edge lengths reveals a large energy because of an increasing exchange energy, whereas the flower state becomes energetically unfavourable at large edge lengths because of an increasing dipolar energy. At intermediate edge lengths the twisted flower state represents the configuration of lowest energy because exchange as well as dipolar energy are kept small.

Figure 13.24 shows spin arrangements in a cube of hard magnetic $Nd_2Fe_{14}B$ of edge length 60 nm [13.51]. As in the preceding case of a soft magnetic material with $Q = 0.1$, in the case of $Nd_2Fe_{14}B$ with $Q = 4.3$ the stray field energy is reduced

magnetic polarization   demagnetizing field

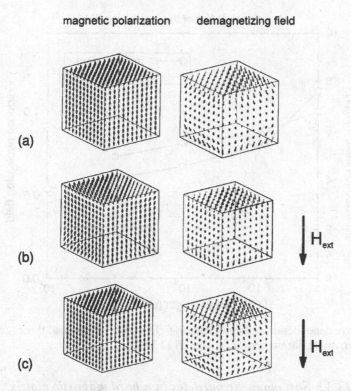

(a)

(b) H_{ext}

(c) H_{ext}

Fig. 13.24. Distribution of the magnetic polarization and of the demagnetization field near the surface of a hard magnetic cube $Nd_2Fe_{14}B$ of edge length 60 nm. Left: J_s; right; H_s; (a) remanent state for $H_{ext} = 0$; (b) undercritical state for $H_{ext} = (2K_1/J_s) \cdot 0.88$; (c) reversed state for $H_{ext} = (2K_1/J_s) \cdot 0.89$ [13.51].

by rotations of M_s near edges and corners. Under the action of an inversely applied field these rotations increase and at a critical field, the nucleation field, H_N, the reversion of magnetization takes place spontaneously. With increasing edge length H_N decreases smoothly and may be described approximately by a logarithmic law,

$$H_N(D) = H_N(\delta_B) - n_{eff}M_s \ln(D/\delta_B), \tag{13.21}$$

with $n_{eff} = 0.086$ and $\mu_0 H_N(\delta_B) = 7$ T. According to Fig. 13.25 the nucleation field determined for an infinite prism above $D = 10^{-1}$ μm is smaller than that of the cube because its demagnetization field is larger than that of the cube. The parameters of eq. (13.21) for the magnetization process in the infinite prism are given by $\mu_0 H_N(\delta_B) = 6.1$ T. The term $n_{eff} \cdot \ln(D/\delta_B)$ corresponds to the N_{eff} value of eq. (6.53). For the cube and the prism $N_{eff} = 0.47$ holds, which agrees rather well with experimental results.

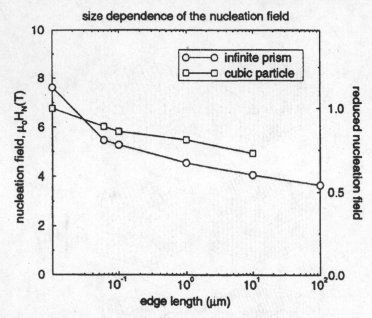

Fig. 13.25. Size dependence of the nucleation field in two (prism) and three (cube) dimensions for the material parameters of $Nd_2Fe_{14}B$ [13.51].

13.3.5 Soft magnetic particles in a hard magnetic matrix

Composite magnetic materials composed of magnetically hard and soft particles are of great interest in developing high-remanence magnetic materials. Using a combination of soft magnetic grains of α-Fe with magnetically hard $Nd_2Fe_{14}B$ grains the remanence may be increased, leading to a permanent magnet with a large maximum energy product. The basic magnetic properties of such an assembly of grains may be studied by embedding an individual spherical α-Fe particle into a cube of $Nd_2Fe_{14}B$ of edge length 150 nm. By the FEM the magnetization curves and the spin configuration of the α-Fe particle have been determined for varying particle diameters [13.54]. Figure 13.26 presents a series of hysteresis loops in the second quadrant showing the decrease of the coercive field with increasing particle diameter. The hysteresis loop remains rectangular up to diameters of 90 nm and then an intermediate demagnetization curve appears before the particle reorients spontaneously into the opposite direction. The dependence of H_c on the particle diameter D_{sphere} is shown in Fig. 13.27. From the logarithmic plot it turns out that the coercive field for $D > 10$ nm decreases according to a $D^{-0.701}$-law. Whereas for $D < 10$ nm the spin configuration remains homogeneous, for larger diameters magnetic vortex-type structures appear which reduce the coercive field drastically. A series of such spin structures is shown in Fig. 13.28 for particle diameters of 20 nm and 90 nm and different inversely applied magnetic fields. It is of interest

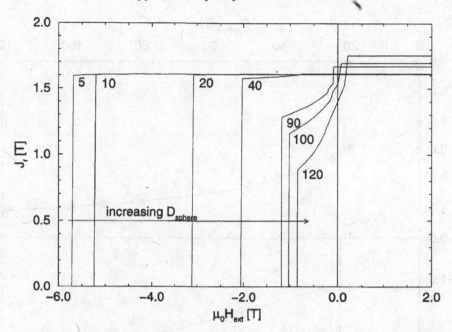

Fig. 13.26. Computed demagnetization curves for the magnetic structure of a soft magnetic α-Fe sphere in a hard magnetic matrix. Diameters of the α-Fe sphere range from 5 nm to 120 nm [13.54].

to note that the simulation of the demagnetization process leads to similar coercive fields assuming α-Fe particles of uniaxial or cubic anisotropy. It is furthermore an important result that magnetically soft particles with diameters $D_{sphere} \ll 2.5\delta_B$ of the hard phase (10 nm) do not reduce the coercive field significantly. Accordingly, remanence enhanced composite magnets with large coercive field require small soft magnetic particles. A reduction of the coercive field by 50% is found for a diameter of 20 nm.

13.3.6 Assemblies of nanocrystalline grains

Many modern magnetic materials used for magnetic recording, permanent magnets and high permeability applications are based on nanocrystalline materials usually produced by melt-spinning, the sputter technique, molecular beam epitaxy or crystallization of amorphous alloys. The characteristic properties of such assemblies of grains depend on the intrinsic material parameters and the microstructure.

The main features of the microstructure are the distribution of easy axes within the assembly of grains, and the magnetic properties of the grain boundaries. Furthermore, the modification of magnetocrystalline energy, exchange energy and spontaneous magnetization within grain boundaries influence the global hysteresis loop of such assemblies of grains. The hysteresis loop is also influenced by the presence

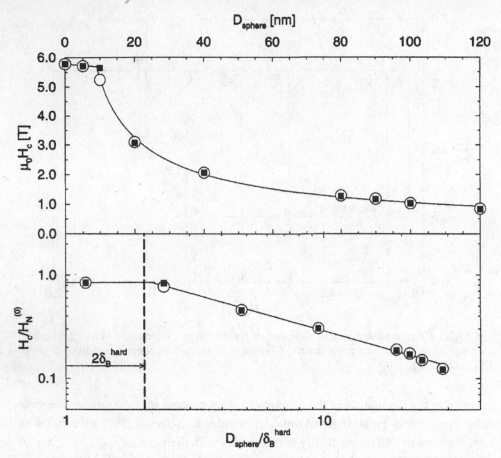

Fig. 13.27. Coercivity H_c and relative coercivity $H_c/H_N(0)$ as a function of the diameter D_{sphere} and $D_{sphere}/\delta_B^{hard}$ of the soft α-Fe phase. δ_B^{hard} denotes the wall width of $Nd_2Fe_{14}B$. The opaque circles belong to an ideally oriented sphere with $\Psi_0 = 0°$ and the filled squares to a misorientation of $\Psi_0 = 45°$ [13.54].

of soft magnetic phases. In the case of composite materials where remanence enhancing is induced by α-Fe grains the reduction of the coercive field determines the optimum amount of the magnetically soft phase acceptable for a permanent magnet. Explicit calculations of hysteresis loops in general are performed within the framework of the Preisach model [13.55]. The parameters used in this model, even in its extended versions, have no direct connection with the real microstructure and therefore correspond to a pure phenomenological theory. Thus the FEM with its ability to treat locally inhomogeneous magnetic systems is superior to the Preisach model.

Fig. 13.28(*opposite*). Magnetization reversal processes for an iron sphere of diameter 20 nm (a) and 90 nm (b) embedded in a matrix of $Nd_2Fe_{14}B$. The numbers give the applied magnetic field in units of the ideal nucleation field $H_N(0)$ of $Nd_2Fe_{14}B$ [13.54].

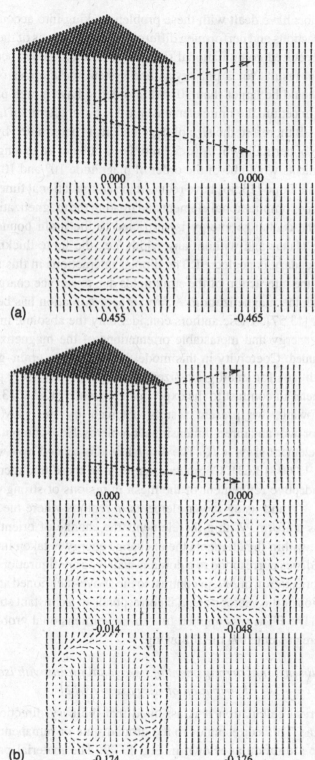

Several authors have dealt with these problems taking into account dipolar and exchange interactions and performing different approximations of the FEM in order to reduce the required computational time. This problem arises because in order to obtain useful results for a polycrystalline system the number of grains taken into account should be larger than 30 and in order to have a high precision in the minimization procedure the finite elements have to be smaller than l_s and l_K. In the case of $Nd_2Fe_{14}B$ therefore the size of the elements is determined by $l_K < 1.3$ nm and in the case of permalloy by $l_s < 5.7$ nm. For a cube of edge length 100 nm the number of elements would be of the order of magnitude 10^6 and 10^4 respectively.

Fukunaga and Inoue [13.56] have reduced the computational time by describing each grain by one computational element and treating the magnetization as uniform. Accordingly, the exchange energy is restricted to the grain boundary, which is assumed to be planar and without any width, i.e., the finite thickness of the dw at the grain boundary is neglected. The dipolar interactions in this approximation are reduced to the interaction of the stray fields due to surface charges at the grain boundaries with the magnetization. A further approximation has been introduced by Otani *et al.* [13.57]. These authors consider only the absolute minimum of the magnetic free energy and metastable orientations of the magnetization within a grain are excluded. Coercivity in this model is due only to grain–grain exchange coupling which fails for small grain sizes because H_c is inversely proportional to the grain diameter in contrast to the experimental results [13.58, 13.59].

In the following, we present results obtained by the FEM described in Section 13.3 for two- and three-dimensional problems. Here computational meshes are used with element sizes $< l_K$. The dipolar field is treated by a vector potential [13.11–13.14, 13.60–13.63]. Computational time in these treatments is saved by applying the adaptive refinement of the mesh in regions of strong changes of the orientation of M_s, and using larger elements in regions where the magnetization is more or less uniform. Within the elements the change of orientation of M_s is linearized, i.e., within the elements the exchange energy is taken into account and the grain boundaries correspond to a domain-wall type configuration with a gradual change of the orientation of M_s, in contrast to the above-mentioned approximations. The linearization of the vector potential corresponds to a constant stray field within each element. The FEM was applied first to two-dimensional problems. Some of these results are presented in the following.

Two-dimensional arrangement of hexagonal grains with isotropic distribution of easy axes [13.12]

For the numerical calculation it is assumed that the easy directions occupy the upper half plane. This corresponds to the ground state or remanent state obtained after magnetic saturation as shown in Fig. 13.29. The material parameters used

easy axes distribution

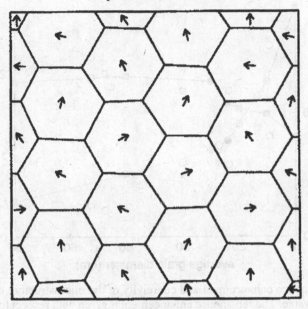

spin arrangements

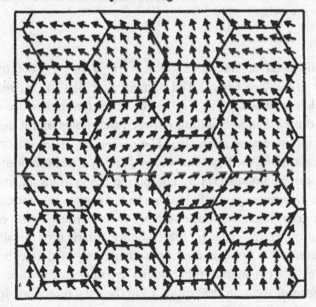

Fig. 13.29. Isotropic distribution of easy axes and the corresponding spin arrangement for zero applied field and for two-dimensional hexagons of diameter 10 nm [13.12].

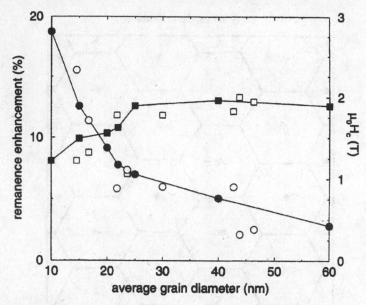

Fig. 13.30. Remanence enhancement and coercivity of the microstructure of Fig. 13.29 as a function of grain size. The remanence enhancement is given with respect to the theoretical limit for noninteracting particles ($0.5 J_s$). ●–● computed remanence enhancement; ○ Ref. [13.12]; ■–■ computed coercive field; □ Ref. [13.12]. Open symbols are experimental values obtained for $Nd_{13.2}Fe_{79.6}B_6Si_{1.2}$ [13.59].

are those of $Nd_2Fe_{14}B$ at 300 K ($K_1 = 4.3 \cdot 10^6$ J/m^3, $K_2 = 0.65 \cdot 10^6$ J/m^3, $A = 7.7 \cdot 10^{-12}$ J/m, $J_s = 1.61$ T. The numerical results for the remanence and the coercive field are presented in Fig. 13.30 showing a remanence enhancement of 20% due to the exchange coupling between the grains and a decrease of the coercive field with increasing grain size. The numerical values are compared with experimental results published by Manaf *et al.* [13.59]. The behaviour of coercivity and remanence can be understood by comparing the spin arrangement of grains of different size in zero magnetic field. Due to the exchange interactions the spin orientation near grain boundaries of one grain is influenced by the easy direction of the neighbouring grains. That is, the resultant polarization parallel to the applied field is increased in comparison to noninteracting particles, thus giving rise to a remanence enhancement or in other words to a magnetic texture. Figure 13.31 gives a comparison of the spin deviations from the easy directions for grain sizes of $D = 10$ nm or $D = 20$ nm. Regions of strong and weak shading deviate by more than 20° and 10° from the easy directions, respectively. In the case of the 10 nm grains the shaded regions extend over a much larger area than in the 20 nm grains, a fact which reflects the increasing remanence enhancement with decreasing grain size.

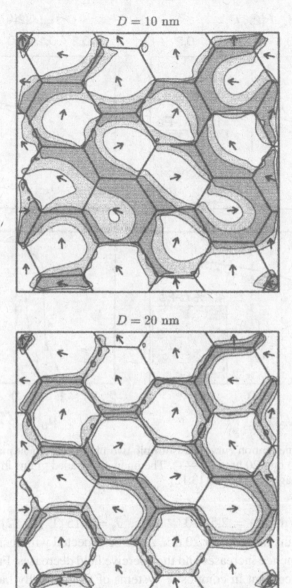

Fig. 13.31. Inhomogenous regions along the grain boundaries. Arrows: easy axes. Shaded areas denote the spin deviations from the easy axes by 10° and 20° (strongly shaded regions) [13.12].

Two-dimensional arrangement of polygonal hard and soft magnetic grains (composite assembly of grains) [13.12]

The demagnetization curves of anisotropic two-phase magnets with increasing amounts of the soft magnetic phase (α-Fe, $T = 300$ K: $K_1 = 4.6 \cdot 10^4$ J/m³,

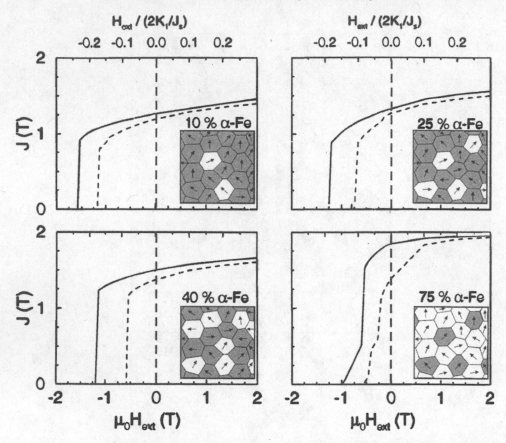

Fig. 13.32. Demagnetization curves of isotropic two-phase permanent magnets for grains of 10 nm (- - - - - -) and 20 nm (———). The open and shaded grains in the inset denote the soft and hard magnetic phases [13.12].

$K_2 = 1.5 \cdot 10^4$ J/m^3, $A = 2.5 \cdot 10^{-11}$ J/m, $J_s = 2.15$ T [13.12] are shown in Fig. 13.32 for grains of diameter 20 nm. As to be expected with increasing amount of α-Fe, the remanence increases and the coercive field decreases. Furthermore, the remanence enhancement in composite systems of magnetically hard (Nd$_2$Fe$_{14}$B) and soft (Fe) grains also increases with decreasing grain size as demonstrated by Fig. 13.33. Here it is obvious that the soft magnetic grains with smaller diameter are aligned much more strongly parallel to the hard magnetic grains.

Three-dimensional assemblies of nanocrystalline grains

A three-dimensional grain configuration and the distribution of magnetization in the remanent state is shown in Fig. 13.34 for a composite system of 35 irregular grains with an isotropic distribution of easy axes [13.63]. The soft magnetic grains

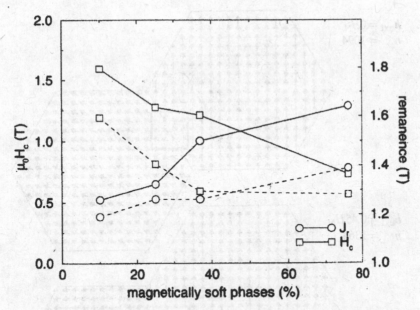

Fig. 13.33. Remanence and coercivity of two-phase permanent magnets as a function of the volume fraction of magnetically soft α-Fe. Remanent polarization for grain sizes of 10 nm (○———○) and 20 nm (○- - - - -○). Coercive field for grain sizes of 10 nm (□———□) and 20 nm (□- - - -□) [13.12].

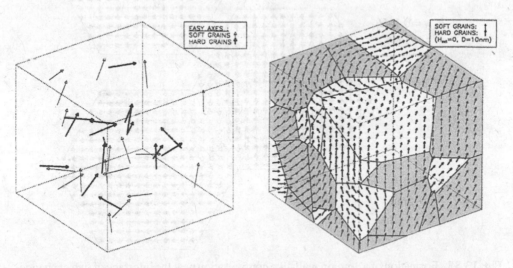

Fig. 13.34. Configuration of the easy axes (left) and the remanent polarization (right) of 35 irregular grains of average diameter 10 nm arranged within the cube with a volume fraction of 51% α-Fe (dark) and 49% $Nd_2Fe_{14}B$ (light).

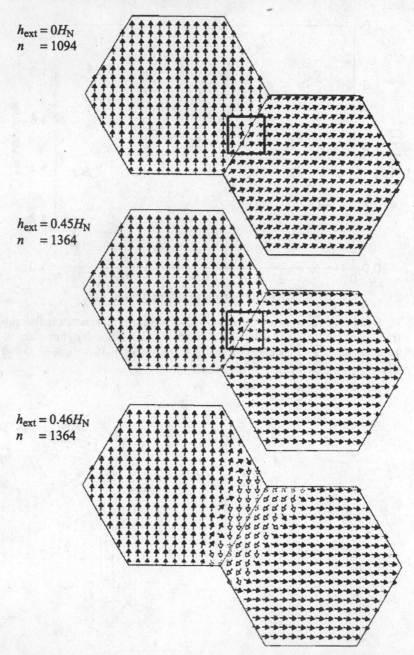

$h_{ext} = 0H_N$
$n = 1094$

$h_{ext} = 0.45H_N$
$n = 1364$

$h_{ext} = 0.46H_N$
$n = 1364$

Fig. 13.35. Formation of a domain wall-like configuration near the interface of two exchange coupled grains for $\mu_0 H_{ext} = 0$ and the undercritical state $H_{ext} = 0.45 H_N$ [13.60].

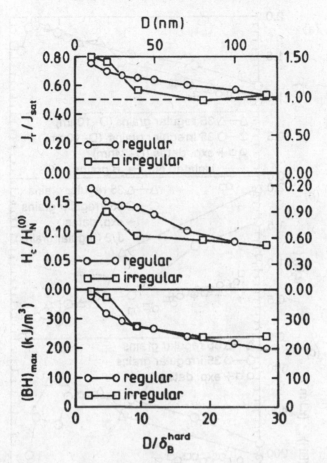

Fig. 13.36. Remanence J_r, coercivity $\mu_0 H_c$ and the maximum energy product $(BH)_{max}$ for the irregular configuration of Fig. 13.34 and a regular tetrahedral grain structure as a function of grain size D [6.24].

of α-Fe (51% volume fraction) are embedded within the hard magnetic matrix of $Nd_2Fe_{14}B$ [13.63–13.65]. At the grain boundaries the magnetization rotates continuously from one easy direction into the neighbouring one. This also is shown in Fig. 13.35 for two grains with perpendicular easy directions for the remanent and the undercritical state. The extension of the domain wall between two grains is of the order of $\delta_B \simeq 4$ nm. Under the assumption of a perfect exchange coupling between the grains, Fig. 13.36 illustrates the dependence of J_r, H_c and $(BH)_{max}$ on the grain size D for the irregular structure shown in Fig. 13.34 as well as for a regular structure composed of 35 dodecahedral grains [13.63]. All three properties decrease with increasing grain size with the exception of H_c of the irregular structure. The regular structure leads to larger values of J_r and $\mu_0 H_c$. The remanence and the

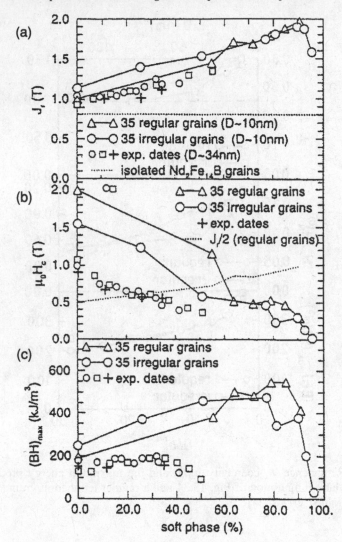

Fig. 13.37. Numerical results for J_r, $\mu_0 H_c$ and the energy product $(BH)_{max}$ of nanocrystalline composite magnets for a regular and an irregular distribution of grains with average grain size 10 nm as a function of the amount of the α-Fe-phase [6.24] (+) [13.66], (○) [13.80], (□) [6.81].

coercive field may be fitted empirically by the following logarithmic laws [13.63]

$$J_r = J_{sat}\left[0.84 - 0.085\ln(D/\delta_B^{hard})\right],$$
$$H_c = H_N(0)\left[0.22 - 0.04\ln(D/\delta_B^{hard})\right], \tag{13.22}$$

with $H_N(0) = 2K_1^{hard}/J_s$ and $J_{sat} = J_s^{hard}\nu^{hard} + J_s^{soft}\nu^{soft}$, where the upper indices refer to the saturation polarization and volume fractions of the hard and soft

magnetic phases. In Fig. 13.37 the numerical results obtained for the three characteristic properties of the hysteresis loop as a function of the α-Fe content [13.63]
are compared with the corresponding experimental data obtained for NdFeB composite magnets (Willcox *et al.* [13.66], Bauer *et al.* [6.80] and Goll *et al.* [6.81]).
The dotted lines represent the case of noninteracting randomly distributed grains.
Whereas the numerical results for J_r agree fairly well with the experimental results, the theoretical predictions for H_c and $(BH)_{max}$ lie appreciably above the
experimental data. In the case of $(BH)_{max}$ this is due to the fact that the condition $\mu_0 H_c > 0.5 J_r$ is experimentally found to be valid for contents of α-Fe $< 30\%$,
whereas the numerical results fulfil this condition up to 50% for α-Fe. The observed
discrepancy is attributed to the role of grain boundaries, which may not be treated by
only considering the change of easy direction. Actually within a grain boundary all
three intrinsic material parameters J_s, K_1, A are submitted to modifications. Due to
atomic disorder K_1 and A in particular may have reduced values. For example, from
amorphous alloys it is well known that the anisotropy is drastically reduced and
also the exchange constant is smaller than in the corresponding crystalline phase,
whereas the spontaneous magnetization is only reduced moderately. In order to take
care of the modification of the material properties it is assumed that within the grain
boundaries of width 3 nm K_1, A and J_s suffer a step-like reduction. Figure 13.38

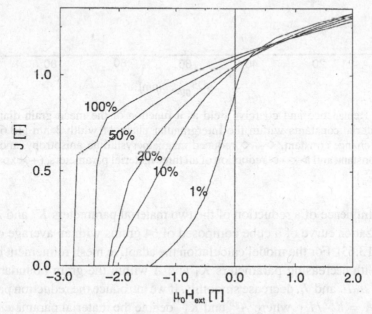

Fig. 13.38. Demagnetization curves for an ensemble of 64 $Nd_2Fe_{14}B$ grains of average grain
diameter 20 nm and an intergranular phase of 3 nm thickness. The percentages refer to the
relative reduction of both the exchange and the anisotropy constants within the intergranular
phase [13.67].

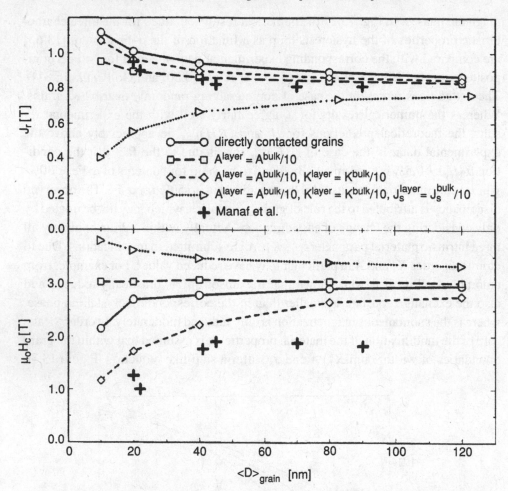

Fig. 13.39. Remanence and coercive field as a function of the mean grain diameter for reduced material constants within the intergranular phase of width 3 nm [13.65]. □--□ Reduced exchange constant, ◇--◇ reduced magnetocrystalline anisotropy, and reduced exchange constant; and ▷ · · · ▷ reduction of all three material parameters, (+) experimental results [13.59].

shows the influence of a reduction of the two material parameters K_1 and A on the demagnetization curve of a cube composed of 64 grains with an average diameter of 20 nm [13.65]. For the model calculation the adaptive mesh refinement has been applied. With decreasing parameters K_1 and A within the grain boundaries both properties, $\mu_0 H_c$ and J_r, decrease smoothly. If we introduce the reduction parameter $f = A^{gb}/A = K_1^{gb}/K_1$, where A^{gb} and K_1^{gb} denote the material parameters within the grain boundary, the coercive field and the remanence, standardized to the nucleation field $H_N(0) = 2K_1/J_s$ and the spontaneous magnetization J_s, respectively, can be written as [13.62, 13.67]

$$H_c/H_N(0) = 0.304 + 0.098f,$$
$$J_r/J_s = 0.646 + 0.036f. \qquad (13.23)$$

A further interesting property is the dependence of the coercive field as a function of the average grain size $\langle D \rangle$. According to measurements of Manaf *et al.* [13.59] the coercive field decreases for grain sizes below 40 nm. Numerical calculations where only the exchange constant A^{gb} is considered do not show a decrease of H_c at small grain sizes (Fig. 13.39). Reducing only K_1^{gb} leads to the experimentally found decrease of H_c; however, the remanence enhancement is found to be too strong (Fig. 13.39). A better quantitative agreement is obtained by reducing A and K_1 simultaneously (Fig. 13.39). A reduction of J_s or of all three material properties does not lead to a satisfactory agreement with the experimental result. Accordingly, H_c and J_r are predominantly determined by the modification of K_1 and A. These results show clearly that for a quantitative agreement between FEM calculations and experimental results the damaging role of grain boundaries has to be taken into account.

References

[13.1] Holz, A., 1967, *Z. Angew. Physik* **23**, 170; 1968, *Phys. Stat. Sol.* **26**, 751; 1968, *Phys. Stat. Sol.* **25**, 567.
[13.2] Hubert, A., 1969, *Phys. Stat. Sol.* **32**, 519.
[13.3] Brown, W.F., Jr., and La Bonte, A.E., 1965, *Z. Appl. Phys.* **36**, 1380.
[13.4] La Bonte, A.E., 1969, *J. Appl. Phys.* **40**, 2450.
[13.5] Holz, A., and Hubert, A., 1969, *Z. Angew. Physik* **26**, 145.
[13.6] Schabes, M.E., and Bertram, H.N., 1988, *J. Appl. Phys.* **64**, 5832.
[13.7] Schabes, M.E., 1991, *J. Magn. Magn. Mater.* **95**, 249.
[13.8] Koehler, T.R., and Fredkin, D.R., 1991, *IEEE Trans. Magn.* **27**, 4763.
[13.9] Fredkin, D.R., and Koehler, T.R., 1990, *J. Appl. Phys.* **67**, 5544; 1990, *IEEE Trans. Magn.* **26**, 1518.
[13.10] Chen, W., Fredkin, D.R., and Koehler, T.R., 1993, *IEEE Trans. Magn.* **29**, 2124.
[13.11] Hertel, R., and Kronmüller, H., 1998, *IEEE Trans. Magn.* **34**, 3922.
[13.12] Schrefl, T., Fidler, J., and Kronmüller, H., 1994, *Phys. Rev. B* **49**, 6100.
[13.13] Kronmüller, H., and Schrefl, T., 1994, *J. Magn. Magn. Mater.* **129**, 66.
[13.14] Schrefl, T., Kronmüller, H., and Fidler, J., 1993, *J. Magn. Magn. Mater.* **127**, L 273.
[13.15] Mansuripur, M., and Giles, K., 1988, *IEEE Trans. Magn.* **24**, 2326.
[13.16] Giles, K.C., Kotiuga, P.K., and Humphrey, F.B., 1990, *J. Appl. Phys.* **67**, 5821.
[13.17] Berkov, D., Ramstöck, K., and Hubert, A., 1993, *Phys. Stat. Sol. (A)* **137**, 207.
[13.18] Rave, W., Fabian, K., and Hubert, A., 1998, *J. Magn. Magn. Mater.* **190**, 332.
[13.19] Grönefeld, M., and Kronmüller, H., 1989, *J. Magn. Magn. Mater.* **80**, 223.
[13.20] Schmidts, H.F., and Kronmüller, H., 1991, *J. Magn. Magn. Mater.* **94**, 220.

[13.21] Schmidts, H.F., and Kronmüller, H., 1994, *J. Magn. Magn. Mater.* **129**, 361.
[13.22] Schmidts, H.F., and Kronmüller, H., 1994, *J. Magn. Magn. Mater.* **130**, 329.
[13.23] Schabes, M.E., and Bertram, H.N., 1988, *J. Appl. Phys.* **64**, 1347; 1991, *J. Magn. Magn. Mater.* **95**, 249.
[13.24] Yan, Y.D., and Della Torre, E., 1988, *IEEE Trans. Magn.* **24**, 2368.
[13.25] Mansuripur, M., 1988, *J. Appl. Phys.* **63**, 5804.
[13.26] Hayashi, N., Nakatumi, Y., and Inoue, T., 1988, *J. Appl. Phys.* **27**, 366.
[13.27] Chantrell, R.W., Hannay, J.D., Wongsam, M., Schrefl, T., and Richter, H.-J., 1998, *IEEE Trans. Magn.* **34**, 1839.
[13.28] Chantrell, R.W., Tako, K.M., Wongsam, M., Walmsley, N., and Schrefl, T., 1997, *J. Magn. Magn. Mater.* **175**, 137.
[13.29] Chantrell, R.W., 1997, 'Interaction Effects in Fine Particle Systems'. In *Magnetic Hysteresis in Novel Magnetic Materials* (Ed. G.C. Hadjipanayis, Kluwer Academic Publ., Dordrecht) p. 21.
[13.30] Tako, K.M., Wongsam, M., and Chantrell, R.W., 1996, *J. Appl. Phys.* **79**, 5767.
[13.31] Tako, K.M., Schrefl, T., Wongsam, M., and Chantrell, R.W., 1997, *J. Appl. Phys.* **81**, 4082.
[13.32] Miltat, J., and Labrune, M., 1994, *IEEE Trans. Magn.* **30**, 4350.
[13.33] Raizer, A., Hoole, S.R.H., Meunier, G., and Coulomb, J.-L., 1990, *J. Appl. Phys.* **30**, 5803.
[13.34] Lewis, D., and Della Torre, E., *IEEE Trans. Magn.* **33**, 4161.
[13.35] Baker, T., 1997, *Finite Elements Anal. Design* **25**, 243.
[13.36] Griffiths, M.K., Bishop, J.E.L., Tucker, J.W., and Davies, H.A., 1998, *J. Magn. Magn. Mater.* **183**, 49; *ibid.*, 2001, **234**, 331.
[13.37] Aharoni, A., 1998, *J. Appl. Phys.* **83**, 3432.
[13.38] Goll, D., Schütz, G., and Kronmüller, H., 2003, *Phys. Rev. B* **67**, 094414.
[13.39] Demerdash, N.A., Nehl, T.W., and Fouad, F.A., 1980, *IEEE Trans. Magn.* **16**, 1092.
[13.40] Hertel, R., and Kronmüller, H., 2002, *J. Magn. Magn. Mater.* **238**, 185.
[13.41] Baibich, M.N., Boto, J.M., Fert, A., Nguyen Van Dau, F., Petroff, F., Etienne, P., Creuzet, G., Friedrich, A., and Chazelas, J., 1988, *Phys. Rev. Lett.* **61**, 2472.
[13.42] Binasch, G., Grünberg, P., Saurenbach, F., and Zinn, W., 1989, *Phys. Rev. B* **39**, 4828.
[13.43] Fert, A., Grünberg, P., Barthélémy, A., Petroff, F., and Zinn, W., 1995, *J. Magn. Magn. Mater.* **140–141**, 1.
[13.44] McMichael, R.D., 1998, Micromagnetic Standard Problems 1–3, http://www.ctcms.nist.gov/~rdm/mumag.html: μMAG – Micromagnetic Modeling Activity Group.
[13.45] Kronmüller, H., and Hertel, R., 2000, *J. Magn. Magn. Mater.* **215–216**, 11.
[13.46] Rave, W., and Hubert, A., 2000, *IEEE Trans. Magn.* **36**, 3886.
[13.47] Kronmüller, H., and Hertel, R., 2001, 'Computational Micromagnetism of Magnetic Structures and Magnetization Processes in Thin Platelets and Small Particles'. In *Magnetic Storage Systems Beyond 2001* (Ed. G.C. Hadjipanayis, Kluwer Academic, Dordrecht) p. 345.
[13.48] Arrott, A.S., and Templeton, T.L., 1997, *Physica B* **223**, 259.
[13.49] Van den Berg, H.A.M., and Vatvani, D.K., 1982, *IEEE Trans. Magn.* **18**, 880.
[13.50] Cowburn, R.P., Kolstov, D.K., Adeyeye, A.O., Welland, M.E., and Tricker, D.M., 1999, *Phys. Rev. Lett.* **83**, 1042.
[13.51] Schrefl, T., Fidler, J., and Kronmüller, H., 1994, *J. Magn. Magn. Mater.* **138**, 15.
[13.52] Rave, W., Fabian, K., and Hubert, A., 1998, *J. Magn. Magn. Mater.* **190**, 332.

[13.53] Usov, N.A., and Peschang, S.E., 1994, *J. Magn. Magn. Mater.* **135**, 111.

[13.54] Fischer, R., Leineweber, T., and Kronmüller, H., 1998, *Phys. Rev. B.* **57**, 10723.

[13.55] Preisach, F., 1929, *Ann. Physik* **3**, 737; 1935, *Z. Physik* **94**, 277.

[13.56] Fukunaga, H., and Inoue, H., 1992, *Jpn. J. Appl. Phys.* **31**, 1347.

[13.57] Otani, Y., Li, H., and Coey, J.M.D., 1990, *IEEE Trans. Magn.* **26**, 2658.

[13.58] Manaf, A., Zhang, P., Ahmed, I., Davies, H.A., and Buckley, R., 1993, *IEEE Trans. Magn.* **29**, 2866.

[13.59] Manaf, A., Buckley, R., Davies, H.A., and Leonowicz, J., 1991, *J. Magn. Magn. Mater.* **101**, 360.

[13.60] Schrefl, T., Schmidts, H.F., Fidler, J., and Kronmüller, H., 1993, *J. Magn. Magn. Mater.* **124**, 251; 1993, *IEEE Trans. Magn.* **29**, 2878.

[13.61] Fischer, R., Schrefl, T., Kronmüller, H., and Fidler, J., 1996, *J. Magn. Magn. Mater.* **153**, 35.

[13.62] Fischer, R., and Kronmüller, H., 1998, *J. Appl. Phys.* **83**, 3271.

[13.63] Fischer, R., Schrefl, T., Kronmüller, H., and Fidler, J., 1995, *J. Magn. Magn. Mater.* **150**, 329.

[13.64] Kronmüller, H., Fischer, R., Seeger, M., and Zern, A., 1996, *J. Phys. D: Appl. Phys.* **29**, 2274.

[13.65] Fischer, R., and Kronmüller, H., 1996, *Phys. Rev. B* **54**, 5469.

[13.66] Willcox, M., Williams, J., Leonowicz, M., Manaf, A., and Davies, H.A., 1994, *13th Int. Workshop on RE Magnets and their Application* (Eds. G.A.F. Manwaring *et al.*, Birmingham, University of Birmingham) p. 443.

[13.67] Fischer, R., and Kronmüller, H., 1998, *J. Magn. Magn. Mater.* **184**, 166.

[13.68] Kronmüller, H., Fischer, R., Hertel, R., and Leineweber, T., 1997, *J. Magn. Magn. Mater.* **175**, 177.

14

Computational micromagnetism of dynamic magnetization processes

14.1 Landau–Lifshitz and Gilbert equations

Computational micromagnetism has been successfully applied for the calculation of magnetic ground states and quasistatic magnetization processes. In these cases the magnetic configurations are determined either by the minimization of the magnetic Gibbs free energy [2.6] or by determining a vanishing torque $[J_s \times H_{eff}]$ exerted by an effective field, H_{eff}, or the polarization J_s. A generalization of the static equilibrium condition (3.1) is obtained from the angular momentum equation

$$dP/dt = L = [J_s \times H_{eff}], \qquad (14.1)$$

by using the magnetomechanical analogue of quantum mechanics

$$J_s = \gamma_0 \cdot P, \qquad (14.2)$$

which relates the mechanical angular momentum to the polarization by the gyromagnetic ratio $\gamma_0 = -g|e|/(2m) = -1.1051 g \cdot 10^5$ (sA/m)$^{-1}$. Inserting eq. (14.2) into eq. (14.1) leads to the undamped dynamic equation

$$dJ_s/dt = \gamma_0[J_s \times H_{eff}]. \qquad (14.3)$$

In order to take care of damping processes Landau and Lifshitz [14.1] introduced an additional term which rotates J_s into the direction of the effective field (Fig. 14.1) and is then written

$$\frac{dJ_s}{dt} = \gamma_L[J_s \times H_{eff}] - \frac{\alpha_L}{J_s}[J_s \times [J_s' \times H_{eff}]]. \qquad (14.4)$$

Following the usual representation of eq. (14.4) we divide by μ_0 and introduce M_s which gives

$$\frac{dM_s}{dt} = \gamma_L[M_s \times H_{eff}] - \frac{\alpha_L}{M_s}[M_s \times [M_s \times H_{eff}]]. \qquad (14.5)$$

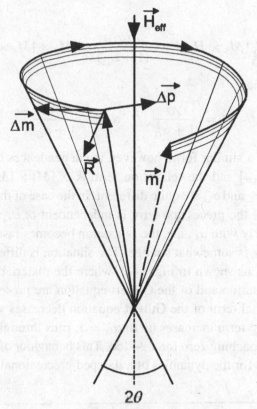

Fig. 14.1. Damped precession of the magnetization M_s around the effective field H_{eff}.

The first term on the right side of eq. (14.5) describes a homogeneous precessional rotation of M_s with frequency $\omega = -\gamma_0 H_{eff}$. Since eq. (14.5) describes a motion of J_s which accelerates with increasing damping parameter α_L, Gilbert [14.2] has proposed an alternative equation which is written

$$\frac{dM_s}{dt} = \gamma_G[M_s \times H_{eff}] - \frac{\alpha_G}{M_s}\left[M_s \times \frac{dM_s}{dt}\right]. \tag{14.6}$$

Here $\gamma_{L,G}$ are the gyromagnetic ratios and $\alpha_{L,G}$ the corresponding damping constants. The subscripts L and G are used in order to differentiate between the Landau–Lifshitz and the Gilbert equations. H_{eff} is defined as in Chapter 3 by the derivation $H_{eff} = -\partial\phi_t/\partial J_s$, or if we introduce the normalized polarization $J_s/J_s = M_s/M_s = m$, $H_{eff} = -(1/J_s)\partial\phi_t/\partial m$. In order to compare the two dynamic equations, in the Gilbert equation on the right side dM_s/dt is replaced by eq. (14.6) itself. This leads to the relation $[M_s \times [M_s \times dM_s/dt]] = -M_s^2(dM_s/dt)$ since $dM_s/dt \perp M_s$ holds for $|M_s| = $ const. Inserting this relation into the substituted Gilbert equation, also denoted the Landau–Lifshitz–Gilbert equation, gives

$$\frac{dM_s}{dt} = \frac{\gamma_G}{1+\alpha_G^2}[M_s \times H_{eff}] - \frac{\alpha_G\gamma_G}{(1+\alpha_G^2)M_s}[M_s \times [M_s \times H_{eff}]]. \quad (14.7)$$

With the replacement

$$\gamma_L = \frac{\gamma_G}{1+\alpha_G^2}, \qquad \alpha_L = \frac{\alpha_G\gamma_G}{1+\alpha_G^2}, \quad (14.8)$$

both equations have a similar form, however, the dependences of the precessional term $P \propto [M_s \times H_{eff}]$ and the relaxation term $R \propto [M_s \times [M_s \times H_{eff}]]$ on the damping constants α_L and α_G are quite different. In the case of the Landau–Lifshitz equation (eq. (14.5)) the precession term is independent of α_L and the relaxation term increases linearly with α_L, i.e., the relaxation becomes faster with increasing α_L, which intuitively is somewhat strange. The situation is different in the case of the Gilbert equation as shown in Fig. 14.2, where the material parameters of the Landau–Lifshitz equation and of the Gilbert equation are presented as a function of α. The precessional term of the Gilbert equation decreases with increasing α_G whereas the damping term increases up to $\alpha_G = 1$, runs through a maximum, and then decreases, approaching zero for $\alpha \to \infty$. This behaviour of both terms agrees with the expectation for the dynamics of a damped precessional movement.

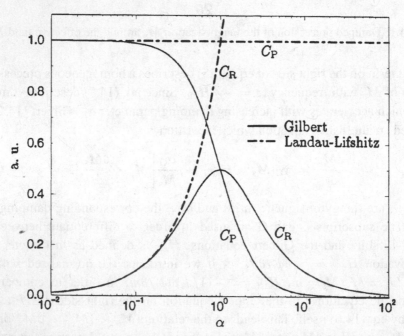

Fig. 14.2. The rates dM/dt due to precession $C_P \propto [M_s \times H_{eff}]$ and relaxation $C_R \propto [M_s \times [M_s \times H_{eff}]]$ in arbitrary units for the Landau–Lifshitz equation and the Gilbert equation as a function of α.

14.2 Characteristic time ranges

An analytical solution of the Gilbert equation has been determined by Kikuchi [14.3] for the case of a sphere with uniaxial anisotropy and a field applied antiparallel to the positive easy direction chosen as the z-axis. In the case of a homogeneous rotation process the demagnetization field $H_s = -(1/3)M_s$ is always oriented strictly antiparallel to M_s and therefore exerts no torque on M_s. Introducing normalized quantities $h_{\mathrm{eff}} = H_{\mathrm{eff}}/H_N^{(0)}$, with $H_N^{(0)} = 2K_1/J_s$, $m = M_s/M_s$ and $\tau = |\gamma_0| t H_N^{(0)}$, the normalized effective field is given by

$$h_{\mathrm{eff}} = \begin{pmatrix} -m_x \\ m_y \\ H_{z,\mathrm{ext}}/H_N^{(0)} \end{pmatrix}, \qquad (14.9)$$

and the Gilbert equation is written (replacing α_G by α)

$$\frac{dm}{d\tau} = \frac{1}{1+\alpha^2}[m \times h_{\mathrm{eff}}] - \frac{\alpha}{1+\alpha^2}[m \times [m \times h_{\mathrm{eff}}]]. \qquad (14.10)$$

In terms of spherical coordinates defined in Fig. 14.3 for M_s with a polar angle, θ, between M_s and the easy axis and an azimuthal angle, φ, the Gilbert equation splits into two differential equations:

$$\frac{d\theta}{d\tau} = -\frac{\alpha}{1+\alpha^2}\sin\theta\,(\cos\theta + h_{\mathrm{ext}}),$$

$$\frac{d\varphi}{d\tau} = \frac{1}{1+\alpha^2}(\cos\theta + h_{\mathrm{ext}}). \qquad (14.11)$$

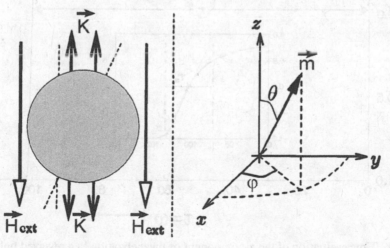

Fig. 14.3. Definition of spherical coordinates θ and φ of the reduced magnetization m. The left part of the figure shows a spherical particle with easy direction K and the magnetization m with angle θ with respect to K.

For the time dependence of $m_z = \cos\theta$ at a field $H_{z,\text{ext}} = H_N^{(0)}$, i.e., $h_{\text{ext}} = -1$, eqs. (14.11) can be integrated giving an implicit solution for the time-dependent z-component $m_z(\tau) = \cos\theta$:

$$\tau = -\frac{1+\alpha^2}{4\alpha}\left[\ln\left(\frac{(m_z(\tau)+1)(m_0-1)}{(m_z(\tau)-1)(m_0+1)}\right) - \frac{2}{m_z(\tau)-1} + \frac{2}{m_0-1}\right], \quad (14.12)$$

where m_0 denotes the z-component of $\boldsymbol{m}$ at $\tau = 0$. In the case of an ideally oriented magnetization, $m_0 = 1$, the torque on $\boldsymbol{m}$ vanishes and no precessional movement of $\boldsymbol{m}$ takes place. The time dependence of $m_z(\tau)$ determined numerically from eq. (14.12) for $m_0 = 0.99$ is shown in Fig. 14.4 for $\alpha = 1$, revealing a behaviour similar to a step function. Here we can distinguish between two time ranges. In range I the magnetization m_z does not change appreciably. The relaxation time, τ_r, elapsing until $m_z(\tau)$ becomes zero is given by

$$\tau_r = \frac{1+\alpha^2}{4\alpha}\left[\ln\frac{1+m_0}{1-m_0} - \frac{2}{m_0-1} - 2\right]. \quad (14.13)$$

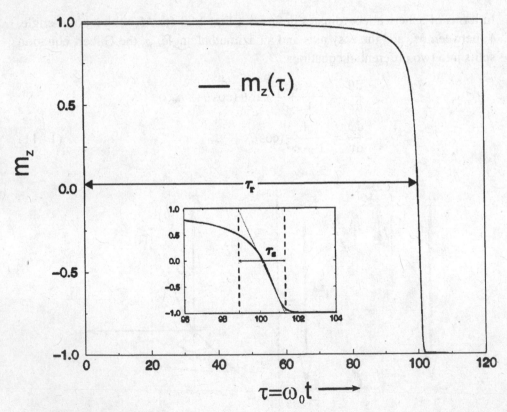

Fig. 14.4. Time evaluation of the z-component of magnetization in a reversed field $H_{\text{ext}} = -H_N^{(0)}$ with $\alpha = 1$ and $m_0 = 0.99$ [14.7].

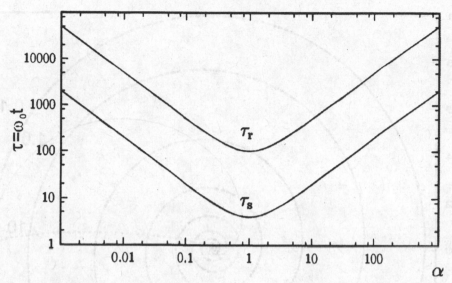

Fig. 14.5. Relaxation time, τ_r, and switching time, τ_s, as a function of α for $m_0 = 0.99$ and $H_{ext} = -H_N^{(0)}$ [14.7].

Furthermore, a range II can be defined where m_z changes its orientation by almost $180°$. This switching time is given by

$$\tau_s = 2\frac{1+\alpha^2}{\alpha}. \tag{14.14}$$

Figure 14.5 shows the characteristic times τ_r and τ_s as a function of the damping constant α for $m_0 = 0.99$. According to eqs. (14.13) and (14.14), τ_r and τ_s have the same functional dependence with a minimum at $\alpha = 1$; however, the values differ by a factor of 20, depending on the choice of m_0. For real materials the α-values are suggested to be in the range of 0.001 to 0.1, i.e., τ_r and τ_s lie on the decreasing branch of the $\tau_{r,s}(\alpha)$ functions. Figure 14.6 shows three characteristic paths of the tip of magnetization depending on the magnitude of the damping parameter. For $\alpha < 1$ a precessional mode with frequency $\omega \approx \gamma_0 H_{eff}$ exists which changes into an aperiodic mode for $\alpha \geq 1$. In the case of $\alpha > 1$ the path of M_s follows the energy gradient which is equivalent to the situation for static calculations. Whereas these analytical solutions for homogeneous precession give a clear insight into the time evolution of $M_z(t)$, the situation for the more general case of inhomogeneous rotation processes is more complex. Inhomogeneous processes are expected for particles larger than the exchange lengths. In this case numerical methods such as the FEM have to be applied.

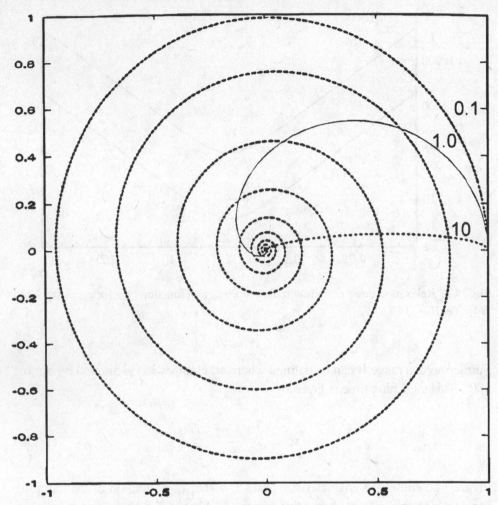

Fig. 14.6. Sketch of the path of the tip of m projected into the plane perpendicular to H_{eff}. In initial conditions at $t = 0$, $M_s \perp H_{eff}$. Three different cases for $\alpha = 0.1$ (damped precession), $\alpha = 1.0$ (limiting case), and $\alpha = 10$ (aperiodic case) are considered [13.68].

14.3 Magnetization reversal in thin films

So far no analytical solution for the precessional reversal of magnetization is available for the case of thin films. The problem was originally treated by Kikuchi [14.3] for the case of a homogeneous precessional movement of M_s on an ellipsoidal cone. He gave approximate solutions for $\alpha \to 0$ and $\alpha \to \infty$. For intermediate values of α he determined the switching time numerically. His numerical results are shown in Fig. 14.7 together with results obtained from an interpolation. The minimum of the relaxation time is now shifted to rather low values of $\alpha^{min} = 0.013$. Accordingly, the reversal of M_s depends in a characteristic way not only on α itself, but also

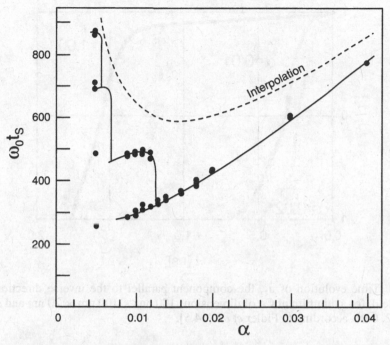

Fig. 14.7. Kikuchi's result for the switching time of thin films as a function of α [14.3] (• computer results), dotted curve: interpolation formula.

on the shape of the particle. It is therefore evident that the growing importance of high-speed magnetoelectronics [14.4] requires an analysis of the dynamics of magnetization reversal in thin films of nanoscale thickness.

Recent micromagnetic simulations of dynamic reversal processes by Fidler *et al.* [14.5] for Co nano-elements of square shape 100×100 nm^2 and thickness 20 nm have confirmed Kikuchi's previous results. The time evaluation of the magnetization component parallel to the negative field direction shown in Fig. 14.8 indicates a minimum of the relaxation time between $\alpha = 0.01$ and 0.02. The switching time does not show a significant variation with respect to the α-parameter. Similar results were obtained by Torres *et al.* [14.6] for nanosquares of thickness $d = 1.5\,l_s$. In this latter paper the role of shape anisotropy and the misalignment between applied field and easy direction have also been investigated.

14.4 Discretization of the Landau–Lifshitz–Gilbert equation

Using a Delaunay triangulation routine a ferromagnetic specimen may be discretized into tetrahedral elements as described in Chapter 13. After a first discretization in space the grid is refined to the degree required by applying the longest edge bisection technique as described in Chapter 13. On the vertices of the finite

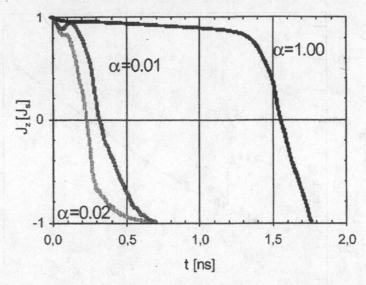

Fig. 14.8. Time evolution of J_z, the component parallel to the inverse direction of the applied field, for a thin film of Co of dimensions $100\,\mathrm{nm} \times 100\,\mathrm{nm} \times 20\,\mathrm{nm}$ and α values 1.00, 0.02, 0.01, according to Fidler *et al.* [14.5].

elements the magnetization, M_s, and the scalar potential, U, of the stray field are calculated by minimization of the total magnetic free energy. Within the volume of the finite elements these functions are linearly interpolated. The discretization with respect to time starts from a splitting of the Landau–Lifshitz–Gilbert equation (see eq. (14.7)) into the precessional term $\Delta m_p = [m \times h_{\mathrm{eff}}]\Delta\tau$ and the relaxation term $\Delta m_r = [m \times [m \times h_{\mathrm{eff}}]]\Delta\tau$ [14.7]. Within the framework of the spherical coordinates φ and θ the changes $\Delta\varphi$ and $\Delta\theta$ after a time step from τ to $\tau + \Delta\tau$ are given by

$$\Delta\varphi = \frac{|\Delta m_p|}{\sin\theta} = |h_{\mathrm{eff}}|\frac{\Delta\tau}{1+\alpha^2},$$

$$\Delta\theta = |\Delta m_r| = |m \times h_{\mathrm{eff}}|\frac{\alpha\Delta\tau}{1+\alpha^2}. \tag{14.15}$$

This technique for integrating the Landau–Lifshitz–Gilbert equation is most suitable because $|M_s|$ remains constant and is independent of the time step, $\Delta\tau$, and also variable time steps can easily be implemented.

14.5 Dynamic nucleation field

In Chapter 6 the coercive field was determined for static conditions. These results apply for magnetization processes up to the μs range, i.e., for soft and hard magnetic materials in conventional applications.

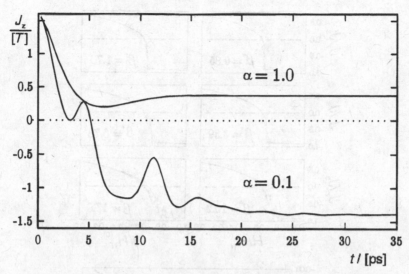

Fig. 14.9. Evolution of the component J_z parallel to the inverse direction of the applied field for a spherical particle and an applied field of $H_{ext} = 0.85 H_N^{(0)}$ and $\alpha = 0.1, 1.0$. Angle between field and easy axis $\psi = 45°$ [14.7].

In modern high-speed magnetoelectronic devices, such as magnetic random access memories (MRAMS), however, magnetization processes take place over times much shorter than µs [14.4]. Here the question arises whether the coercive field depends on the field rate dH_{ext}/dt. The evolution of the z-component of the polarization J_s is shown in Fig. 14.9 for a field of $H_{ext} = 0.85 H_N^{(0)}$ applied at an angle of 45° with respect to the negative easy direction for two damping constants $\alpha = 0.1$ and $\alpha = 1$ [14.7]. Even though the initial state for both calculations is exactly the same, the final stationary states are different. For $\alpha = 1$ the system remains in a metastable state on the upper branch of the hysteresis, whereas for $\alpha = 0.1$ the system changes in a ringing mode into a stable state with minimal energy. Figure 14.10 shows demagnetization curves for $\alpha = 1.0$ and $\alpha = 0.1$ and different reduced field rates

$$\beta^* = \frac{1}{\omega_0 H_N^{(0)}} \left| \frac{dH_{ext}}{dt} \right| = \frac{\beta}{\omega_0 H_N^{(0)}}, \qquad (14.16)$$

with

$$\omega_0 = -\gamma_0 H_N^{(0)} \quad \text{and} \quad H_N^{(0)} = 2K_1/J_s.$$

For the calculations the material parameters of $Nd_2Fe_{14}B$ have been used and a sphere of radius $\delta_B = 4.2$ nm is considered. According to Fig. 14.10 the coercive field decreases with the lower α-value. Furthermore, with increasing field rate the reduction of H_c increases. The reduction of the coercive field $\Delta H_c = H_c^{0.1} - H_c^{1.0}$

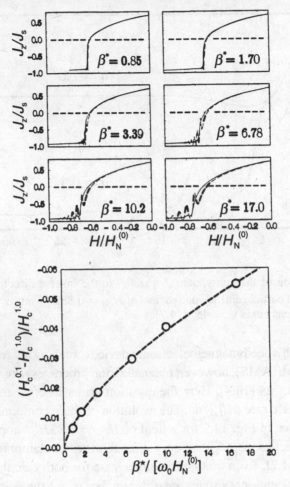

Fig. 14.10. Demagnetization curves with the applied field increasing linearly for different field rates β^* and $\alpha = 1.0$ (dashed line). Angle between field and easy direction $\psi = 45°$. The lower plot shows the reduction of the coercive field $H_c^{0.1}$ for $\alpha = 0.1$ with respect to the coercive field $H_c^{1.0}$ for $\alpha = 1.0$ as a function of the field rate β^* [14.7] ($\beta^* = \beta/(\omega_0 H_N^{(0)})$, $\beta = \mathrm{d}H_{\mathrm{ext}}/\mathrm{d}t$).

is shown at the bottom of Fig. 14.10. A dynamic nucleation field may be defined by the minimum pulse height that is sufficient to reverse the magnetization. This dynamic nucleation field, H_N^{dyn}, is a function of the field rate β^*, of the orientation of H_{ext} with respect to the easy direction and of the damping constant α, i.e., $H_N^{\mathrm{dyn}} = H_N^{\mathrm{dyn}}(\beta, \psi, \alpha)$ [14.7].

Varying all the relevant parameters of H_N^{dyn} would be a rather time-consuming task. Therefore the calculations have been restricted to a Stoner–Wohlfarth particle [6.63] with the material parameters of $Nd_2Fe_{14}B$ assuming homogeneous

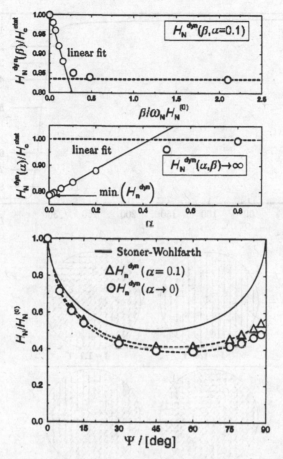

Fig. 14.11. Dynamic nucleation field of a Stoner–Wohlfarth particle. Top: dynamic nucleation field for $\alpha = 0.1$ and $\psi = 45°$ versus field rate β^* (open circles). Centre: H_N^{dyn} for $\beta \to \infty$ versus α (open circles) and $\psi = 45°$. Bottom: H_N^{dyn} in the limit of $\beta^* \to \infty$ for $\alpha = 0.1$ (open triangles) and $\alpha \to 0$ (open circles) versus orientation ψ of the field with respect to the easy direction. The full curve represents the static Stoner–Wohlfarth result [14.7].

precession of M_s. Figure 14.11 shows in the upper part $H_N^{dyn}/H_N^{(0)}$ as a function of β^* and of α. For large values of β and α H_N^{dyn} approaches constant values becoming independent of β or α, respectively. The lower part of Fig. 14.11 shows the dependence of H_N^{dyn} on the angle ψ of H_{ext} with respect to the negative easy axis for $\beta \to \infty$ and $\alpha = 0$ and $\alpha = 0.1$. As compared with the conventional Stoner–Wohlfarth results for $H_N(\psi)$, H_N^{dyn} is reduced giving a minimum value of $H_N^{min,dyn} = 0.39 H_N^{(0)}$, which corresponds to a reduction of H_N by 20%. These results are in perfect agreement with analytically determined dynamic nucleation fields of Porter [14.8].

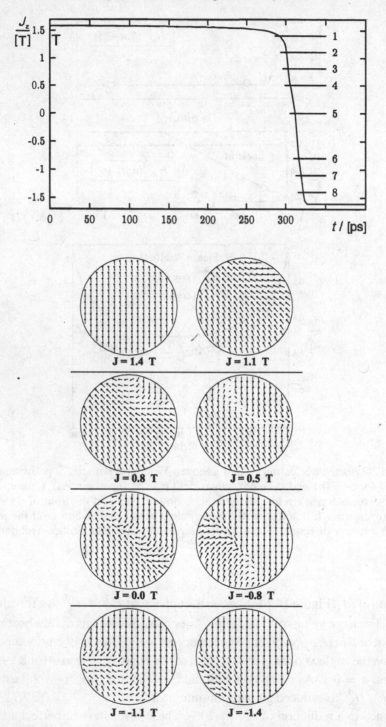

Fig. 14.12. Evolution of the z-component of magnetization with time for a sphere of Nd_2Fe_4B and $\alpha = 0.5$. The numbers 1–8 indicate the points where the distribution of magnetization, shown at the bottom, has been determined.

The dependence of the reversal process of magnetization on the damping constant α is caused by the different paths the system is taking toward the energy equilibrium. For large damping constants M_s follows the energy gradient on a discrete path as shown in Fig. 14.4. For low damping the path of M_s is governed by a precessional rotation perpendicular to the energy gradient (H_{eff}). This results in large relaxation times and, as a consequence of fast changes of the applied field, leads to nonstationary magnetization states with high energy densities (see Fig. 14.6). The distribution of magnetization within the cross-section of a sphere as determined for $\alpha = 0.5$ and different times during the switching process are shown in Fig. 14.12. In the upper part the numbers 1–8 indicate the positions where the magnetic states have been determined [14.7]. Here it becomes obvious that the reversal takes place by an inhomgeneous rotation process at the surface. Up to this starting process the demagnetization process proceeds reversibly and then becomes irreversible if the coercive field is approached. Here the system does not have enough time to relax into the local momentary local minimum of the energy. In particular, for low damping, precession with large opening angles occurs. To describe this process a dynamic calculation is necessary because the nonstationary states cannot be calculated statically. Nevertheless, the static calculation leads to approximately the same hysteretic properties, like coercive field, remanence or maximum energy product, as the dynamic calculation.

Due to the nonuniform demagnetization process the switching time, τ_s, for low α-values deviates from the analytical results given in eq. (14.13) and Fig. 14.5. As shown by Fig. 14.13, for $\alpha < 1$ the numerical values of τ_r and τ_s are larger by a factor of 10 as compared to the analytical result for homogeneous precession.

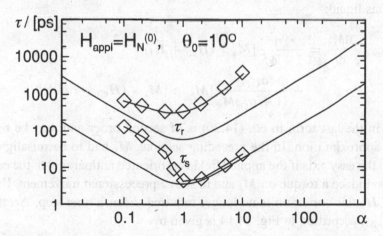

Fig. 14.13. Numerically calculated switching times, τ_s, for varying damping constants α (–◇–◇–) compared with the analytically calculated τ_s of eq. (14.14).

14.6 Dynamics of thermally activated reversal processes

14.6.1 Thermal fluctuations

In the preceding sections all calculations have been performed by neglecting thermal fluctuations. Previous investigations by Néel [14.9] and Brown [14.10, 14.11] gave a first insight into the reversal of magnetization of small particles by thermal fluctuations.

Following Brown [14.10, 14.11] thermal effects can be included by adding a stochastically fluctuating field, $H_f(t)$ to the effective field. The statistical properties of this field are the following:

$$\langle H_f(t) \rangle = 0, \tag{14.17}$$

$$\langle H_{f,i}(t) \cdot H_{f,j}(t + t') \rangle = \mu \delta_{ij} \delta(t'), \tag{14.18}$$

where eq. (14.18) is a consequence of the fluctuation-dissipation theorem [14.12, 14.13] and guarantees that the fluctuating field components $H_{f,i}$ are uncorrelated and for $i = j H_{f,i}(t)$ and $H_{f,i}(t + t')$ are also uncorrelated. μ corresponds to a temperature-dependent constant which corresponds to the variance of the statistical variable H_f, which is Gaussian and is given by

$$\mu = \frac{2k_B T \alpha}{|\gamma_0| M_s V}, \tag{14.19}$$

where V denotes the volume of the particle or the discretization volume of the finite elements, and k_B corresponds to Boltzmann's constant. Inserting H_f into the Landau–Lifshitz–Gilbert equation we obtain a micromagnetic equation of the motion of M_s similar to the Langevin equation [14.14–14.16] of a Brownian particle in a viscous liquid:

$$\frac{dM_s}{dt} = \frac{\gamma_G}{1 + \alpha_G^2} [M_s \times (H_{eff} + H_f)]$$

$$- \frac{\alpha_G \gamma_G}{(1 + \alpha_G^2) M_s} [M_s \times [M_s \times (H_{eff} + H_f)]]. \tag{14.20}$$

Here H_f in the last term in eq. (14.20) is of second order and may be neglected in a first approximation. In the preceding sections M_s had to be misaligned with respect to the easy axis if the applied field was oriented antiparallel to the easy axis, in order to induce a torque on M_s and to start a precessional movement. Under the action of H_f this condition is no longer required. After a time step, Δt, the polar angle, $\theta_{t+\Delta t}$, according to Fig. 14.14 is given by

$$\theta_{t+\Delta t} = \theta_t - \Delta\theta \pm \Delta\vartheta, \tag{14.21}$$

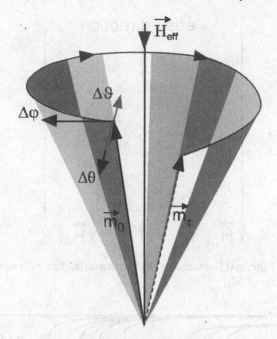

Fig. 14.14. Movement of magnetization during a time step Δt leading to $\Delta\theta$ and $\Delta\varphi$ due to the Gilbert equation and $\Delta\vartheta$ due to thermal fluctuations.

where $\Delta\theta$ describes the change of θ_t due to the Gilbert equation and $\Delta\vartheta$ that due to the thermal fluctuation, H_f, at $T > 0$. The fluctuation takes place self-consistently, not changing the total energy of the system. Changes of energy are exclusively due to the dissipation described by the Gilbert equation. In the case of $\theta_t = 0$, naturally $\Delta\theta = 0$ holds and the $\Delta\vartheta$ then describes the temperature dependence of the spontaneous magnetization, which for $T < T_c/2$ follows Bloch's $T^{3/2}$ law:

$$M_s(T) = M_s(0)\left(1 - (T/T_0)^{3/2}\right). \tag{14.22}$$

The average deviation $\langle|\Delta\vartheta|\rangle$ is related to the temperature by $\cos(\langle|\Delta\vartheta|\rangle) = \sqrt{2}\,(T/T_0)^{3/4}$. The thermal fluctuations of $M_s(T)$ are related to thermally excited spin waves with frequencies larger than the homogeneous precession frequency, $\omega_0 = -\gamma_0 H_{ext}$. The role of the thermal fluctuations in a first approximation can be described by the average misalignment $\langle|\Delta\vartheta|\rangle$ assuming that these misalignments in neighbouring finite elements are uncorrelated and at random orientations, as shown in Fig. 14.15. For numerical calculations the frequency of the fluctuations is assumed to be $\omega = 50\,\omega_0$. Figure 14.16 shows a series of simulated thermally activated reversals of magnetization for different applied fields and temperatures, i.e., different strengths of the elementary fluctuations $\langle|\Delta\vartheta|\rangle$. For this calculation, at each vortex uncorrelated deviations $\Delta\vartheta$ are produced by a statistical random

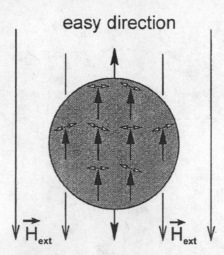

Fig. 14.15. Model of thermal fluctuations of the magnetization with stochastic distribution of angular fluctuations of M_s.

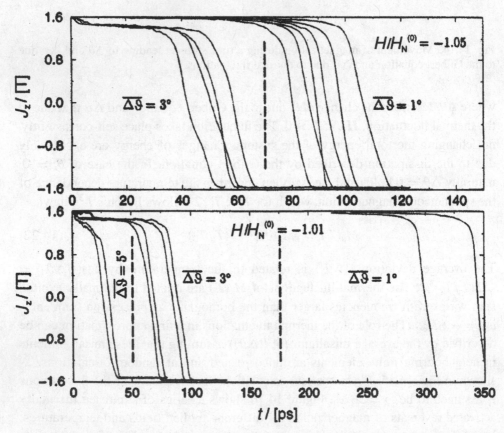

Fig. 14.16. Time dependence of the z-component of M_s of a sphere of $Nd_2Fe_{14}B$ for different fields H_{ext} and different strengths $\overline{\Delta\vartheta}$ of thermal fluctuations [14.21].

number routine, where the probability $P(\Delta\vartheta)$ follows a Boltzmann distribution, $P(\Delta\vartheta) \propto \exp(-\Delta\vartheta/\langle\Delta\vartheta\rangle)$. The demagnetization curves for different numerical runs for constant H_{ext} and $\langle\Delta\vartheta\rangle$ show moderate changes of the relaxation time, τ_r, and of the switching time, τ_s. For constant external field and different $\langle|\Delta\vartheta|\rangle$-values, increasing $\langle|\Delta\vartheta|\rangle$-values lead to a decrease of τ_r and increasing applied fields also lead to a decrease of τ_r.

In the case of an overcritical field, $H_{ext} = 1.05 H_N^{(0)}$ and $\langle|\Delta\vartheta|\rangle = 1°$ we observe two types of switching modes. A type I switching mode with $\tau_s \approx 10\,\text{ps}$ and a type II mode with switching time $\tau_s \approx 20\,\text{ps}$. An investigation of the distribution of M_s of these two modes reveals differences in the type of nucleation process. In the case of the type I mode the nucleation takes place at the surface of the sphere, whereas the type II mode nucleates in the bulk of the sphere. Figure 14.17 presents the development of the demagnetization process for these two demagnetization modes.

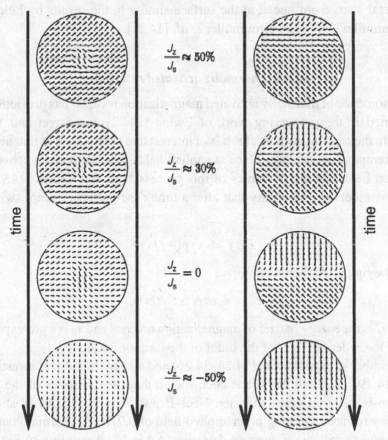

Fig. 14.17. Evolution of reversal of magnetization starting in the centre or on the surface of a sphere.

The dynamics of thermally activated reversal processes on the basis of eq. (14.20) has been studied recently by several authors. Boerner and Bertram [14.17] investigated the thermally activated reversal of elongated particles. The probability of not switching has been determined for different applied fields.

Torres *et al.* [14.6] also considered the dependence of the relaxation times on applied field and temperature. Chantrell *et al.* [14.18] discussed the high-speed switching of magnetic recording media taking care of thermally activated processes. For short field pulses the coercivity is shown to increase in the ns range.

Zhang and Fredkin [14.19] studied the switching of tiny particles. The probability of not switching is shown to be described by a delayed stretched exponential. The magnetization reversal of micro-sized thin films has been investigated by Koch *et al.* [14.4]. These authors show that the reversal takes place inhomogeneously starting from the edges of the films. Scholz *et al.* [14.20] investigated the reversal of magnetization in cubic and spherical particles. It is shown that in particles with low anisotropy the magnetization rotates coherently whereas for large anisotropies the reversal starts from nuclei at the surface similar to the results by Leineweber and Kronmüller [14.7] and Kronmüller *et al.* [14.21].

14.6.2 Thermally activated relaxation

The phenomenon of thermally activated magnetization reversal has previously been investigated in the pioneering work of Ewing [14.22] and Street and Woolley [14.23]. In the case of tiny particles it is of interest that the reversal of magnetization at finite temperature can take place at applied fields smaller than the conventional nucleation field. For the dynamics of this process Néel and Brown [14.9–14.11] have shown that the probability that after a time t the particle has not switched is given by

$$P(t) = \exp(-t/\tau), \tag{14.23}$$

with τ obeying an Arrhenius law,

$$\tau = \tau_0 \exp(\Delta E/(kt)), \tag{14.24}$$

where ΔE is the energy barrier of magnetization reversal and τ_0 is a pre-exponential factor of the order 10^{-9} (s) of the order of the Larmor frequency.

In a number of experiments [14.24–14.26] and also by numerical investigations [14.17–14.19, 14.27, 14.28] it has been shown that the dynamics of the thermal relaxation is more complex than the Néel–Brown model; Fig. 14.18 shows the probability for not switching in an applied field of $0.97H_N^{(0)}$. Thermal fluctuations are assumed to induce an average deviation $\overline{\Delta\vartheta} = 5°$. According to Fig. 14.18,

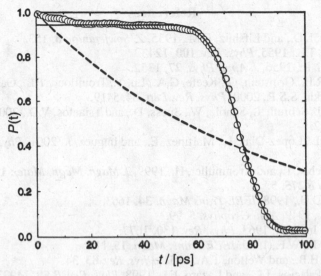

Fig. 14.18. Probability, $P(t)$, of nonswitching of a $Nd_2Fe_{14}B$ sphere under an applied field of $H_{ext} = -0.97H_N^{(0)}$ and $\alpha = 0.5$. Strength of thermal fluctuations $\langle \Delta \vartheta \rangle = 5°$. ($\circ \circ \circ$) calculated values of $P(t)$; ($-----$) Brown–Néel law; (———) Weibull distribution.

before switching we deal with a stationary state with an average magnetization below the 1.61 T of the spontaneous magnetization. The Néel–Brown model obviously cannot describe the numerical results. However, the Weibull [14.28] distribution gives a rather nice description of the probability $P(t)$ as proposed by Zhang and Fredkin [14.19]. The numerical results are fitted by

$$P(t) = \begin{cases} 0.945, & t < 44 \text{ ps} \\ 0.92 \exp\left[-\left(\frac{t[\text{ps}]-44}{33.5}\right)^{4.5}\right] + 0.25, & t > 44 \text{ ps}. \end{cases} \quad (14.25)$$

The Weibull distribution corresponds to a delayed stretched exponential of general form

$$P(t) = A \exp\left[-\left(\frac{t - \tau_r'}{\tau}\right)^\kappa\right] + B. \quad (14.26)$$

As in the case of nonthermal reversal processes there exists a relaxation time τ_r' which determines the waiting time until the switching process takes place. The relaxation time τ corresponds to the time after which the probability that the grain has not switched has become $1/e$. The parameter κ influences the switching time. Increasing values of κ lead to a reduction of the switching time. Its effect on the relaxation time, τ_r, is less significant.

References

[14.1] Landau, L.D., and Lifshitz, E.M., 1935, *Z. Sowjetunion* **8**, 153.

[14.2] Gilbert, T.L., 1955, *Phys. Rev.* **100**, 1243.

[14.3] Kikuchi, R., 1956, *J. Appl. Phys.* **27**, 1352.

[14.4] Koch, R.H., Grinstein, G., Keefe, G.A., Lu, Y., Trouilloud, P.L., Gallagher, W.L., and Parkin, S.S.P., 2000, *Phys. Rev. Lett.* **84**, 5419.

[14.5] Fidler, J., Schrefl, T., Scholz, W., Suess, D., and Tsiantos, V.D., 2001, *Physica B* **306**, 112.

[14.6] Torres, L., Lopez-Diaz, L., Martinez, E., and Iñiguez, J., 2001, *Physica B* **306**, 216.

[14.7] Leineweber, T., and Kronmüller, H., 1999, *J. Magn. Magn. Mater.* **192**, 575; 2000, *Physica B* **275**, 5.

[14.8] Porter, D.G., 1998, *IEEE Trans Magn.* **34**, 1663.

[14.9] Néel, L., 1949, *Ann. Geophys.* **5**, 99.

[14.10] Brown, Jr., W.F., 1963, *Phys. Rev.* **130**, 1677.

[14.11] Brown, Jr., W.F., 1979, *IEEE Trans. Magn.* **15**, 1196.

[14.12] Callen, H.B., and Welton, T.A., 1951, *Phys. Rev.* **83**, 34.

[14.13] Garcia Palacios, J.L., and Lazaro, F.J., 1998, *Phys. Rev. B* **58**, 14937.

[14.14] Chandresekhar, S., 1943, *Rev. Mod. Phys.* **15**, 1.

[14.15] Wang, M.C., and Uhlenbeck, G.E., 1945, *Rev. Mod. Phys.* **17**, 323.

[14.16] Øksendal, B., 1992, *Stochastic Differential Equations*, 3rd Edition (Springer Verlag, Berlin–Heidelberg–New York).

[14.17] Boerner, E.D., and Bertram, H.N., 1997, *IEEE Trans. Magn.* **33**, 3052.

[14.18] Chantrell, R.W., Hannay, J.D., Wongsam, M., and Lyberatos, A., 1998, *IEEE Trans. Magn.* **34**, 349.

[14.19] Zhang, K., and Fredkin, D.R., 1999, *J. Appl. Phys.* **85**, 5208.

[14.20] Scholz, W., Schrefl, T., and Fidler, J., 2001, *J. Magn. Magn. Mater.* **223**, 296.

[14.21] Kronmüller, H., Leineweber, T., and Bachmann, M., 2001, *16th IMACS World Congress, Lausanne* (2000 IMACS).

[14.22] Ewing, A., 1885, *Philos. Trans. R. Soc.* **176**, 569.

[14.23] Street, R., and Woolley, J.C., 1949, *Proc. Phys. Soc. A* **62**, 562.

[14.24] Ledermann, M., Schultz, S., and Ozaki, M., 1994, *Phys. Rev. Lett.* **73**, 1986.

[14.25] Wernsdorfer, W., Dudin, B., Mailly, D., Hasselbach, K., Benoit, A., Meier, J., Ansermet, J.-Ph., and Barbara, B., 1996, *Phys. Rev. Lett.* **77**, 1873.

[14.26] Wernsdorfer, W., Hasselbach, K., Sulpice, A., Benoit, A., Wegrowe, J.-E., Thomas, L., Barbara, B., and Mailly, D., 1996, *Phys. Rev. B* **53**, 3341.

[14.27] Lyberatos, A., and Chantrell, R.W., 1993, *J. Appl. Phys.* **73**, 6501.

[14.28] Nakatani, Y., Uesaka, Y., Hayashi, N., and Fukushima, H., 1997, *J. Magn. Magn. Mater.* **168**, 347.

[14.29] Weibull, W., 1951, *J. Appl. Mech.* **18**, 293.

Appendix

Scaling laws of the statistical pinning theory

•

A.1 Basic equations

The coercive field determined by the statistical pinning theory according to the results of Section 7.2 depends on the microstructural parameters such as defect density, N, characteristic parameters of the defects, such as Burgers vector or dislocation length, and micromagnetic parameters such as δ_B, magnetostriction and dw areas. If we characterize the interaction force between a defect and a dw by its maximum value P_0 we obtain for rigid dws the following relation

$$\mu_0 H_c \propto \frac{N^{1/2} P_0 \delta_B^{\pm 1/2}}{M_s L_x^{1/2}}, \tag{A.1}$$

where the $(+)$-sign holds for long-range interactions decreasing as $1/r$ and the $(-)$-sign holds for short-range interactions decreasing as $1/r^3$ (dipoles), L_x denotes the lateral dimension of the mobile dw area. If it is taken into account that dws are flexible and they accommodate to the potential valleys, the coercive field may be represented as [A.1]

$$H_c = \overline{H}_c + \left[c \overline{(\Delta H_c)^2} \right]^{1/2}, \tag{A.2}$$

where the first term results from the average pinning forces and the second term is due to the fluctuations of the pinning forces. $\overline{(\Delta H_c)^2}$ denotes the average of the statistical variance of the fluctuations and c corresponds to a constant of order 1. The accommodation of the dw leads to a finite value of $\overline{H}_c$, whereas in the case of rigid dws $\overline{H}_c$ vanishes. In the case of bowing dws the repulsive interaction forces are increased because they now interact with a larger number of defects.

The calculation of the coercive field of flexible dws starts from the energy due to the curvature of the dw. The deformation of the dw wall is described by its elongation along the z-axis given by $z(x, y)$. The shape-dependent part of the dw wall energy may then be written as

$$E_{dw} = \int\limits_0^{L_x} \int\limits_0^{L_y} dx \, dy \left\{ \frac{1}{2} \gamma_B \left[(dz/dx)^2 + (dz/dy)^2 \right] \right.$$

$$\left. + \frac{\mu_0 M_s^2}{2\pi} dz(x, y)/dy \int\limits_0^{L_x} \int\limits_0^{L_y} \frac{dz(x', y')/dy'}{|\mathbf{r} - \mathbf{r}'|} dx' \, dy' \right.$$

$$+ \sum_{\nu} \delta(x - x_\nu)\, \delta(y - y_\nu)\phi\,[(z_\nu - z(x, y))/\delta_0]$$

$$- (J_1 - J_2)H_{\text{ext}} \times z(x, y)\Bigg\}. \tag{A.3}$$

L_x and L_y are the dimensions of the wall in the x- and y-directions. The first two terms in eq. (A.3) represent the increase of wall energy and of the stray field energy of the bowed dw. The stray field results from the magnetic charges induced if the dw bows out along the direction of the y-axis ($\parallel J_1 - J_2$). The third term represents the interaction energy with defects with coordinates (x_ν, y_ν, z_ν) and the last term corresponds to the magnetostatic energy of the bowed out region in the external field. As shown previously [A.1, A.2] the behaviour of dws depends on the quality parameter $Q = 2K_1/M_s J_s$, which describes the effect of the stray field on the rigidity of the dw. In soft magnetic materials with $Q \ll 1$ the dw remains rigid in the x-direction and bowing occurs stray field free along the y-direction. In hard materials with $Q \gg 1$ two-dimensional bowing out of the dw is expected [A.1, A.2].

A.2 One-dimensional bowing in soft magnetic materials, $Q \ll 1$

Omitting the curvature dz/dx the dw may be projected onto the yz-plane. The number of defects per unit area is then given by $N \cdot L_x$. From eq. (A.3) we obtain by minimization with respect to the elongation $z(x, y)$ the following one-dimensional equation [A.1] for a 180°-wall:

$$d^2z/dy^2 + (P_0/\gamma_B L_x) \sum_{\nu} \delta(z - z_\nu)p(z(y) - z_\nu)/\delta_B) + 2J_s H_{\text{ext}}/\gamma_B = 0, \tag{A.4}$$

where P_0 is the maximum interaction force of an individual defect at position z_ν and the function $p(z)$ describes the normalized profile of the interaction force. From eq. (A.4) we derive the generalization of eq. (A.1) in the form[1]

$$J_s H_c/\gamma_B = \text{const.}(L_x N)^\alpha (P_0/\gamma_B L_x)^\beta \delta_B^\gamma. \tag{A.5}$$

Relations for the exponents α, β, γ may be derived by scaling down the field of defects in the z-direction, e.g., by a factor of 2. The factors of eq. (A.5) then transform in the following way: $(L_x N)' = 2(L_x N)$; $(H_c J_s/\gamma_B)' = \frac{1}{2}(H_c J_s/\gamma_B)$, $(P_0/L_x \gamma_B)' = \frac{1}{2}(P_0/L_x \gamma_B)$, $\delta_B' = \frac{1}{2}\delta_B$. In order that eq. (A.5) remains invariant on inserting these relations into eq. (A.5) the following scaling law has to be fulfilled:

$$1 = -\alpha + \beta + \gamma. \tag{A.6}$$

Scaling the y-direction similarly yields $(L_x N)' = 2(L_x N)$, $(H_c J_s/\gamma_B)' = 4(H_c J_s/\gamma_B)$, $(P_0/L_x \gamma_B)' = 2(P_0/L_x \gamma_B)$ and gives

$$2 = \alpha + \beta. \tag{A.7}$$

Equations (A.6) and (A.7) have to be satisfied for one-dimensionally bowed dws. For an explicit calculation of the exponents we make use of another physically motivated scaling relation. In the case of large defect densities the average interaction force cancels out. Only spatial fluctuations of the defect density result in finite pinning centres. The strength of

---

[1] The exponents should not be intermixed with the critical exponents of the magnetic phase transition in Chapter 12.

these fluctuations increases as the square root of the defect density [2.8, 7.3, A.3] leading to the following relation:

$$\beta = 2\alpha. \tag{A.8}$$

For large defect densities we now obtain

$$\alpha = 2/3, \qquad \beta = 4/3, \qquad \gamma = 1/3,$$

and

$$\mu_0 H_c = \text{const.} \frac{N^{2/3} P_0^{4/3} \delta_B^{1/3}}{\mu_0 M_s \gamma_B^{1/3} L_x^{2/3}}. \tag{A.9}$$

Equation (A.9) has also been derived from computer simulations [A.1, A.4] and from detailed theories of dw bowing [A.1, A.5].

The special case of rigid dws as given by eq. (A.1) is obtained from the above scaling relations for the limit $\gamma_B \to \infty$, i.e., for rigid dws H_c does not depend on γ_B which can only be fulfilled for $\beta = 1$ (see eq. (A.5)). With this condition eq. (A.6) and eq. (A.8) give the rigid wall exponents $\alpha = 1/2$, $\beta = 1$, $\gamma = 1/2$ as derived previously [2.8, 7.3, 7.6] and according to eq. (A.1). The coercive field of the dipole configuration is obtained by replacing P_0 by P_0/δ_B in eq. (A.5).

A.3 Two-dimensional bowing in hard magnetic materials, $Q \gg 1$

Minimizing E_{dw} of eq. (A.3) the following integro-differential equation is obtained:

$$\frac{d^2 z(x, y)}{dx^2} + \frac{d^2 z(x, y)}{dy^2} + \left(\frac{\mu_0 M_s^2}{\pi} / \gamma_B\right) \int_0^{L_x} \int_0^{L_y} \frac{dz(x', y')/dy'}{|\mathbf{r} - \mathbf{r}'|} dx' dy'$$

$$- (P_0/\gamma_B) \sum_{\nu=1} \delta(x - x_\nu) \delta(y - y_\nu) p[(z(x, y) - z_\nu)/\delta_B] + 2J_s H_{ext}/\gamma_B = 0. \tag{A.10}$$

Due to the stray field term in eq. (A.10) the treatment of the double curved dw becomes complicated and the wall rigidity becomes anisotropic. The coercive field now depends on a fourth parameter, $\mu_0 M_s^2/\gamma_B$, and is written

$$H_c = \text{const.} \, N^\alpha (P_0/\gamma_B)^\beta \, \delta_B^\gamma (\mu_0 M_s^2/\gamma_B)^\delta. \tag{A.11}$$

The stray field term reduces the amplitude of the bowing and results in an elliptical deformation of its shape. Therefore independent scaling in the x- and y-directions is not possible. The obstacle array therefore is scaled down by a factor of 2 in both directions x and y: $N' = 4N$, $(J_s H_c/\gamma_B)' = 4(J_s H_c/\gamma_B)$, $(P_0/\gamma_B)' = (P_0/\gamma_B)$, $(J_s^2/\gamma_B)' = 2(J_s^2/\gamma_B)'$. This leads to the scaling relation

$$2 = 2\alpha + \delta \tag{A.12}$$

and to eq. (A.6), $1 = -\alpha + \beta + \gamma$, for the scaling in the z-direction. Because of the parameter J_s^2/δ_B, eqs. (A.12) and (A.6) are not sufficient to determine the exponents. For $\mu_0 M_s^2 \delta_B \ll \gamma_B$ the stray field term in (A.10) and (A.11) may be neglected, i.e., $\delta = 0$. Equations (A.8), (A.12), and (A.6) then give

$$\alpha = 1, \quad \beta = 2, \quad \gamma = 0 \tag{A.13}$$

and

$$\mu_0 H_c = \text{const.} \frac{N P_0^2}{M_s \gamma_B}. \tag{A.14}$$

In the other limit, $\mu_0 M_s^2 \delta_B \gg \gamma_B$, the rigidity of the dw is highly anisotropic impeding the curvature in either the x- or the y-direction. Assuming $d^2 z/dx^2 \ll d^2 z/dy^2$ an independent scaling in all three dimensions is possible. Equation (A.12) thus splits into

$$\left. \begin{array}{l} 2 = \alpha + \beta + 3\delta \\ 0 = \alpha - \beta - 2\delta \end{array} \right\}. \tag{A.15}$$

Together with (A.8) the exponents are given by

$$\alpha = 4/3, \quad \beta = 8/3, \quad \gamma = 1/3, \quad \delta = -2/3.$$

The coercive field is now written

$$\mu_0 H_c = \text{const.} \frac{N^{4/3} P_0^{8/3} \delta_B^{1/3}}{\mu_0 M_s^2 \gamma_B^{5/3}} \left(J_s^2 / \gamma_B \right)^{-2/3}. \tag{A.16}$$

From the scaling relations derived so far the dependence of H_c on the micromagnetic and the microstructural parameters could be derived for three different regions as shown in Fig. A.1:

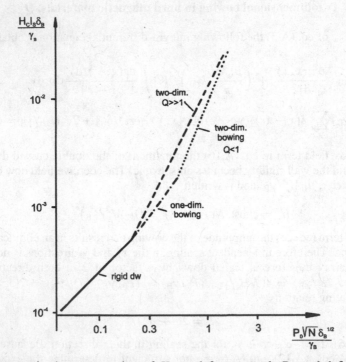

Fig. A.1. Summarized representation of H_c on defect strength P_0, and defect density N for small and large stray field effects. $L_x = L_z = 10^3 \delta_B$; (—) rigid dw model, (– · – · –) one-dimensional dw bowing, (– – –) two-dimensional dw bowing accomplished ($Q \gg 1$); (· · · · · ·) onset of two-dimensional dw bowing ($Q = 0.2$).

1. The rigid domain wall range.
2. One-dimensional dw bowing for $Q \ll 1$.
3. Two-dimensional dw bowing for $Q \gg 1$.

The transition between the rigid dw range and the bowing range is determined by the parameter $P_0\sqrt{N}/(\gamma_B \delta_B^{1/2})$ and the extensions L_x and L_y of the dw. From a numerical calculation Hilzinger and Kronmüller [A.1] found for the crossover point

$$\frac{200\gamma_B \delta_B^{1/2}}{P_0\sqrt{N}} = \frac{L_x}{\delta_B} \simeq \frac{L_y}{\delta_B}. \tag{A.17}$$

As shown in Fig. A.1 at low defect densities or low pinning forces the rigid dw pinning holds. At a parameter $P_0\sqrt{N}/(\gamma_B \delta_B^{1/2}) > 0.25$ dw bowing takes place. One- and two-dimensional bowing depends on the Q-parameter and the strength, P_0, of the pinning forces.

Comparing the different relations obtained for H_c it becomes obvious that the parameter β increases with increasing tendency of bowing from $\beta = 1$ to 4/3, 2 and 8/3. Similarly the parameter α increases from $\alpha = 1/2$ to 2/3, 1 and 4/3.

References

[A.1] Hilzinger, H.-R., and Kronmüller, H., 1976, *J. Magn. Magn. Mater.* **2**, 11.
[A.2] Hilzinger, H.-R., 1977, *Philos. Mag.* **36**, 225.
[A.3] Kronmüller, H., 1997, in *Magnetic Hysteresis in Novel Magnetic Materials* (Ed. G.C. Hadjipanayis, Kluwer, Dordrecht) p. 85.
[A.4] Hilzinger, H.-R., 1976, *Phys. Stat. Sol. (A)* **38**, 487.
[A.5] Bertotti, G., 1998, *Hysteresis in Magnetism* (Academic Press, San Diego) p. 357.

Index